W9-DJH-411

CHICAGO PUBLIC LIBRARY
HAROLD WASHINGTON LIBRARY CENTER

R0016338039

.VISION

The Ch   Library

Nuclear Navy

**United States Atomic
Energy Commission
Historical Advisory
Committee**

Chairman, Alfred D. Chandler, Jr.
Harvard University

John T. Conway
Consolidated Edison Company

Lauchlin M. Currie
Carmel, California

A. Hunter Dupree
Brown University

Ernest R. May
Harvard University

Robert P. Multhauf
Smithsonian Institution

Nuclear Navy

1946-1962

Richard G. Hewlett
and
Francis Duncan

The University of Chicago Press   Chicago and London

VA
58
.H43
Cop. 2

The University of Chicago Press
Chicago 60637
The University of Chicago Press
Ltd., London

Published 1974
Printed in the United States of America
International Standard Book Number:
0–226–33219–5
Library of Congress Catalog Card
Number: 74–5726

RICHARD G. HEWLETT is chief historian
of the U. S. Atomic Energy Commission.
He is coauthor, with Oscar E.
Anderson, Jr., of *The New World,
1939–1946* and, with Francis Duncan, of
Atomic Shield, 1947–1952.

FRANCIS DUNCAN is assistant historian
of the U.S. Atomic Energy Commission.
He is the coauthor of *Atomic Shield.*

[1974]

BUSINESS/SCIENCE/TECHNOLOGY DIVISION
THE CHICAGO PUBLIC LIBRARY

OCT 19 1977

Contents

Illustrations

Foreword

The members of the Historical Advisory Committee of the United States Atomic Energy Commission have closely followed the writing of this volume and find the completed study an honest, scholarly, and balanced history of the Navy's nuclear propulsion program. We enjoyed the opportunity to review the draft and final chapters and to discuss them at length with the authors. At our meetings we had access to all the information they used, both classified and unclassified, and also had the opportunity to inspect the plants, laboratories, and nuclear-powered vessels whose development they were describing and analyzing. In the reviews of the draft and the final chapters we did not, of course, attempt to verify the accuracy of the details, based as they were on voluminous files of documents, many of which had been opened for historical research for the first time. Nor did we try to influence the authors' interpretations of the documentary record. The review did, however, permit us to say with certainty that this study in all respects meets exacting canons of historical scholarship.

The story told here has significance for men of affairs as well as scholars. It says much about the innovation and development of a basic new technology under the guidance of the federal government. It describes the complex relationships among the scientists who handled the basic research, the civilian and military officials (usually technically trained engineers), who were responsible for carrying out the programs, and the contractors (usually private corporations), who built the plants, equipment, components, and ships. The study suggests both the problems raised in the process of putting a new technology to work and the techniques and procedures devised to solve these problems. In this way it provides a rare insight into the inner workings of the military and civilian governmental offices carrying out the task. Above all this history emphasizes the critical role played by individual personalities in the execution of a highly sophisticated, impersonal technological program within a large and sometimes impersonal bureaucracy.

> Alfred D. Chandler, Jr.
> Chairman, Historical Advisory Committee
>
> June 25, 1973

Preface

This book had its origins in a series of discussions with Admiral Hyman G. Rickover beginning in the spring of 1962. Having read *The New World,* the first volume in the Atomic Energy Commission's historical series, Admiral Rickover urged the authors to undertake a history of the naval nuclear propulsion program. Such a study, he believed, would reveal for the first time the truly significant aspects of the development of nuclear technology in the United States, a subject which, in his view, *The New World* had merely skirted. Although the authors of *The New World* found Admiral Rickover's suggestion an exciting possibility, work had already started on the second volume in the series and it was not feasible to take on another book. However, discussions with the admiral continued over the next six years with growing interest on both sides.

By 1968 the authors of the present book were completing *Atomic Shield,* the second volume in the Commission's historical series. Our research had reinforced our earlier impression that the Navy project deserved careful study. More than ever we were intrigued by the suggestion that Admiral Rickover and his group might have devised some especially effective approach to reactor development which others had not found. If Rickover had such a "magic formula," would it not be sensible to find out what it was so that others could use it?

The chance to write history that might have practical as well as intellectual value was certainly attractive, but we could foresee problems. The first was the obvious difficulty of defining Rickover's "formula." The challenge of trying to elucidate something Rickover and his own staff were unable to define was reason enough to hesitate. Even more serious in our view was the stress on administrative methods and engineering practices which such a study would seem to require. We were not specialists in public administration, management, or engineering. We could bring to the project only our talents and experience as historians. Rickover himself discounted this objection with the observation that the task required generalists rather than specialists. In his opinion the only person better qualified for the job would be a sociologist with exceptionally broad intellectual interests and experience.

These reservations still troubled us, but we were now fascinated with the idea of writing the history of the naval nuclear propulsion program. Finally, in October 1968, we agreed to write the book if: (1) we had complete and unrestricted access to all the records of the project and to all persons who had participated in it; (2) we would be free to determine the scope, content, and approach of the book; and (3) review of the manuscript would be

xi

limited to matters of security classification and factual accuracy. Admiral Rickover accepted these terms and added only one of his own: that we would not use our access to the project for any purpose other than writing this book.

Admiral Rickover and his staff have honored his agreement both in letter and in spirit. The admiral ordered his staff and the principal contractors to open all their files related to the nuclear propulsion project, to answer all our questions, to show us anything we wanted to see, to make available any personnel we wanted to interview. The result was a freedom of access, an openness, a degree of cooperation which historians seldom enjoy and cannot usually expect. This open access and freedom gave us the opportunity to check personal recollections against the record, to compare conflicting opinions, and to get beyond the legends and myths which had grown around the project. There we found an underlying consistency which gave us confidence that we were approaching the truth. Any failure to reach that goal must be attributed to our own limitations as historians and not to our sources.

The writing of this book thus became a challenging intellectual experience in which we found with increasing confidence that we could probe the thoughts and opinions of the principal protagonist in our study without fear of compromising our integrity. The entire manuscript was completed before the admiral or any member of his staff saw it. Then, true to his promise, the review was confined to points of factual accuracy. We evaluated each comment on its merits and accepted or rejected it accordingly. The final version, as it appears in this book, represents the authors' opinions and conclusions alone.

Before beginning our research, we reached a firm decision that, for better or worse, our product would be a historical analysis. That is, we would not attempt to use the analytical methods of the political scientist or sociologist, disciplines in which we have little competence. Rather, we proposed to use as best we could our abilities as historians to study the development of the Navy project as a historical process. We would attempt to place events in the larger historical context of the Atomic Energy Commission, the Navy, the Department of Defense, other parts of the executive branch, and the Congress. Because we did intend to write history in the sense of presenting a reasonably complete and well-rounded account of selected events and topics, we knew that we would have to terminate our study far short of the present. We decided that we would use the earliest possible cut-off date that would permit us to describe the Navy project in its fully evolved, if not final, form.

As a result of these decisions, the opening chapters of this book take the form of a historical narrative. Beginning with chapter 5 we begin to shift from an almost purely narrative approach to a more analytical study. In the latter chapters we have selected those elements which seem to us to illustrate the principles of the Rickover approach to technological innovation. Some participants may complain that we have omitted themes which dominated their attention for months and years. Others will surely claim that we have not attributed appropriate credit to many individuals who gave all of their professional lives to this project. We have tried to be conscientious about such matters, but we have felt constrained to place a higher priority on our primary goal, which was to define the principles of the Rickover approach.

Similarly, some will complain that in cutting off the book at the end of 1962, we have excluded some of the most pertinent issues in evaluating the Rickover approach. We are not able, for example, to present Rickover's running battle with the Department of Defense over the use of nuclear power in surface ships or the bitter controversy which was carried on with Secretary of Defense Robert S. McNamara and his aides over the application of systems analysis in the decision-making process. Acknowledging these omissions, we contend that reliable historical analysis simply is not yet possible for the years after 1962. Too many of the protagonists, including Rickover and his key staff, are still active; too many of the issues are still alive and in contention; too little hindsight is available to provide historical perspective. As these words are being written, Admiral Rickover is still in charge of the organization, and the project as he created it continues to grow and evolve. We hope we have been able to capture its essential characteristics from the limited perspective we enjoy. We must leave the final judgments to another generation of historians.

Because this book has been sponsored by the Atomic Energy Commission and because we did our research and writing as government employees, we were granted unrestricted access to the records of the Commission and the Navy Department as well as those of the Division of Naval Reactors. Many of these records are still classified for reasons of national security and cannot be made available to the public, but we were able to convey the substance of these records in the text of this book. Although we believe that we have been able to give a balanced and accurate account of our subject within the constraints of security classification, those constraints have affected the text in subtle if not always important ways. We have, for example, been unable to present the technology of nuclear propulsion with the kind of engineering detail available in the classified records. Furthermore, in discussing the impact of nuclear propulsion on fleet operations, particularly in chapter 11, we have not been able to present all the issues which historians operating without classification restraints would want to present to their readers. Given the problems of writing contemporary history of classified subjects, we know of no solution other than to warn our readers that such discrepancies exist. We stand by our original contention, however, that these discrepancies are minor and do not impair the fundamental integrity of our narrative or conclusions.

So many people have given us assistance and encouragement that it is impossible to name them all, but we do wish to thank individually some who went far beyond their professional or official duties to help us. We are especially indebted to the members of the Commission's historical advisory committee. Serving without compensation, the members were willing to read and criticize successive drafts of the manuscript and to subject themselves to the agonies which historians always suffer in trying to clarify their thinking. Many of the better qualities of this book are the result of the committee's efforts, but we the authors assume responsibility both for the final judgments and the errors that may appear.

We are also grateful to the Atomic Energy Commission and its staff for making it possible for us to write this book as a part of the agency's history program. Both the members of the Commission and the staff understood our needs, made all records available, and gave us the freedom to draw our own conclusions. We particularly express our appreciation to Chairman Dixy Lee Ray and her predecessors, James R. Schlesinger and Glenn T. Seaborg. For administrative support and protection we depended upon Woodford B. McCool, the secretary of the Commission, and his successor, Paul C. Bender. Robert E. Hollingsworth, the general manager, and his deputy, John A. Erlewine, assured us unstinting support from the staff.

Literally hundreds of individuals from high-ranking government officials to anonymous shipyard workers and seamen gave us their impressions of the project. Those whose comments were recorded by name in our notes are in the section on sources. We feel obliged, however, to single out for special mention here a few persons whose assistance went far beyond what we would expect to receive in a normal interview. Admirals Arleigh A. Burke and Robert B. Carney, both former Chiefs of Naval Operations, and Admiral James L. Holloway, Jr., former Chief of Naval Personnel, not only were generous in their time for interviews but also permitted us to use their personal files and memoirs in the Navy History Division. We are also deeply in debt to several present and former members of Admiral Rickover's senior staff, including William Wegner, David T. Leighton, Lawton D. Geiger, Louis H. Roddis, Jr., and James M. Dunford, for giving us almost countless hours of their time to explain activities during their years with the project.

We cannot begin to express the debt we owe to our own staff. John V. Flynn, our research assistant during the early years of the project, not only did yeoman's service in reviewing hundreds of boxes of records but also brought his mastery of naval nomenclature and specialized technical subjects to bear on many portions of the draft. Alice L. Buck completed several long-term research projects which helped us decide how to treat a number of subjects which lay outside our specialized knowledge and experience. We were also fortunate to obtain for some months the services of L. Robert Davids, a historian with experience in the Navy, who helped us to understand some of the intricacies of naval administration. Roger M. Anders served ably as our research assistant during the last two years of the project. Betty J. Wise typed the entire manuscript in more drafts than we care to remember and checked editorial style and references. Somehow she also found time to carry on the essential administrative activities of the office so that we could concentrate on research and writing. Without her skill and understanding of our needs we could not have completed this book.

Seldom have historians had a more challenging assignment than the one we faced in writing this volume. During a period of sharply increasing awareness of the implications of technological innovation we were privileged to trace the development of a technology which has profoundly affected both the civilian and military spheres of our society. We have also had the exceptional advantage of being able to observe some of that development in the making and to question those who directed the project. Our hope in undertaking this volume was to throw some light on how technological innovation was accomplished in a major government program. How well we have met our goal is for others to say.

> Richard G. Hewlett
> Francis Duncan
>
> Germantown, Maryland
> April 25, 1973

First to sign the surrender document were Foreign Minister Mamoru Shigemitsu and General Yoshijiro Umezu for Japan. General of the Army Douglas MacArthur signed for the allied nations and Fleet Admiral Chester W. Nimitz for the United States. Then, one after another, representatives of the other states which had been at war with Japan came forward to the green-covered table on board the battleship *Missouri* and affixed their signatures.

For Nimitz that moment on September 2, 1945, in Tokyo Bay was the climax of a distinguished career. He had become commander in chief of the United States Pacific Fleet within a month after the disaster at Pearl Harbor. Starting with that shattered force, he had organized in the next four years one of the most powerful battle fleets in history. By the time the atomic bomb was dropped, American battleships lay off Japan's home islands, bombarding shore installations while planes from carriers ranged freely inland.[1]

Triumphal Tour

A few weeks after the memorable ceremonies on the deck of the *Missouri*, Nimitz returned to the continental United States to receive a hero's welcome. Thousands cheered him in San Francisco as he rode to the city hall to receive official greetings from Governor Earl Warren. He told the assembled throng that despite atomic bombs or any other new weapons, "our Navy today is a guarantee of peace for tomorrow." He admitted that new weapons might change the character of battle, but the prerequisite for military success would be "control of the sea."[2]

In Washington, on October 5, Nimitz received an accolade comparable only to that accorded earlier to General Dwight D. Eisenhower. Hundreds of thousands of Washingtonians lined the streets as Admiral Nimitz rode to the Capitol to address a joint session of Congress. During a parade to the Washington Monument grounds, a thousand naval aircraft—fighters, torpedo bombers, and dive bombers—flew overhead. The Admiral, wearing blues, gold braid, and a stiff white collar in place of the rumpled khakis which had been his customary uniform in the Pacific, recounted for the crowd that filled the monument grounds the achievements of American and allied forces in the Pacific. He did not, however, belittle the importance of the atomic bomb. "The introduction of atomic power," Nimitz said, "has given new importance to seapower. . . . Our defense frontiers are no longer our own coast lines. . . . Today our frontiers are the entire world."[3]

It had been a day of triumph, not just for Nimitz but, in a far more important sense, for the Navy. Seldom in the nation's history had the exploits of the Navy so completely captured the attention of official Washington; seldom again would the Navy have such sweeping command of its own destiny. Nimitz had not missed the opportunity. He had spoken out clearly and effectively for the Navy as a vital part of the balanced defense forces of the future.

The next day, before going on to New York City, Nimitz stopped at the Main Navy Building to see Secretary James V. Forrestal. Rumors were already circulating in Washington that Fleet Admiral Ernest J. King, Chief of Naval Operations, expected to retire by the end of the year. On October 8, King wrote Forrestal that he considered Nimitz "the officer clearly and definitely indicated" to be the new Chief of Naval Operations.[4]

Two days later Forrestal told Nimitz that he could have the assignment for a period of not more than two years.[5] Forrestal did not propose to announce his decision for several weeks, but Nimitz could return to Pearl Harbor with the knowledge that he would be guiding the Navy's destiny in the critical years ahead.

The Assignment

Receiving the thanks of a grateful nation was a pleasant if arduous task. During the parades, banquets, and speeches, Nimitz could have given little thought to an assignment he had received from King on August 30, 1945. In a brief formal letter—copies of which he sent to other key organizations in the Navy—King asked for the recommendations of the Pacific Fleet on future developments in gunnery and ships, in fact on all the forms of endeavor that had gone into the defeat of the Japanese. No other American fleet in the course from defeat to victory had fought so many types of combat —carrier duels in the Coral Sea and at Midway; destroyer attacks in the East Indies; amphibious assaults on Pacific atolls; submarine raids in enemy waters; and battleship engagements at Surigao. In the wide-ranging questions that he posed, King was asking Nimitz and his officers to place no constraints upon their views. Not only would they concern themselves with material and equipment, but also with the ships themselves. How had they stood up under combat? What types could be discontinued, modified, or added? Because King intended to disseminate the information throughout those parts of the Navy concerned with postwar developments, he wanted Nimitz to express his own views in the report.[6]

The size of the postwar Navy was not a new subject. Shortly after the death of Franklin D. Roosevelt, Forrestal had offered President Truman a brief presentation on the matter. King had outlined his own views to Forrestal on April 27, 1945. For some time, perhaps beginning before the end of the war, the Navy would be able to reduce its strength. After peace arrived, the United States would have to have land, sea, and air components and the overseas bases from which to deploy them. Control of the seas would be necessary so that the United States could move its forces into areas where hostilities threatened. As King now saw it, the Navy would have to have the strength and the bases to control the Western Atlantic, the entire Pacific, and their approaches.[7]

Within a few days King had made ready a more detailed study. He recognized that in many instances its conclusions could only be tentative, but underlying his thoughts was a single principle: the idea of a balanced force. The Navy afloat would be divided into Atlantic and Pacific Fleets, each with ships in reserve. The active fleet was to be balanced as to type of ship. Further, the active fleet would be divided into five carrier task forces, two in the Atlantic and three in the Pacific. Although these forces would not be identical in composition, each would be built around large carriers supported by battleships, cruisers, and destroyers. The reserve fleet, too, was to consist of several types of ships so that it also would be a balanced force.[8]

If King did not question the idea of a balanced fleet, he was less certain of the characteristics each type should possess. A study group of officers on his staff reported on August 22, 1945, that the new Navy would have to build on the hard-won combat experiences of World War II. It was clear that in peacetime neither funds nor personnel would be plentiful, but there would be an opportunity to correct certain design deficiencies that had been accepted only under the stress of war. Yet these lessons had to be combined with plans for new weapons. The group thought the next ten or fifteen years should see revolutionary developments in controlled missiles, explosives, and the utilization of energy. The officers proposed that one part of the problem was at least subject to immediate study: ships on hand and under construction should be analyzed to see what improvements could be made.[9] For King there was no better source of information on the strengths and weaknesses

1. Fleet Admiral Ernest J. King congratulates Fleet Admiral Chester W. Nimitz after the announcement on November 21, 1945, that Nimitz would succeed King as Chief of Naval Operations.
U.S. Navy

of the Navy's ships than Nimitz and the Pacific Fleet. With these thoughts in mind, King had asked Nimitz to take on the study.

The Chief of Naval Operations

In drafting recommendations for King, Nimitz and his staff could automatically take into account the complex organizational hierarchy by which the Navy was administered. At the head of the structure was the civilian Secretary of the Navy, a cabinet position established in 1798. During the early years of the republic when the number of ships and personnel was small, the Secretary had little difficulty in administering the Navy. As technology advanced, reorganization became necessary. In 1842 Congress established a system under which various entities called bureaus had jurisdiction over large segments of the Navy, including shipbuilding and outfitting. The new system had some advantages, but relations between the bureaus and the Secretary proved difficult.[10] Another complication was that officers in the fleet were often dissatisfied with the ways of the bureaus. When the nation was building a modern navy at the end of the nineteenth century, many experienced officers were convinced that ships were being built more in accordance with the ideas of the bureaus than in response to the practical requirements of ships in battle.

The struggle by a number of officers to establish a military chief of the Navy came to a head during the first administration of Woodrow Wilson. The officer group, with the help of a few key congressmen, introduced legislation directly charging an officer as Chief of Naval Operations with the responsibility of seeing that the fleet was prepared for war. After a hard fight Secretary Josephus Daniels was successful in watering down the legislation so that the new military head would clearly be acting under the authority of the Secretary. Established in 1915, the position of Chief of Naval Operations soon became one of the most powerful in the Navy, although the bureaus still remained directly under the Secretary.[11]

As Assistant Secretary of the Navy from 1913 to 1920, Franklin Roosevelt had witnessed the struggle to establish the position of Chief of Naval Operations. As president in 1941 after the defeat at Pearl Harbor, Roosevelt decided to concentrate even further the military direction of the Navy. To Admiral King, who was already Chief of Naval Operations, he gave authority to coordinate and direct the efforts of the bureaus, an extension of power long sought by earlier Chiefs of Naval Operations and given almost without notice during the crisis atmosphere early in the war.[12]

The flow of power to King accentuated other changes in Navy administration, particularly in diminishing the importance of the General Board. Established in 1900, the board was intended to provide the Secretary with advice on the size, composition, and disposition of the fleet. Even further, the board drew up recommendations on such matters as the speed, armor, and armament of new ships. The functions of the General Board declined as the responsibilities of the Chief of Naval Operations grew, and King failed to find an effective way to use the board during World War II. Before the end of the war King set up the ship characteristics board under his control. Its full-time members analyzed the characteristics proposed by the bureaus for new ships, and decisions were voted by the full board, which included representatives from the bureaus concerned with the type of ship under consideration. Very soon the General Board had little left to do, save for such projects the Secretary assigned to it.[13]

Bureau of Ships

Any plans Nimitz might develop for new combat vessels would inevitably involve the Bureau of Ships. A relatively recent organization going back only to 1940, the bureau was the child of the union of two powerful units in the prewar Navy: the Bureau of Construction and Repair and the Bureau of Engineering. Originally all shipbuilding had been in the province of the Bureau of Construction and Repair, but as ship design and construction became more sophisticated with advancing technology, the Bureau of Engineering had come to have an equally important role in fixing the design of new naval vessels. As the Navy began to expand before World War II, Congress accepted arguments that constructing large numbers of new ships demanded a single agency to design and build them. In 1940 the two entities were combined into a new Bureau of Ships.[14]

The Bureau of Ships had emerged from the war a formidable, united, and effective organization, although its roster still showed a conscious balance of assignments between those officers who were naval architects and those who were engineers. At the peak of its operation during the war the bureau had a staff of more than 6,000 officers and civilians in Washington and operated 465 shipyards employing over one million people. During the course of the war the bureau had spent $17 billion building more than 110,000 ships.[15]

The extraordinary accomplishments of the Bureau of Ships during the war reflected in part the competence of the officers and civilian engineers who manned the prewar organization. The bureau successfully used the wartime emergency to attract capable and energetic young engineers to join the ship-building effort. Bringing a fresh breeze to the Navy bureaucracy, these young officers and civilians used imagination both in technical and administrative areas to get the job done. The Bureau of Ships had been equally fortunate in the caliber of its career officers. Through a carefully planned post-graduate training program which involved advanced studies in the leading engineering schools such as the Massachusetts Institute of Technology and extensive engineering duty both at sea and in Navy yards, the bureau had developed a cadre of officers whose professional talents in ship design and construction were virtually unsurpassed in any other nation. Typical of this technical excellence were the two leaders of the bureau during the war, Edward L. Cochrane and Earle W. Mills. Cochrane, who had graduated at the top of his Annapolis class in 1914, had taken a graduate degree in naval architecture at MIT and had worked as a junior officer in several Navy yards. Most of his service had been in the old Bureau of Construction and Repair. Mills, an Annapolis graduate of 1917, had served as a regular line officer on battleships, destroyers, and cruisers before taking a graduate degree in naval engineering at Columbia University. With experience as an engineering officer at sea and in the Bureau of Engineering before the war, Mills, like Cochrane, had become a key member of the design group in the Bureau of Ships in 1940. Two years later, when new leadership was needed to take on the enormous burdens of the wartime shipbuilding effort, Cochrane and Mills were promoted to rear admiral over the heads of many other officers and were appointed chief and deputy chief, respectively, of the Bureau of Ships. They were to serve in these positions until 1946.[16]

Although the organization of the Bureau of Ships shifted occasionally during the war, the bureau's role in ship design remained relatively constant. New requirements for ships generally came from the Chief of Naval Operations through the General Board. A preliminary design group within the bureau worked with the General Board in arriving at those characteristics which seemed best suited for the ship's intended mission. Every ship design was a compromise of many factors such as size, speed, and armament. Often, if the new ship was to be radically different from earlier types, several studies were required, sometimes as many as fifty.

Once the Chief of Naval Operations had approved the preliminary design, the General Board issued a directive. This document authorized the Bureau of Ships to establish the lines and body plan, develop the general arrangement plan, and make necessary strength calculations. At the David Taylor Model Basin along the Potomac in nearby Maryland, engineers towed hull models to estimate the shaft horsepower needed to drive the vessel at the required speed. When all the calculations and checks had been completed, another organization in the bureau prepared contract plans and specifications. Before construction could begin, the Secretary of the Navy had to approve the design. From the contract plans the shipbuilder, either a naval shipyard or a private contractor, made the thousands of detailed drawings needed for actual construction. The bureau, however, exercised full control over all plans and purchase orders. To insure compliance with specifications and to inspect the work, the bureau had representatives stationed in the field and at shipyards.[17] As the construction program grew during the war, the bureau in Washington could no longer follow all the details of the work and came to rely more and more on field representatives. This system, however, had produced the fleet which had spearheaded the victory in the Pacific.

Research in the Navy

As Nimitz had made clear in his October speeches, the future of the Navy would depend heavily on the vigor and quality of research on new weapon systems and ships. Research was certainly not a new idea in the Navy, but not all bureaus had pursued it with equal intensity. Several of the bureaus had their own installations devoted to solving practical problems in such areas as engineering, hull design, and ordnance. Research which did not fall within the cognizance of any one bureau did not fit easily into the Navy structure. In 1915 Secretary Daniels tried to create closer ties with American scientists and engineers as a preparedness measure when World War I showed no signs of ending. The naval consulting board he organized with Thomas A. Edison as chairman was not particularly successful. Its main legacy was the Naval Research Laboratory, which began operation in 1923 in the District of Columbia. The laboratory had done notable work in investigating radio phenomena and had played an essential role in developing radar.[18] From time to time the laboratory was under the Secretary of the Navy and the Bureau of Ships.

Impressed by Vannevar Bush's efforts in 1940 to mobilize the nation's scientific manpower and resources by creating the National Defense Research Committee and the Office of Scientific Research and Development the following year, Secretary Frank Knox gave special attention to ways of increasing the Navy's effectiveness in research and development. During the war, the Navy benefited greatly from its close contacts with scientists. Determined that this experience should not be lost, Forrestal established on May 19, 1945, the Office of Research and Inventions, which was to report to him rather than to one of the bureaus. Under the new office came the Naval Research Laboratory. The function of the larger organization was two-fold: to continue some of the wartime research of interest to the Navy and to encourage and coordinate new efforts in areas not being covered by any of the bureaus.[19]

Forrestal had chosen Rear Admiral Harold G. Bowen to head the new office. Bowen was an aggressive officer who as chief of the Bureau of Engineering in the 1930s had fought for the use of high-temperature and high-pressure steam in the Navy. When the Bureau of Engineering became a part of the Bureau of Ships, Bowen lost out in the struggle to head the new organization. Instead he became director of the Naval Research Laboratory in 1939 and supported preliminary research in atomic energy. Under his leadership, the laboratory expanded rapidly. Bowen also had the confidence of Forrestal, a factor of no mean importance in the shifting organization of the postwar Navy.[20]

Bowen faced a more uncertain future than Nimitz or Cochrane. The Chief of Naval Operations was perhaps the most powerful individual in the Navy, rivaling even the Secretary. Cochrane had in the Bureau of Ships a cumbersome structure, but it had built ships during the war and would continue to do so. Bowen, on the other hand, headed an office that had as its bailiwick research, an amorphous term at best, but never more so than in 1945 when political leaders and scientists were debating the relations of science and the federal government. In his new office, Bowen had the advantage of being independent of the bureaus, but he also lacked the protection of a time-honored organization. It was by no means clear in the autumn of 1945 whether Bowen could realize his hopes for a consolidated research organization in the Navy.

The Postwar Fleet

While Nimitz was on his victory tour, a board of officers at Pacific Fleet headquarters had begun drafting the report which Admiral King had requested in August. The vast scope of the assignment made for a bulky document containing recommendations for various types of ships.[21]

At the end of the war the Navy had three large *Midway*-class aircraft carriers under construction. These were larger than other carriers and incorporated the British feature of the armored flight deck, which had proved itself against Japanese suicide attacks. The review board, however, was not certain about some of the features and recommended extensive operation of these ships under conditions as near to combat situations as possible. The *Essex*-class carrier had done admirably, and its engineering plant had out-performed expectations. Weaknesses of this class were its lack of protection against *kamikaze* attacks, inadequate antiaircraft batteries, and vulnerability while rearming and refueling planes. The board, however, considered both the *Midway* and *Essex* classes fundamental to a balanced fleet.

The *Iowa* class of battleships—of which the *Missouri* was one—had given invaluable support in fast carrier operations. They were maneuverable and fast, and carried the world's longest-range ship-mounted gun. The antiaircraft armament of the *Iowa*s was the best in the world. The ships were rugged, could stay at sea during the most prolonged and severe operations, and could provision smaller units of a task force. The class, which had been designed before Pearl Harbor, would be effective in the postwar period with only relatively minor modifications. The board was convinced that these battleships were indispensable to the postwar fleet.

The board found cruisers and destroyers the most difficult to analyze because these ships had been so versatile in World War II operations. The board saw future employment of cruisers and destroyers in fighting enemy surface ships, destroying enemy commerce, conducting shore bombardments, protecting bases, patrolling and scouting, waging antisubmarine campaigns, and fulfilling the traditional function of "showing the flag" in foreign ports. Whatever the mission of the ships, the board urged that the design of cruisers and destroyers be kept simple so that they would serve as prototypes for ships which might be built later for actual combat. These ships would operate with carrier forces, but the multitude and variety of these missions meant that no fleet would be balanced without them.

Of all the ship classes covered in the report, the board gave its greatest attention to submarines—not that the Navy considered submarines the backbone of the fleet, but rather because the role of the submarine had changed drastically during the war. Operations in World War II, and particularly the German experience, had posed a monumental dilemma for submarine designers: with limited undersea endurance the submarine had to be designed for efficient operation on the surface, but these same features impaired the ship's performance as an undersea craft. The dilemma had existed since the first practical submarines had been built at the turn of the century, but it had become more severe as tactical demands increased for a submarine with more endurance while submerged.

For efficient surface and submerged operations the submarine had to have two propulsion systems. Diesel engines could provide high speed and long range for surface propulsion, but, submerged and shut off from the earth's atmosphere, the vessel had to depend on battery-powered electric motors. These drove the submarine at a much lower speed. Furthermore, battery capacity limited the submarine's endurance—the faster a submarine traveled on electric motors, the quicker its batteries were exhausted. When the ship resurfaced, the diesels could be used to recharge the batteries, but this operation could require as long as six hours.

Although the submarine inflicted heavy losses upon surface shipping during World War II, it was a weapon with severe limitations. Below the surface the submarine was slow and dependent upon her periscope for an accurate determination of her own or the enemy's position. Most often the submarine used her surface speed to gain a position from which to launch a torpedo attack and then submerged to wait for her unsuspecting quarry to come within range. In the face of attack by aircraft or surface ship, the submarine usually sought to conceal herself below the surface. Once below periscope depth, the vessel was blind and, although she could still hear a surface enemy, almost any ship was fast enough to escape.

In the latter years of the war, the Germans tried desperately to improve their submarines, first by adapting the Dutch-invented snorkel. This device consisted essentially of two tubes which, extending from the submerged vessel to the surface, served as air intake and exhaust for the diesels. The device was only a palliative, however; once below snorkel depth, the submarine was as limited as ever. Even with the snorkel, operation was noisy and it was difficult for the vessel to use her own detection gear. As another approach, the Germans had in operation a few submarines which carried three times the usual number of batteries as well as the snorkel. These had a submerged range of 30 nautical miles at 15 knots, 110 miles at 10 knots, and 285 miles at 6 knots. The war ended before these vessels could enter combat. More advanced than either the snorkel or the improved battery-powered units was the closed-cycle system in which oxygen for the engines was released from chemicals. One of the several closed cycles had been developed to the point where its feasibility had been established by the time Germany collapsed.[22]

Improved propulsion systems for submerged operations not only gave the submarine a new defensive advantage but also increased its offensive poten-tial. During World War II the Navy had been successful in attacking German submarines either by destroying the vessels on the surface or by driving them under. Once the submarine had dived, its speed was greatly reduced and sound detection gear on surface vessels could get a good fix for launching depth charges. New submarines capable of relatively high submerged speeds would seldom have to surface and could evade sonar detection by destroyers even if the submarines could not outrun them. A capability for high sub-merged speeds promised to make the submarine an important weapon in any future war.[23]

As a veteran submariner, Nimitz was aware of the submarine's dilemma. The ultimate solution was nuclear power, which would make possible both surface and submerged operation on a single propulsion system. The goal was a true submarine, one capable of operating at high speeds for extended periods below the surface. The idea was a fascinating one, but in the fall of 1945 it seemed to Nimitz and others to be far in the future.

By early November Nimitz was back in Honolulu and had a chance to study the review board's draft report. On the draft pages he added comments on those sections he thought needed elaboration or correction. Most of his criticisms applied to technical evaluations of wartime performance; on the general assumptions and conclusions of the report he had no important reser-vations. In its final form on November 8, the report represented Nimitz's premises for building the postwar Navy. Coming from one of the Navy's most experienced admirals and a prospective Chief of Naval Operations, Nimitz's endorsement gave the report more than ordinary significance.

Stance for the Future

Soon after Nimitz completed his "balanced fleet" report for King, new de-velopments began to draw him back to the mainland once again. Early in November new rumors of his appointment as King's successor began to ap-pear in the press. These reports were of special interest to the Senate Military Affairs Committee, which had been making headlines during the autumn of 1945 by pointing up the bitter controversy between the Army and the Navy over unification of the armed services. The committee had discovered that the prospective Chief of Naval Operations in 1944 had prepared a statement in favor of unification. In view of Forrestal's and King's opposition to that idea, the committee decided to ask Nimitz to testify.[24]

Nimitz told the Senate committee in Washington on November 17 that he had changed his mind since making his 1944 statement. He argued that the final year of the Pacific war had demonstrated the effectiveness of unified command without a merger of the Army and Navy. He also maintained that American naval power, rather than the atomic bomb, had been responsible for the defeat of Japan. The bomb, Nimitz said, was a force to be reckoned with, but its implications for the future were not clear. "Pending further knowledge and experience, those charged with the protection of our country have the duty of maintaining adequate naval, air, and ground forces for the security of our people and the peace of the world." He concluded that the existing organization of the military establishment offered "an adequate basis for further progressive development and improvement."[25]

Statements such as these tied Nimitz as closely to the balanced fleet as King had ever been. That both men, and most senior officers in the Navy for that matter, should accept that principle was understandable. The balanced fleet concept summarized almost four decades of naval experience; it had produced victory in a global conflict. Furthermore, the concept of the balanced fleet entailed a general principle, not a fixed doctrine; it was neither universal nor inflexible. In adopting it, King and Nimitz were not excluding the possibility of innovations in naval strategy or ship design. As one of his last acts as Chief of Naval Operations, King had established a new section in his office to deal with atomic weapons, nuclear propulsion, and guided missiles.

Nimitz, even more than King, seemed aware of the potential of wartime technology. By the end of 1945, as the hitherto secret products of American science and engineering came to light, the full dimensions of a revolution in military technology had begun to appear. Sensing some of this, Nimitz, in his homecoming address at the Washington Monument on October 5, had declared: "Perhaps it is not too much to predict that history will refer to this present period not as the ending of a great conflict but as the beginning of a new atomic age."[26]

Such words had a ring of the future about them, but it was hard to tell how seriously Nimitz meant them. In the first few months after Hiroshima such phrases as "a new atomic age" had gained currency in American newspapers, on radio programs, and in the halls of Congress. Much more often during these same months did Nimitz speak out for the balanced fleet and attack proposals for unification of the armed services. The idea of the balanced fleet was capable of adaptation; but, like the Navy's resistence to unification, it tended to look backward rather than forward. The lessons of the past more than the challenge of the future seemed to dominate the thinking of King, Nimitz, and most of the Navy. Compared to many of his predecessors, Nimitz was a forward-looking, progressive officer, but perhaps Forrestal had been right in wanting a younger, more versatile man than Nimitz to lead the Navy into the postwar world.

If, as Nimitz and others had suggested, the Navy was entering a new age of unprecedented technological change, was it wise to entrust its destiny to a senior officer approaching the end of his career? Would it have been more prudent to bring in a younger man, perhaps one who had a solid understanding of the new technology? Should the new Chief of Naval Operations be more concerned during his tenure with traditional fleet problems or with the application of new technologies such as electronics and nuclear power? Forrestal, perhaps in the face of overwhelming pressure, accepted experience over creativity.

In this sense the issue Forrestal faced in the fall of 1945 transcended the question of personalities. It epitomized the fundamental question which the Navy and all the services faced in the years after World War II: how could the armed services incorporate into the nation's defense the startling technical discoveries which the war had produced? A quarter of a century later there is still some question about the most effective way of accomplishing technological innovations, but we now face an additional, much more troubling, question. So greatly has the rate of change accelerated, so rapidly has the complexity of technology increased, that the pace of technological development which knowledgeable men were tempted to call revolutionary in 1945 would not be considered so today. Now we are inclined to ask: Given the extraordinary complexity of modern technology and of the political and economic institutions upon which we depend to control its development, can we reasonably hope that effective systems of technological control can be devised? And more to the point, does the recent history of technology suggest any clues to a practical solution?

No one volume could presume to answer such questions in their totality. In the following pages we have limited ourselves to the historical approach. Beyond that we have focused our attention on the Navy and its struggles with the adoption of nuclear power for ship propulsion. That story is important in itself; its wider implications may suggest tentative answers to the broader issues raised above.

2

The Idea and the Challenge

Although, as Admiral Nimitz had suggested, the world appeared to be on the threshold of the atomic era in 1945, the United States Navy had had little opportunity during the war to prepare for that kind of a future. Excluded by President Roosevelt from the wartime project, the Navy had been able to send only a few of its officers and civilian engineers to the Manhattan District laboratories. In all top echelons of the Navy, only a handful of officers had the slightest conception of what a nuclear reactor was, and none of them could have begun to direct the design of one. While the Army had been spending $2.5 billion in building a nation-wide complex of nuclear laboratories, production plants, and reactors, the Navy was permitted to do little more than preliminary development of a secondary process used to produce fissionable material for the atomic bomb.

The idea of using nuclear power to propel naval vessels had been in fact one of the earliest uses envisioned. Because a nuclear chain reaction required a very small amount of fuel and no oxygen for combustion, it offered at least the theoretical possibility of a naval fleet with unprecedented range and submarines with the incomparable advantage of unlimited operations while submerged. But new ideas are not necessarily pursued simply because they are obvious. The transformation of scientific principles into practical engineering designs is usually difficult. The cost of the potential application often seems unreasonably high or at least not worth foregoing other equally useful and perhaps more immediately promising ideas. Ultimately the issue may come down to whether the idea is practical or desirable. Even if the application is obvious and the need for it compelling, those who seek it may lack the imagination, technical knowledge, management skills, and resources necessary to accomplish it. If the need is great enough, the idea imposes a challenge upon those who pursue it. Indeed, the challenge may be as important as the idea itself.

In the nine years after the discovery of nuclear fission in 1939, the Navy faced that kind of a challenge. How the challenge of nuclear power developed and how the Navy responded to it is the subject of this chapter.

The Beginnings

Of all the agencies of the United States Government the Navy had been the first to seize upon the possible application of nuclear power when the fission process first became known to the world in January 1939. Late that month Niels Bohr, the great Danish physicist, and Enrico Fermi, the young Italian who had recently won the Nobel Prize for his research on nuclear reactions, attended the fifth Washington Conference on Theoretical Physics at the George Washington University. Bohr and Fermi had fascinated the group by discussing the startling report that Otto Hahn and Fritz Strassmann, two German scientists at the Kaiser Wilhelm Institute in Berlin, had succeeded in splitting the nucleus of the uranium atom. The most exciting result of the experiment was that the fission process had released a significant amount of the energy in the atomic nucleus. At least in theory man had now gained access to the energy of the atom.[1]

The news of the Hahn-Strassmann experiment made a special impression on Ross Gunn, a physicist at the Naval Research Laboratory. Gunn had studied electrical engineering and had earned a doctorate in physics at Yale in 1926. After twelve years at the naval laboratory he had become superintendent of the mechanical and electrical division and, more recently, technical advisor to the director. As soon as Gunn heard reports of the Washington conference, he called his friend Merle Tuve of the Carnegie Institution of Washington, which had sponsored the meeting with the university. Like physicists at several eastern universities, Tuve and his associates had already confirmed the reports from Berlin.

The results were interesting in a theoretical sense, but they had little practical import as long as the energy release was on the submicroscopic scale of the atomic nucleus. Gunn may not yet have known what Fermi and others suspected—namely, that the fissioning uranium nucleus released one or more high-energy neutrons which might be used to start additional fissions and thus lead to a chain reaction. Fermi mentioned this possibility at a meeting which George B. Pegram, the venerable dean of the physics department at Columbia University, arranged at the Navy Department in Washington on March 17, 1939. In his usual conservative way, Fermi was reluctant to predict the possibility of the chain reaction without more data, but Gunn knew enough to make allowance for that. While most of the Navy personnel present concentrated their attention on a nuclear weapon, Gunn was already turning over in his mind the idea of using nuclear power to drive the world's first true submarine.[2]

Interest at the Naval Research Laboratory

Gunn wasted no time. Because the laboratory's budget offered no prospect of funds, he turned for help to Rear Admiral Bowen, who was then chief of the Bureau of Engineering, which at that time was responsible for the laboratory. The best Bowen could do was to provide $1,500, which Gunn then allotted to Tuve and his associates at the Carnegie Institution for studies of the fission process. On the chance that only the rare 235 isotope of uranium was susceptible to fission, Gunn also approached Jesse W. Beams, a physicist at the University of Virginia, whose knowledge of the centrifuge might lead to a practical way of separating uranium 235 from the much more common 238 isotope.[3]

Gunn's dependence on private research institutions and the universities was typical of the predicament facing scientists who were seeking financial support for basic research in the 1930s. Since the end of World War I the federal government had spent little on scientific research even for military projects. Virtually the only government agency engaged in research in physics in 1939 was the National Bureau of Standards, and that organization had to pinch pennies to meet even its primary responsibilities.[4]

The Naval Research Laboratory itself was a small organization concerned more with applied than basic studies. Despite the efforts of Admiral Bowen and others, the laboratory's budget was small, a fact reflecting, in Bowen's opinion, the Navy's lack of interest in research. For that matter, no government agency had funds for the kind of work that would be necessary to explore the new technology which the discovery of fission had suggested. Only the extraordinary promise of the Hahn-Strassmann experiment enabled Gunn to muster as much support as he did in the summer of 1939. Even had the funds been available, Gunn thought that such administrative barriers as the restrictions on government contracting would have prevented him from launching any large-scale investigations.

Other scientists were coming to the conclusion that traditional means of support were insufficient. Leo Szilard and Eugene P. Wigner, two refugee physicists from Nazi Europe, were so alarmed at the prospects of a German nuclear weapon that they took affairs into their own hands and prevailed upon Albert Einstein to sign a letter calling the dangerous potential of atomic energy to President Roosevelt's attention. Even after reading Einstein's letter, the president was painfully slow to react. Perhaps for reasons of security Roosevelt decided to restrict consideration of a policy for nuclear research to a small government committee. Lyman J. Briggs, director of the National Bureau of Standards, served as chairman of the Advisory Committee on Uranium. The Navy representative was not one of the officers from the Bureau of Engineering, which was responsible for new propulsion systems, but rather was Commander Gilbert C. Hoover of the Bureau of Ordnance. For the moment at least, military interest seemed to be centered on the use of nuclear energy for weapons, not for propulsion.[5]

With so little authority the Briggs committee could do little more than prepare reports. On November 1, 1939, the committee wrote to Roosevelt that the chain reaction was a possibility but that it was still unproved. "If it could be achieved and controlled, it might supply power for submarines. If the reaction should be explosive, it would provide a possible source of bombs with a destructiveness vastly greater than anything now known." The reference to submarines was perhaps calculated to appeal to a president who had once been Assistant Secretary of the Navy, but the absence of any White House response until the spring of 1940 was disappointing. The Briggs committee proceeded cautiously with only vague assurances of presidential interest.

Rumors of Nazi activity in uranium research and reports from American physics laboratories did more than anything else to demonstrate the need for government support. Research at Columbia University indicated a good chance for a chain reaction with low-energy or slow neutrons in a mass of uranium 235. Such a system appeared feasible as a power source, but it would require development of a satisfactory isotope separation process to provide uranium 235 and the selection of a light element to serve as a moderator in slowing down neutrons. In May Fermi and Szilard announced that graphite appeared to have a low appetite for neutrons and thus might be a good moderator.

To Ross Gunn these developments suggested the need for broad and effective cooperation between the government and the scientists in the universities. Fortunately Gunn had better prospects for Navy support in the spring of 1940 than he had had a year earlier. By that time the Bureau of Ships had been established, and Bowen had become director of the Naval Research Laboratory, reporting directly to the Secretary of the Navy. Bowen considered his designation as technical aide to the secretary merely a face-saving device, but it could prove valuable to Gunn.[6]

Now free of conflicting responsibilities, Bowen could help Gunn launch some nuclear research. First Bowen asked Harold C. Urey, a world authority on isotope separation, to organize a group of scientists to advise the President's Committee on Uranium. A less direct but more effective channel for Gunn's concern was Tuve at the Carnegie Institution. Tuve told his chief, Vannevar Bush, that submarine propulsion appeared more practical at the moment than an atomic bomb, but he favored government support of isotope separation studies, which would be the first step toward a weapon.

Gunn's access to Bush through Tuve could prove extremely important to his hopes for nuclear power. Bush, formerly vice-president of the Massachusetts Institute of Technology, was not only president of the Carnegie Institution but also chairman of the National Advisory Committee for Aeronautics. One of the most influential scientists in the nation, Bush was working closely with James B. Conant, the distinguished president of Harvard University and supporter of the administration's mobilization efforts in marshalling the nation's scientific talent for defense. In June 1940 Bush, Conant, and others had persuaded Roosevelt to establish a National Defense Research Committee. No longer dependent upon Army or Navy requests for new projects, the scientists in the NDRC could start work they thought important. In the uranium project this meant replacing the ordnance officers, who were providing token representation of the military services, with a new uranium committee which included scientists like Tuve and Gunn.[7]

Gunn's influence was prominent in the decisions which the new committee reached before the end of June 1940. At the group's request the War and Navy departments approved a thorough study of isotope separation and allotted $100,000 for the work, which the Naval Research Laboratory would administer with the committee's help. The members urged Bush to set aside an additional $140,000 to study fundamental physical constants and to explore neutron multiplication in a small assembly containing about one-fifth the amount of uranium judged necessary for a chain reaction.[8] Although modest by later standards, these grants represented a significant step toward government support of research and development.

The Navy in Isolation

During 1941 Gunn kept tabs on the research projects supported by the Navy contracts, especially Beams's efforts to develop the centrifuge method for producing uranium 235. Scientists at Columbia were using the funds Gunn had obtained from the National Defense Research Committee to investigate the gaseous-diffusion process for the same purpose.

At the Carnegie Institution in Washington, Philip H. Abelson was exploring uranium isotope separation by the thermal-diffusion process. Abelson, a former student of Ernest O. Lawrence at the University of California, had been on the verge of discovering fission in 1939 when news of the Hahn-Strassmann experiment reached the United States. Abelson had proceeded with others at Berkeley to discover neptunium, the first man-made element. In 1940 he was one of the most promising young physicists in the nation.

Impressed by his work, Gunn arranged in the summer of 1941 to bring Abelson and his thermal-diffusion experiment to the Naval Research Laboratory,
where higher steam pressures and superior shops were available. Thus the
Navy was not only supporting nuclear research contracts but also had a small
isotope-separation experiment in its own laboratory.[9]

Gunn's efforts, however, did not guarantee the Navy a strong voice in the
government's uranium project. One of the deficiencies which Bush had detected in his new research committee was that it operated on the same level
with the government laboratories and thus had difficulty in exerting control
over the increasing number of research projects on a variety of subjects.
Bush's answer was the Office of Scientific Research and Development, which
President Roosevelt established on June 28, 1941. Under the new organization, Briggs's uranium committee became the S-1 Section and, as Bush tact-
fully explained in a letter to Gunn, Army and Navy personnel would no
longer be members of the sections. Technically Gunn would continue to
serve as a liaison officer and as a consultant on isotope separation processes,
but in fact he had little contact with the S-1 Section after the reorganization.[10]

Despite the Navy's growing isolation from atomic energy development as
the nation moved toward war in the fall of 1941, Abelson continued to pursue his work on the thermal-diffusion process at the Naval Research Laboratory. Unfortunately for him the first results of his research were not available until February 1942, and by that time President Roosevelt had decided
to rely on the Army to build the necessary plants for producing fissionable
materials and the atomic bomb.[11]

Bush, probably hoping to avoid a squabble, did not inform the Navy of
the president's decision. While the Navy's voice in the S-1 project gradually
faded, the Army quickly took over the task of translating laboratory experiments into huge production plants. By September 1942, the Army had established the Manhattan project under the firm hand of General Leslie R.
Groves. Under the circumstances, it was not surprising that Gunn and others
at the Naval Research Laboratory tended to equate their growing isolation
with the rise of Army control.[12]

Gunn kept fighting for full access to data on nuclear reactions, but he now
found the Naval Research Laboratory almost completely cut off from the
Manhattan project. Probably in response to Navy pressure, General Groves
ordered two inspections of Abelson's experiments, once in February and
again in September 1943. Because it did not then seem that the thermal-
diffusion process could produce significant amounts of uranium 235 in time

for use in weapons during the war, Groves decided not to encourage the Navy project.[13]

Even under these difficult conditions Abelson persisted in his research on thermal diffusion. Now ready to test large equipment, Abelson obtained approval to build a new plant in the Naval Boiler and Turbine Laboratory at the Philadelphia Navy Yard. The plant was under construction in the spring of 1944 when J. Robert Oppenheimer, director of the weapon laboratory at Los Alamos, learned that Abelson's plant would be producing small amounts of slightly enriched uranium by July. There was enough steam capacity at the Philadelphia site for a plant three times the size Abelson was building.

Aware of the obstacles which the Manhattan District had encountered in developing other isotope-separation processes for the huge plants under construction at Oak Ridge, Oppenheimer suggested to Groves that Abelson's plant might be the best way to produce uranium 235 quickly. Groves immediately reestablished contact with the Navy. Time was so critical that he decided to gamble on building a full-scale thermal-diffusion plant at Oak Ridge without further experiments. On June 26, 1944, Admiral King ordered the blueprints for Abelson's plant sent to the Manhattan District. Within three months the first columns of the Oak Ridge plant were in operation. During the critical days in the spring of 1945, when Oak Ridge was producing uranium 235 for the Hiroshima weapon, the thermal-diffusion plant advanced by about a week the delivery of the first material to Los Alamos. Largely on its own resources the Navy had made a small but measurable contribution to the development of the atomic bomb.[14]

Postwar Considerations

More important than the role played in the production of the bomb by the small amount of material produced in the thermal-diffusion plant was the claim it gave the Navy for a share in atomic energy development after the war. The plant had demonstrated not only the competence of the Naval Research Laboratory but also the Navy's determination to pursue the goal of nuclear power. All this was to the good, but the wartime experience bred in some parts of the Navy a distrust of the Army and Groves that died hard. Quite likely both Bowen and Gunn exaggerated the hostility they saw in Groves's decision to exclude the Navy from the Manhattan project. In fact, Groves was more than ready to give the Navy information under appropriate conditions. In the autumn of 1944 he invited the Navy to name two officers to serve on a committee which would study postwar policy for the development of atomic energy. Fully aware of the importance of the committee, Groves had selected as chairman Richard C. Tolman, a physicist on his personal staff.[15]

Under the circumstances it was reasonable that the two officers should come from the Bureau of Ships, which would probably be responsible for any development of nuclear propulsion systems for the Navy. Rear Admiral Cochrane, the wartime chief of the bureau, could hardly have done better than to choose his deputy, Rear Admiral Mills, as one of the officers for the assignment. Mills, with his wide experience both in the fleet and in Washington, would be in a good position to appraise the potential of nuclear power for the Navy. Cochrane selected as the second member Captain Thorvald A. Solberg. Like Mills, Solberg was a graduate of Annapolis and the Columbia University engineering school. In addition to having served at sea as an engineering officer, Solberg had become an expert in research on boiler water treatment both in the Navy's engineering experiment station in Annapolis and the Naval Boiler and Turbine Laboratory in Philadelphia. Early in the war he had distinguished himself as a liaison officer with British scientists and engineers in London. After he returned to the Bureau of Ships in Washington as chief of the research and standards branch in the shipbuilding division in February 1944, he learned something of Abelson's work on thermal diffusion at Philadelphia. Of all the officers in the bureau at that time, Solberg probably was the only one who had been exposed to any details about the Manhattan project.

Mills and Solberg joined the Tolman committee early in November 1944 for a series of interviews with scientists and engineers from all parts of the Manhattan project. On November 8 the committee held one session at the Naval Research Laboratory, where Gunn and Abelson had a chance to express their interest in nuclear propulsion. The incident seemed to have no special significance at the time, but the committee's final report to General Groves did propose postwar development of nuclear power for the Navy. In the spring of 1945, perhaps to stress the Navy's interest, Mills sent Tolman an appendix which explored the advantages of nuclear propulsion. The greatest benefits, in Mills's opinion, would be the vastly increased range of Navy ships at all speeds and the freedom from the dangers of refueling under combat conditions or during sudden storms at sea. The appendix, which Tolman sent on to Groves, made a good impression on the general, but, by the very nature of the report, no action could be expected until the war ended.[16]

The Navy after Hiroshima

With the Hiroshima attack the existence of atomic energy burst upon mankind as an almost incredible reality. From the Smyth report[17] the world learned that uranium fission could be controlled in a reactor and that the process released vast amounts of energy, a fact demonstrated by the huge cooling-water facilities supporting the production reactors at Hanford, Washington. One of the most obvious potential uses of this new form of energy was for ship propulsion, a fact which the Navy could now proclaim in public. In testifying in an open hearing before the Special Senate Committee on Atomic Energy on December 13, 1945, Gunn declared that the main function of atomic energy should be "turning the world's wheels and driving its ships."[18]

If the end of the war made possible a public appeal for a nuclear project by the Navy, it also proved a difficult time for new military proposals for use of atomic energy. General Groves, who had successfully directed one of the greatest engineering feats in history, came under attack by many Manhattan project scientists who feared military limitations on postwar research. Weeks of public hearings on the Pearl Harbor disaster tarnished the bright image of a victorious Army and Navy with an unflattering picture of ineptness and even incompetence at the highest levels of command.

There was perhaps no better measure of the swift public reaction against the military than the passion for demobilization which swept the nation. The Navy faced not only the task of transporting the flood of veterans home from Europe and Asia but also a drastic reduction in its own personnel. Using battleships and aircraft carriers as well as troopships, the Navy brought home more than two million men between October 1, 1945, and May 1, 1946. Personnel on active duty in the Navy dropped to fewer than one million in June 1946 from more than three million during the last month of the war. In the months after V-J day the Bureau of Ships canceled the construction of more than 9,800 combat vessels and small craft, amounting to a reduction of more than $1 billion in expenditures. More than two thousand vessels were assigned to the reserve Sixteenth and Nineteenth Fleets for inactivation and almost seven thousand ships were declared surplus to the needs of the post-war Navy.[19]

The sudden shift in public opinion and the precipitous pace of demobilization were demoralizing enough for those responsible for the future of the Navy. The elimination of the German and Japanese fleets made it difficult to justify the need for a large fleet in the national defense budget. Even more threatening was the uncertainty of the future. The atomic bomb with its awesome power seemed to invalidate all traditional military doctrines. The Navy felt itself particularly vulnerable to the charge that the bomb and airpower had made ships and seapower obsolete. There was doubt whether the carriers, battleships, and host of smaller craft that made up the once-proud task forces were still necessary. Almost as an act of desperation the Navy began planning a test of the effects of the atomic bomb upon naval vessels.

Furthermore, the assertion that the Navy had a legitimate interest in nuclear power had to be qualified. In the final analysis, the availability of uranium ore determined the course which nuclear power development would follow. In 1946 the supply was dangerously small, perhaps barely enough to meet minimum requirements for nuclear weapons, to say nothing of nuclear propulsion. The Navy was interested in both nuclear weapons and nuclear power. In November 1945, within the office of the Chief of Naval Operations, Admiral King had established a division of special weapons which reflected this dual interest. Under the direction of Vice Admiral William H. P. Blandy, the new division was responsible for keeping abreast of research and development on guided missiles, atomic power, and nuclear weapons.

The division of special weapons was tied to General Groves and the Manhattan District through Rear Admiral Solberg, who had served as the Bureau of Ships liaison officer with the district during the war, and through Commodore William S. Parsons, who had worked at Los Alamos. In the new division Solberg was in charge of the atomic power section, and Parsons led the sections on guided missiles and atomic weapons.[20] Both officers had seen enough of the Manhattan project to sense some of the difficulties the Navy would encounter in trying to transplant nuclear science and technology from the wartime laboratories to the Navy. In late 1945 it was still common to regard atomic energy as something which only physicists and chemists of Nobel prize stature could master. It did not seem likely that a mere transfer of technical reports could give the Navy an effective atomic energy laboratory without at least some of the people who had worked in the wartime project. Solberg and Parsons could see a role for nuclear propulsion in the Navy eventually, but they were convinced it would take time for the Navy to build proficiency in the new technology.

Within the Bureau of Ships, Cochrane and Mills looked upon nuclear power as only one of the many possibilities for improving the performance of ships to be built for the postwar fleet. Following the balanced fleet concept, the bureau was designing a variety of new ships, including submarines, heavily armored aircraft carriers, submarine-killer ships, and destroyers. Of all these the submarine seemed to offer the greatest challenge. What the Navy needed was a new propulsion plant and a new hull design capable of high speed at substantial depth. In the spring of 1946 the Bureau of Ships was not at all certain how it should use its limited funds. It seemed likely that a closed-cycle system could be brought into operation within a few years. Nuclear power offered enormous advantages, but development would be long and difficult.[21]

Only in Admiral Bowen's office and in the Naval Research Laboratory was there any live expectation over the immediate development of nuclear power. The period of isolation from the Manhattan project had dampened none of the initial enthusiasm of Bowen and Gunn, and both men were in a good position to make their views felt. Gunn appeared to have more influence than ever before at the laboratory, and Bowen seemed to be making good on his desire to centralize all Navy research and development in one office. During the war he had made the most of the direct line to the Secretary of the Navy granted him in 1939. The assistance he gave Under Secretary James V. Forrestal early in the war stood him in good stead when Forrestal became Secretary. Bowen had not been able to obtain independent bureau status for the Naval Research Laboratory, but in October 1944 he had convinced Forrestal to create the Office of Research and Inventions. Under Bowen's direction, the new office took over the Naval Research Laboratory from the Bureau of Ships and the special devices division from the Bureau of Aeronautics. Bowen also acquired authority for Navy policy on patents and research contracts. With this charter Bowen hoped the new office would provide a focus for all nuclear research and development in the Navy.[22]

The Chimera of Independence

Bowen assumed that only by controlling all activities related to the naval use of atomic energy could the Navy be certain that nuclear propulsion would be vigorously pursued. To Bowen this meant that the Navy would have to develop its own capabilities, not only in propulsion but also in the basic nuclear sciences. A proposal from the Naval Research Laboratory late in 1945 called for transferring the personnel from Abelson's Philadelphia project to the Washington laboratory. There Abelson's group would study isotope separation processes, nuclear physics as it applied to various aspects of reactor design, and nuclear chemistry, including the processing of reactor materials and developing metals and ceramics for reactor use. Eventually this Navy group would need a new laboratory consisting of several large buildings on a remote site of several square miles and capable of accommodating a hundred scientists and engineers.[23]

Bowen realized that an independent Navy effort would require broad access to technical data, almost all of it classified, which the Manhattan project had generated. He decided to ask Secretary Forrestal for support in obtaining clearances to all Manhattan data for a dozen people in his office. This request was the subject of a discussion late in December 1945, with Captain Parsons, who represented the special weapons division, and Solberg, who spoke for the Bureau of Ships. Parsons immediately raised the question of security. The Manhattan District had always prohibited the circulation of technical data between sites, and General Groves was not likely to find reasonable a Navy request for general access. Solberg had similar reservations and doubted the immediate necessity for an independent Navy effort. He thought it would be wiser to work within the Manhattan project until the larger policy issues were settled. Not only had Bowen failed to get support from stronger organizations within the Navy, but he had also aroused old suspicions going back to 1939. Solberg later wrote Mills that he considered Bowen's approach too aggressive and that Bowen was obviously trying to take over all atomic energy work within the Navy.[24]

Among the many unsettled policy issues in early 1946, the role of the federal government in atomic energy development was probably foremost in Solberg's thinking. The extraordinary impact of this new force, particularly as it was demonstrated at Hiroshima and Nagasaki, made complete monopoly by the government a certainty, at least until some form of international control could be devised. But three months of public debate over atomic energy legislation had confused rather than clarified the question of how the government monopoly would be managed. Atomic scientists who had manned the Manhattan project during the war had launched the first attack on the legislation drafted by the Army. By the time Senator Brien McMahon had introduced a new bill in late December 1945, the issue had become one of "civilian" versus "military" control of atomic energy. As hearings before McMahon's special Senate committee dragged on into 1946, the chances of establishing a new atomic energy commission during that session of Congress diminished steadily. In the meantime, Groves was trying to hold the nation's atomic energy program together on the slender basis of his wartime authority.[25]

Bowen had reason to complain that in the six months since the end of the war the Navy had accomplished almost nothing to advance the use of nuclear power, but his frontal assault on the Manhattan District in December and on the McMahon bill in January seemed quixotic. As Parsons had predicted, Groves felt no compunction about refusing the clearance request, and Bowen's eagerness to attack the McMahon bill as a threat to a nuclear Navy seemed to play into the hands of the McMahon forces. After the Senate special committee adopted the Vandenberg amendment, which assured the armed forces a voice in the new Commission through a military liaison committee, Secretary of War Robert P. Patterson and Groves decided to accept the McMahon bill as the best the Army could hope for under the circumstances. Parsons, Mills, and Solberg agreed, and they were not pleased when Bowen persuaded Forrestal to adopt his statement attacking the bill as the official Navy position.[26]

Mills and Solberg were no less interested in nuclear power than was Bowen, but they saw the realities of the situation. Groves and the Manhattan District still held tight control of all technical data on atomic energy. Alienating Groves and his staff would not help, whether the McMahon bill passed or not. It would have been prudent to wait a few months until Congress had decided the fate of the bill, but Bowen and the Naval Research Laboratory were pushing ahead under full steam.

In March 1946, the laboratory distributed a report by Abelson and two assistants proposing construction of a nuclear-powered submarine to be in operation within two years. Because such a ship would operate underwater at high speed, Abelson suggested that the Navy use the most advanced hull which the Germans had developed for a closed-cycle system. In May 1944 the Germans had awarded a contract for construction of one hundred of these submarines, designated as Type XXVI, but none was ever built. Like other German submarines, the Type XXVI design used two concentric hulls, an inner pressure hull and an outer hull containing fuel and ballast tanks. In the Type XXVI the hydrogen peroxide for the closed-cycle system would be placed in large plastic bags which would collapse under seawater pressure as the fuel was consumed. Abelson claimed that a nuclear-powered ship built on this design would require only minor hull changes and could retain intact most of the machinery. The reactor and the primary heat-transfer system would be mounted in the space previously occupied by fuel tanks under the main pressure hull.[27]

Many features of the proposal were questionable or vague. It was doubtful whether the reactor could be located under the main pressure hull, where it would be completely inaccessible while the ship was at sea. Furthermore, the plan contained essentially nothing about the reactor. The only feature Abelson mentioned was that the reactor would use a sodium-potassium alloy as the heat-transfer material between the reactor and the propulsion turbine, and Abelson himself admitted that this alloy had never been used in such an application. Rather than describe the reactor, Abelson and his associates concentrated on approximating the specifications for conventional submarine equipment. The proposal was admittedly nothing more than an effort to operate a reactor in a submarine hull. The report also suggested that the use of nuclear power constituted only a modification of existing submarine propulsion equipment and did not require a completely new technology. The proposal, in short, did not advance the cause of the independent Navy project.

Over the years since 1939, Admiral Bowen had waged a hard-fought and courageous battle for a nuclear Navy. As an engineer he had the kind of practical approach necessary to produce results, but his strong convictions and tenacity bred an inflexibility that misled him. The idea of an independent Navy project was a chimera. The Naval Research Laboratory possessed neither the personnel nor the facilities for such an effort. Abelson was already making plans to return to his prewar post at the Carnegie Institution of Washington, and the Bureau of Ships had taken the first steps toward a cooperative effort with the Manhattan District. Bowen had the drive and the intelligence needed to establish a nuclear project, but without a solid base in nuclear technology all his energy and enthusiasm were in vain.

The Bureau Takes Command

By the end of March 1946 it was clear that any action the Navy might take on nuclear power would have to come from the Bureau of Ships. Through Mills and Solberg the bureau had good liaison with the Manhattan District and, through Parsons, with the Chief of Naval Operations. What the bureau lacked, however, was a strong advocate of nuclear power, such as Admiral Bowen. Mills, Solberg, and Parsons were convinced that, for better or worse, the Navy would have to rely on the Manhattan District. They were prepared to adopt Groves's suggestion that the Navy assign a small number of engineering officers full-time at Oak Ridge to learn the fundamentals of nuclear technology.

The idea really came into focus on March 26, when Charles A. Thomas of the Monsanto Chemical Company proposed to the Bureau of Ships that the Navy participate in a joint government-industry project to build an experimental power reactor at the Clinton Laboratories, which the company operated for the Army at Oak Ridge. The plan was to build a small power reactor proposed by Farrington Daniels, a chemist who had been director of the Chicago Metallurgical Laboratory during the final months of World War II. Daniels' idea was not necessarily to achieve a practical or economic power reactor, but rather to build an experimental unit quickly by selecting a design which would require the smallest possible extrapolation from existing technology. Daniels contended that by making this a cooperative effort involving the Manhattan project laboratories, American industry, and the armed forces, each group would soon have the basic technology needed for specific applications. Solberg, who represented the Navy at a meeting with Daniels in New York City on April 11, 1946, noted that the reactor would not apply directly to the Navy's needs, but he told Daniels and the industrial representatives that the Navy would be glad to cooperate.

For Cochrane and Mills the Daniels proposal could not have come at a better time. During the spring of 1946, perhaps stimulated by some of Bowen's actions, the higher echelons in the Navy had begun to think more seriously about nuclear power. In response to a request from Forrestal, the General Board had undertaken an investigation of various possibilities for advanced propulsion systems in the Navy. In March Cochrane had received a request from the board for a study on the subject, but before the bureau could complete its reply, the board had recommended to Forrestal on April 4 that "active comprehensive study and development of atomic power for utilization in propulsion of Naval units be initiated without delay."[28]

In drafting his reply to the General Board, Cochrane could now refer to the bureau's decision to assign a group of officers to Oak Ridge to work on the Daniels project. "It is the Bureau's opinion," Cochrane wrote, "that the action being taken by the Manhattan District to develop an experimental power pile is the soundest possible approach to this problem and will produce the fastest results." Contrary to public opinion that nuclear power was just around the corner, Cochrane and his staff believed that "at least 4–5 years will elapse before it will be possible to install atomic energy in a naval ship for propulsion purposes."[29]

Within the bureau, one of the officers interested in nuclear propulsion was Captain Albert G. Mumma, chief of the machinery design division. A graduate of Annapolis in 1926, Mumma had early distinguished himself in engineering and had been the first Navy officer in several decades to be sent to Europe for postgraduate studies. After two years at L'Ecole D'Application du Génie Maritime in Paris, Mumma had returned to the United States in 1936 with a new respect for French naval engineering and a strong conviction that sound technical training would be a key to American naval strength in any future war.[30]

During World War II Mumma had specialized in machinery design and had been a member of the Alsos mission which had moved into Germany with the forward units of the allied invasion armies in 1945 to intercept any German atomic energy activities. Poised and intelligent, with a breadth of intellectual interests unusual in engineering officers in the Navy, Mumma had become by 1946 one of the most promising officers in the Bureau of Ships and a close advisor to Admiral Mills.

Mumma's experience on the Alsos mission had given him an opportunity to obtain some information on nuclear technology, and he was convinced that nuclear power would provide an incomparably superior energy source for ship propulsion, especially in submarines. He agreed that the Navy should begin to develop some competence in nuclear technology, and he supported the proposal to send some bureau personnel to Oak Ridge for training. Although Mumma believed it would be several years before the Navy could begin to build a nuclear propulsion system, he wanted to launch on a broad scale the kind of technical development which might eventually help that effort. One promising idea was to use liquid metals as the heat-transfer medium in steam generating systems. The high thermal conductivity of these materials suggested certain theoretical advantages in steam plants. Mumma probably had seen Abelson's study proposing the use of a liquid sodium-potassium alloy in a nuclear propulsion plant. Whatever the prospects for nuclear propulsion might turn out to be, liquid metals seemed worthy of investigation.

During the closing days of June 1946 Mumma arranged two contracts which would use up research funds unexpended during the fiscal year. One with the Mine Safety Appliances Company provided for research on the chemical and physical properties of sodium-potassium alloys. A second contract with Babcock & Wilcox Company covered the possibility of using sodium-potassium as a heat-transfer fluid in a gas-turbine generator. Neither contract mentioned anything about nuclear propulsion, but that application was prominent in Mumma's mind when he approved the studies. The next step would be to arrange a contract with General Electric to design a nuclear propulsion plant using a liquid-metal coolant.

Mumma also took an active part in recommending personnel for nuclear training. With his own broad academic background, he appreciated the importance of choosing officers and civilians who would be able to cope with the complexities of nuclear physics and then find ways to apply these principles in Navy projects. In Mumma's mind, it was really more important to build a base of sound technical competence in the bureau than to train men for a specific short-term project.

In considering engineering duty officers for the Oak Ridge assignment, Mumma thought first of Lieutenant Commander Louis H. Roddis, Jr. The son of a naval medical officer, Roddis had grown up in the Navy. He had graduated first in the Annapolis class of 1939 and had achieved equivalent academic distinction in graduate engineering studies at the Massachusetts Institute of Technology. Just twenty-eight years old, Roddis had already gained the reputation of being one of the most promising younger engineering duty officers in the Navy. At the moment he was serving on Admiral Solberg's staff, which was organizing Navy participation in the forthcoming nuclear weapon tests at Bikini. If selected, Roddis would have to report to Oak Ridge after completion of the Bikini tests during the summer.

Lieutenant Commander James M. Dunford was Mumma's second recommendation. A classmate of Roddis's both at Annapolis and MIT, Dunford had an academic record only a shade less distinguished. Driving personal ambition and unwavering confidence in his ability as an engineer had led Dunford to apply to the Bureau of Ships for any special or unusual assignments. His transfer to Oak Ridge shortly after his arrival at the Pearl Harbor Naval Shipyard in the spring of 1946 would be a personal inconvenience, but the assignment seemed to be the kind of opportunity Dunford was seeking.

Like Roddis and Dunford, Miles A. Libbey, Mumma's third recommendation, was also a lieutenant commander, an Academy and MIT graduate, and an officer looking for new ideas. Libbey was already investigating the use of radioisotopes, particularly in determining the wear characteristics of materials. This interest made him a natural choice for the Oak Ridge assignment.

The fourth officer recommended by Mumma had a different background. Raymond H. Dick had always been sensitive about the fact that he was not a graduate of the Naval Academy. Pugnacious, strong-willed, and intellectually sharp, Dick resented the condescension of Annapolis officers. His determination to outperform Academy men resulted in an exceptional combat record and a spot promotion to lieutenant during World War II. At Ohio State University, Dick had done graduate work in metallurgy, a specialty unusual for engineering officers in the bureau. His knowledge and experience promised to be especially useful at Oak Ridge.

In proposing names for Admiral Mills's consideration, Mumma did not overlook the many talented civilian engineers and physicists who worked in the bureau. Although the top policy positions in the bureau at that time were reserved for career officers, the officers relied heavily upon professionals in civil service positions for specialized technical knowledge and experience. From his own machinery design section, Mumma recommended Alfred Amorosi, an engineer who had been studying advanced propulsion systems for submarines. Also from his section Mumma proposed George B. Emerson, who had been following the design of steam power plants for naval ships. The third civilian on Mumma's Oak Ridge list was Everitt P. Blizard, a physicist who had spent the war working on degaussing systems for the Navy.

To head the project Mumma thought it was important to have a senior officer with broad engineering experience in ship design and development. For this position he proposed Captain Harry Burris, who had done an outstanding job in expediting the production of steam propulsion plants for destroyer escorts during World War II. With the approval of other senior officers in the bureau, Mumma sent his list to Mills.

Mills had no trouble approving Mumma's suggestions, except for one. Without discounting Burris's capabilities, Mills thought he had a better candidate to head the Oak Ridge group in Captain Hyman G. Rickover. Just forty-six years old, Rickover had a good technical background. An Annapolis graduate in 1922, he had earned a master's degree in electrical engineering at Columbia University in 1929 and was qualified to command submarines. Following several assignments to sea duty as chief engineer of the battleship *New Mexico* and as commanding officer of the U.S.S. *Finch,* a mine sweeper on the Asiatic station, Rickover had applied to become an

"Engineering Duty Only" officer. Men with this designation were still line officers, but specialized in such areas as electrical engineering and propulsion. To become an EDO was a mark of achievement. Those who were chosen, however, were barred from exercising command afloat. As an EDO Rickover had served as assistant planning officer at the Cavite Navy Yard in the Philippines. In the fall of 1939 he had been assigned to the rapidly growing electrical section in the Bureau of Ships in Washington.[31]

What really distinguished Rickover from his colleagues was his performance as head of the electrical section. Driven by a passion to produce the electrical equipment needed by the fleet, Rickover had insisted on retaining in his section the full engineering design capabilities that had characterized most technical units in the bureau before World War II. Under the pressure of building the thousands of ships needed during the war, most bureau sections had delegated the design function to its officers in the field and had limited the headquarters task to administering contracts, inspections, and procurement schedules. Rickover, however, had followed a distinctive and much more difficult approach. He had assembled in his section a group of the best officers and civilian engineers he could find. He personally sifted through battle reports and inspected every battle-damaged ship he could reach to see for himself how electrical equipment performed under combat conditions. Working with his staff, he decided what changes in equipment were required. Then through close supervision of contractors he saw to it that the equipment was produced on time and, more important, to the required specifications.

Rickover's severely practical approach, his tireless energy, and his refusal to compromise on technical excellence paid off handsomely during the war. His own inspections of the fleet revealed electrical equipment of poor reliability and obsolete design: circuit breakers that would pop open when the ship's guns were fired, cable that would leak and carry water through bulkheads to control switchboards, new electrical motors built according to specifications dating back to the 1920s, and junction boxes that would emit poisonous gases in submarines when fires occurred. In addition to correcting scores of such deficiencies, the electrical section under Rickover developed fundamental engineering data on such subjects as shock-resistance and took the lead in designing new and improved equipment such as motors, generators, lighting systems, power distribution systems, circuit-breakers, relays, cable, and infrared detection gear.

Although the electrical section initiated, directed, and evaluated all these activities, private industry did the actual technical work. Rickover personally selected the contractors and worked directly with the responsible official in each company. He and his staff worked directly with the contractors in designing the equipment, and once the plans and specifications were established, he insisted that the manufacturers follow them to the letter. In the process Rickover established close working relationships with the major electrical equipment contractors such as General Electric and Westinghouse and earned the reputation of being a tough-minded, exacting, but reliable customer. On Commander Rickover's word alone contractors were willing to start work on a new project even before they had been offered a letter contract. By 1945, when he left the bureau to set up a ship repair base at Okinawa, Rickover had built the most creative, productive, and technically competent section in the Bureau of Ships.

This accomplishment alone was enough to convince Mills that Rickover was the officer to head the Oak Ridge group, but Mills knew that many officers in the bureau would oppose the assignment. Rickover had anything but an ingratiating personality. He remorselessly pointed out flaws in Navy equipment even when they were outside his own responsibility. He could speak with devastating frankness, never put personal feelings above his mission, and did not try to conceal his contempt for such military traditions as captain's inspections or full-dress parades.

These predilections had sometimes antagonized Rickover's fellow officers, but within the Bureau of Ships there was a more fundamental source of opposition. In insisting upon personal and firm technical direction over whatever activity he had under his command, Rickover took what often seemed to others a narrow, proprietary, and almost obsessive view of his responsibilities. During the war officers like Mumma and Burris had witnessed the development of an administrative system which gave the bureau general supervisory control over a vast empire of shipyards and contractors. The very size and technical complexity of the bureau's mission appeared in their minds to preclude the kind of personal attention which Rickover gave to technical details. Instead these officers advocated what Rickover was to call "the systems approach," which would provide the bureau with leaders who were not primarily technical specialists but rather officers with broad administrative experience in managing a variety of bureau activities. From the point of view of an officer like Mumma, giving the development of nuclear propulsion to Rickover would be a mistake. Mills's action would place the development of the bureau's most advanced and potentially revolutionary technical effort in the hands of an officer who did not accept the bureau system and who would fight for nuclear power with a single-mindedness that would ignore the bureau's other responsibilities.

Mills understood these arguments, but he also saw the need for prompt investigation of nuclear technology. Even if he accepted all these arguments against the Rickover appointment, Mills still saw it as the best way of getting a firm fix on the engineering possibilities of nuclear propulsion. In assigning Rickover to Oak Ridge, Mills's only concession to his fellow officers was that Rickover was not to be in charge of the group. The officers would report to the Army colonel who served as the Manhattan District engineer, and the civilians would be assigned to the scientist directing the Daniels reactor project at Oak Ridge.

The Oak Ridge Assignment

Rickover arrived in Oak Ridge before the end of June 1946 with mixed feelings about his assignment. After twenty-seven years in the Navy he seemed near the end of a career which, despite his demonstrated competence, he believed would never bring him flag rank in the ordinary course of events. But Rickover possessed a driving ambition and a sense of history. He was convinced that nuclear power would revolutionize the Navy, and in this new technology he saw the seeds of opportunity. On the debit side, he was painfully aware of his ignorance in the nuclear sciences, and he did not need many days at Oak Ridge to discover that there was little in existence there which would be of any help to him. The situation was unpromising enough to suggest the false impression that Mills had sent him to Oak Ridge to get him out of Washington.

The Navy officers soon found that they were not the only newcomers at Oak Ridge. As a part of the cooperative effort to build what was hoped to be the world's first power reactor, a number of American corporations and the Army Air Force had also sent some of their most promising younger engineers to work on the Daniels project. In the barracks and laboratory at Oak Ridge, Rickover discussed reactor technology with several men who were later to have a role in the nuclear submarine project: John W. Simpson and Philip N. Ross of Westinghouse, Harry E. Stevens of General Electric, and Harold Etherington of Allis-Chalmers.

The apparent aimlessness of much of the activity at the Clinton Laboratories bothered Rickover. The scientists wanted to continue the research projects started during the war, but they had little confidence in the future of the laboratory under the direction of an industrial contractor like Monsanto. Discouraged and restless, the scientists were mildly contemptuous of Daniels' project to build a power reactor. The Daniels group was drafting technically ambitious plans for the reactor, but there was little evidence of a systematic effort to define engineering problems. Without a general plan or definite assignments, the Navy, Air Force, and industrial representatives were presumably to find ways of making themselves useful and to pick up what information they could on their own initiative.

The casual way of doing things at Oak Ridge suited Rickover's purposes, not because he wanted to be casual himself but because he made good use of the freedom from assignments or routines that would distract him from his mission. The organization chart carried Rickover as deputy to Colonel Walter J. Williams, director of operations for the entire Manhattan project. An experienced engineer who had managed one of the isotope separation plants during the war, Williams coordinated production activities at Oak Ridge and other sites. To Williams the appointment of Rickover as his deputy meant little more than that the naval officer would share his office. In fact, Rickover soon found a private office for himself at the laboratory, where he could avoid administrative chores and devote himself entirely to technical reports.

Dunford, Libbey, and the three Navy civilians had found desk space in a large office nearby and frequently discussed their work with Rickover. Amorosi, Emerson, and Blizard spent most of their time on the Daniels project, but the officers had no specific assignments. They were free to study documents and attend informal lectures on nuclear physics. The more formal courses beginning in the fall were originally intended only for scientists with doctoral degrees, but the officers hoped that by working hard during the summer they could hold their own with the scientists.

The Navy Team

By September 1946 Dick and Roddis had joined the others at Oak Ridge, and Rickover began to use his rank to assert some leadership over the officers in the Navy group. He had already established himself as a hard worker, a good engineer, and a man with an obsession about nuclear power. Although Rickover did not have the advantages of the training in advanced mathematics and physics which they enjoyed, the younger officers had to admit that Rickover's industry more than made up for any deficiencies in formal training. Furthermore, there was no way to suppress his intense desire to pull the group into an effective task force, even if the younger men had wished to do so. The ineffectual performance of other groups at Oak Ridge reinforced Rickover's conviction that the kind of strong leadership he had exercised in the electrical section during the war was necessary to give the Navy the information it needed. Rickover neatly circumvented the restrictions placed on his authority in the Bureau of Ships by obtaining from his Army superiors at Oak Ridge permission to prepare for each of the officers the periodic "fitness" reports upon which their chances for promotion would depend. This authority simply formalized the leadership he had already established by force of his own personality.[32]

Experience, particularly in the Bureau of Ships during the war, had given Rickover some definite ideas about method. Creating hardware, whether it was a simple electrical component or something as complicated as a reactor, required technical accuracy. He did not believe that technical mastery of anything as complex as atomic energy could be acquired by "osmosis," simply by casual exposure to engineering activity. A few weeks of concentrated study had enabled him to follow the esoteric terminology which the scientists at Oak Ridge used. His method was direct—to read the technical reports and abstract the data needed to design a power reactor.

This approach quickly set the operating pattern for the group. Beginning with himself, Rickover required each of the officers to master the new technology. He knew it was possible to distill from the jargon of physics and chemistry the hard data the engineers needed. One or more of the group signed up for every course given at the Clinton Laboratories, attended every lecture, and investigated every project. These activities alone filled much of the working day; the rest of the educational process, and perhaps the most important part, was relegated to the remaining working hours, evenings, and weekends.

The results of these studies were summarized in written reports. These were not informal notes for the personal use of the writer, but were expected to be reliable technical information for the whole group and perhaps even for the Navy at large. Rickover insisted that the reports be clear and concise, written in good English, correct in technical detail, and relevant in some way to the central mission of the group. No activity was too big to be covered by a report—if a technical symposium included a dozen papers, all were dutifully summarized—and no relevant technical detail was too small to be recorded in writing. Preparation of these reports often required further study and this in turn led to new information. Slowly but steadily the Navy group amassed a compendium of reports which clearly depicted the status of nuclear technology.[33]

In general, the Navy group did not see any hope for quick development of a naval propulsion reactor. In a report in November 1946 Rickover was even more pessimistic than Admiral Cochrane had been in April about the time required to develop a shipboard system. Rickover estimated that it would take five to eight years to build such a plant with existing resources, and he warned that the work would involve some difficult engineering. One of the most obvious needs was to design an effective shield to protect personnel from the enormous amounts of radiation generated in a reactor. This would require original research because the production reactors built during the war were planned with a comfortable margin of safety; minimum amounts of shielding in terms of volume and weight had not then been important.[34]

Scarcely less vital in Rickover's estimation would be new materials. Metals that would withstand high temperatures were available, but to meet reactor specifications they would have to have a low attraction for neutrons and be capable of resisting prolonged and intense neutron bombardment. Other problems Rickover foresaw were the selection of a coolant to transfer heat from the reactor to the propulsion equipment and the design of the heat exchangers, pumps, and valves which would be leak-proof and trouble free. Even if all the necessary funds and talent were available, Rickover thought it would take at least three years to build the first propulsion reactor.

The Role of Industry

The Bureau of Ships recognized through its participation in the Daniels project the potential importance of industry's role in developing nuclear power. The government, through the Navy, might have to supply the funds, but private industry would bear the burden of actual design and construction of any naval propulsion plant. Large equipment manufacturers like General Electric, Westinghouse, Babcock & Wilcox, and Allis-Chalmers had been supplying equipment for the Navy for decades. Navy contracts had been an important part of their business. For companies in the electrical equipment industry, nuclear power had an additional attraction. If, as some engineers had predicted, uranium would soon become an important fuel for power generation, companies like General Electric and Westinghouse could not

begin too soon to learn the fundamentals of the new technology. The eagerness of such companies to participate in the Daniels project showed that they
understood the potential of nuclear power.

Perhaps more than most large American companies, General Electric had
a keen eye to the future. Its research laboratory—one of the few large scientific institutions established by American industry before World War II—
demonstrated the company's dedication to new ideas. Already a leading supplier of power equipment for the Navy, General Electric found the prospects
of nuclear propulsion intriguing. Harry A. Winne, the company's vice-president in charge of engineering, had caught a glimpse of the future of atomic
energy while serving as a member of the Acheson-Lilienthal committee early
in 1946. Winne would never forget Robert Oppenheimer's fascinating descriptions of a new world of industrial development. Others in the company
were equally excited about the prospects for nuclear power. The idea of using a liquid metal as the heat-transfer medium had caught the attention of
Cramer W. LaPierre and others in the company's general engineering and
consulting laboratory. In May 1946 LaPierre sparked a company proposal
to the Navy for a preliminary study of a nuclear-powered destroyer.

In August, a few weeks after President Truman had signed the act establishing the Atomic Energy Commission, General Groves had approved a
contract with General Electric for a paper study of a liquid-metal–cooled
reactor plant for a destroyer, and Admiral Mills had assigned two officers
from the Bureau of Ships to work with LaPierre and his group at Schenectady. The timing of this action suggested that the Navy was anxious to
establish a working relationship with an experienced contractor while the
Manhattan District was still in existence rather than to wait until the new
Commission could organize itself. The stress upon civilian control in the legislative struggle over the act may have caused the Navy to expect a less
sympathetic response from the Commission than from the Army.[35]

Subsequent events would reveal some foundation for these fears. Some
months before General Electric signed the Navy contract, the company had
also accepted General Groves's request that it operate the plutonium production plant at Hanford, Washington, in exchange for a promise that the
government would provide a nuclear development laboratory for the company at Schenectady, New York. This decision would later threaten the Navy
project in two ways. After taking over the atomic energy program in January
1947, the new Commission would become deeply concerned about the production of plutonium for weapons and less than enthusiastic about other

activities that might distract General Electric's attention from the enormous task facing the company at Hanford. At the same time, acceptance of the Hanford project gave General Electric a solid claim on government funds for the new nuclear research installation, which was to be called the Knolls Atomic Power Laboratory. The very name pointed out the company's primary interest—to develop atomic energy as a power source for civilian purposes.

Just as the Commission would fear the new laboratory as a distraction from the company's main task at Hanford, so the Navy could suspect it as a diversion from what it saw as the more important and immediate goal of building a nuclear ship. Whatever manpower and resources General Electric could spare from the Hanford project would go into development of a power reactor at Knolls, not to the Navy work. It was true that LaPierre's group was studying liquid-metal power systems under the Navy contract, but this work was firmly under the control of Captain Mumma and the Bureau of Ships. Rickover did not see much prospect of picking up this contract as part of an independent development project.

Despite these complications, General Electric still seemed the best single hope for early development of a nuclear ship. The company was eager to participate whenever the Navy and the Commission straightened out their priorities. In the autumn of 1946 General Electric was ready to talk to Mills, Mumma, Rickover, or anyone else who had an idea about a feasible nuclear project.

No survey of engineering resources for a nuclear ship could overlook the Westinghouse Electric Corporation in Pittsburgh. Like General Electric, Westinghouse was a large corporation in the electrical equipment business and had been a major supplier of propulsion equipment to the Navy. Both companies had participated indirectly in the wartime atomic energy program as suppliers of electrical equipment for the Manhattan project. General Electric's greater success in establishing a position in atomic energy after the war stemmed from the fact that General Electric was a larger company than Westinghouse and from its reputation as a company strongly oriented to forward-looking scientific research. The existence of the General Electric research laboratory seemed an asset of overriding importance in 1946, when atomic energy was still considered the exclusive province of the scientist.[36] Westinghouse, however, had also earned a high reputation for scientific research, and the company was respected for its solid engineering capacity. Another Westinghouse asset from the Navy's perspective was its new president, Gwilym A. Price. A lawyer and banker, Price had joined Westinghouse to help negotiate the settlement of war contracts. In this capacity he had demonstrated his ability to lead the company through the difficult transition from military to civilian production. In May 1946 Price still did not know much about the company's peacetime products and even less about atomic energy. But after a conversation with Rickover, who was about to go to Oak Ridge, Price was pretty well convinced that nuclear power was a field the company could not overlook in the postwar years.[37]

The True Submarine

During the first six months at Oak Ridge the Navy group had properly taken a broad view of the application of nuclear power to naval propulsion. Although Rickover and his associates fully appreciated the special advantages such a propulsion system would have in a submarine, they had not narrowed their focus to undersea craft alone. Nuclear power would also have advantages in surface ships, and it seemed likely that installation of a power reactor would be easier in a surface vessel than in a submarine.

Developments within the Navy during the autumn of 1946, however, were strengthening the inclination of the Oak Ridge group to concentrate its attention on the submarine. In a series of conferences since September 1946, submarine officers had been discussing antisubmarine techniques and new submarine designs. These men had concluded that "we cannot expect surface and near-surface detection to long remain in their present states of development. When the snorkelling submarine becomes readily detectable, nothing short of a deep-running true submarine will be acceptable." This event would mark the end of air-breathing engines for submarines and would make nuclear power "most attractive."

On January 9, 1947, the submarine officers recommended a broad effort to improve the nation's submarine forces, including a gradual replacement of existing submarines with new diesel models capable of greater submerged speed and endurance. They assigned high priorities to the development of the closed-cycle system to replace the diesel engine for still greater submerged speed, and the design and development of "nuclear power plants for eventual installation in submarines to give unlimited submerged endurance at high speed." Such a ship would be the world's first true submarine. The following day Admiral Nimitz approved these recommendations.[38]

Just how the Bureau of Ships would carry out this new assignment was

2. Rickover with Commission and General Electric officials at Schenectady, probably in December 1946. Left to right: C. Guy Suits, John J. Rigley, Hyman G. Rickover, Leonard E. Johnston, and Harry E. Winne.

2

not at all clear. Mills, who had now succeeded Cochrane as chief of the bureau, had just established on his staff the positions of Coordinator and Deputy Coordinator for Nuclear Matters. Mills gave these officers complete authority over nuclear matters "whether the nuclear energy [was] to be used as an explosive or source of power." In other words, the new office would be responsible for changes in ship design necessary to accommodate nuclear armaments as well as for nuclear propulsion plants. With this broad charter, it made sense to Mills to give this responsibility to the bureau's director of ship design (a position soon to be filled by Captain Armand M. Morgan) and his deputy (Captain Mumma).[39]

However sensible this appointment may have appeared to most officers in the Bureau of Ships, Rickover found it a discouraging development. To him it meant that the bureau would attempt to integrate nuclear propulsion into the regular organization of ship design and construction activities. Rather than enjoying the separate status of an independent project with special priorities and attention, nuclear power would be handled, along with the closed-cycle plant, as just another approach to a better submarine. From his single-minded perspective Rickover could not believe such an arrangement could produce a nuclear submarine in the near future.

Another complication facing Mills was the role of the new Atomic Energy Commission, which had taken over the entire Manhattan project from the Army on January 1, 1947. Until the Commission could organize itself and establish some policy, it would be difficult to make any plans for the submarine reactor or any other atomic energy project. The Atomic Energy Act made it clear that the Commission had exclusive authority over nuclear research and development. Any proposals for the submarine reactor would now have to clear the Commission as well as the Navy, and no one knew when or how the Commission would respond to any Navy proposal.[40]

General Electric

Whatever the Navy accomplished on nuclear power in the immediate future would depend largely on General Electric. The company was already heavily engaged in the atomic energy program at Hanford and was making plans for the new Knolls Atomic Power Laboratory which the government was committed to build. LaPierre's group had already started work on liquid-metal systems under the Manhattan District contract which would run until the end of the fiscal year on June 30. One of the bureau's first tasks in 1947 was to discuss with General Electric its plans for the coming year.[41]

The way Mills handled this requirement revealed a weakness in his new organizational directive. Instead of ordering Morgan and Mumma to Schenectady, he permitted Rickover and Roddis to open discussions with General Electric. Whatever the responsibility assigned on paper to Morgan and Mumma, the fact remained that Rickover and his group were the only naval officers who could evaluate General Electric's proposals for nuclear propulsion. The assignment brought Rickover and his group for the first time into a position from which they could influence bureau policy on nuclear power.

Rickover used the occasion to drive home his conviction that the Navy should concentrate on the nuclear submarine. The bureau's original proposal had been to design a propulsion reactor for a destroyer, where space requirements would not be so rigorous. Preliminary studies by General Electric, however, now suggested that it might be possible to build a liquid-metal–cooled reactor small enough to fit in a submarine hull. A submarine plant not only represented the optimum application of nuclear power in the Navy but also had the advantage of requiring less power and therefore less fissionable material than would a destroyer. Rickover believed that with sufficient effort it might be possible to have such a reactor operating in a submarine by the end of 1950. Such a schedule would preclude a long search for an optimum design. Rickover's idea was to aim for a full-scale operating installation at the earliest possible time. A reactor using slow neutrons, as did all but one existing model, and a liquid-metal coolant for greater efficiency, could first be installed in a destroyer escort. Then as that effort proceeded, General Electric could see what changes would be needed for use in a submarine. The proposal was that the Bureau of Ships would attempt to negotiate a contract with General Electric by July 1, 1947, to design and build the reactor, shielding, controls, heat exchangers, and associated equipment for both the destroyer escort and the submarine and to provide all the main propulsion machinery for the latter.[42] Rickover carried the proposal to Mills, who considered it far too ambitious.

A Question of Priorities

Even if Mills had favored the idea, the possibility of negotiating a new contract with General Electric was by no means certain. The existing contract would expire on June 30, and any extension or new contract would have to have the Commission's approval. Until the Senate confirmed President Truman's appointments to the Commission, the new agency would scarcely be able to organize its headquarters staff, and even then there would be no one within the Commission prepared to evaluate the Navy proposal.

To a large extent, the fate of the nuclear submarine rested with the Navy. If the Navy made a strong appeal for nuclear power, the new Commission might offer support. Parsons, now a rear admiral and director of atomic defense in the office of the Chief of Naval Operations, thought the obvious application of nuclear power to submarines would seem to "justify a conclusion that submarine propulsion by atomic power should be assigned national priority number one." But Parsons was not ready to make such a recommendation. He thought it might be better for the moment to focus on power units generally rather than on naval reactors specifically. He feared that work on a nuclear submarine might distract the Navy from much-needed improvements on conventional ships. Furthermore, Parsons asserted, the engineering problems of building a Navy reactor were more difficult than those faced in creating the first nuclear weapon. Premature engineering solutions might actually delay rather than advance the development of a nuclear submarine. Parsons also believed that until more uranium ore was available, the nation should put breeders (which would produce more fuel than they would consume) ahead of all power reactors, including those for naval propulsion. It seemed reasonable to Parsons that within five years there would be enough talent and information to provide a sound basis for a submarine reactor. In the meantime, he suggested that the Navy assign a few high-caliber engineering specialists to work on reactor projects at the Commission's laboratories.[43]

Parsons did not make explicit another consideration which must have colored his attitude toward nuclear power. As an ordnance expert and a member of the Los Alamos staff, Parsons was thoroughly familiar with atomic weapons. He had personally witnessed the effect of the bomb on Hiroshima. He was convinced that in the postwar struggle between the military services, the atomic bomb more than anything else would guarantee the Navy a prominent place in national defense plans. Nuclear propulsion as a long-range possibility should not be permitted to divert the Navy from its primary goal, the establishment of its capability to deliver nuclear weapons. This was an opinion which many line officers in the Navy shared.

Both as a high-ranking officer on Nimitz's staff and as an authority on atomic energy, Parsons could expect his views to dampen whatever enthusiasm might exist in the Navy for nuclear power. Certainly he had not helped the cause of those who favored priority development of a nuclear submarine. Rickover especially disagreed with Parsons, but as a relatively unimportant engineering officer on detached service at Oak Ridge, he did not have much voice in the Navy. The best he could do was to put his opinions in a memorandum to Mills.[44]

If private industry had no economic motivation for developing nuclear power and if the Commission, for a time at least, would have to concentrate on weapons, the Navy would have to provide the drive and inspiration for nuclear propulsion. How soon the Navy would have such a power plant would depend almost entirely, in Rickover's opinion, on how much effort the Navy invested. With existing support, it might take eight or ten years; with greater investment in engineering (as opposed to scientific research) it might take only three to five years.

Whatever the priority, Rickover thought the engineering would be difficult. He had not changed his estimate of the most important targets for engineering studies; they were still shielding, materials for construction, reactor controls, coolants, and heat-exchanger equipment. Solving these problems would require a large number of engineers trained in nuclear technology. Although there had been some progress since the war in training nuclear engineers, no more than seventy-five were yet available, and 20 percent of these were products of the Navy program at Oak Ridge. Engineering resources were still so small that Rickover believed it essential to keep his group together when the Oak Ridge assignment ended in September.

Rickover had little opportunity to follow up his memorandum to Mills. He and his associates were about to begin a tour of the Commission's major installations. It was a trip he had been planning since January as the last and perhaps most important part of the year's training. From the middle of July until late in August 1947, he would be largely out of touch with Mills and the Bureau of Ships in Washington. Although there was some interest in a nuclear submarine in Washington, it did not approach the intensity which Rickover now felt. Mills saw nuclear power as something the Navy had to pursue, but, like Parsons, he was not ready for full-scale development. On his staff in the Bureau of Ships Mills now had five Navy captains[45] who were serving as consultants on atomic power. In approving a timetable, he was more likely to rely on them than on Rickover, who he believed had a tendency to demand the highest priorities for any project he led.

From Rickover's point of view the situation was discouraging because no one in Washington seemed to understand the real obstacles and opportunities in developing power reactors. Neither Mills nor his consultants had ever studied the details of reactor technology. The new Commission had on its Washington staff only one man who had any experience with reactors, and he was not an engineer but a physicist strongly oriented toward research. The Commission was still struggling during the summer of 1947 to find itself and, until the staff had formulated some general plans, was reluctant to commit itself to any project. In other words, there were people within the Navy and in the Commission's laboratories who saw the potential of nuclear power for the Navy, but few if any of them were willing to support the kind of effort Rickover was proposing.

Mills's sincere but cautious interest in nuclear power was soon to pervade the military establishment. In July 1947 he arranged to discuss his plans for a nuclear project with the atomic energy committee of the Joint Research and Development Board. This complicated title accurately reflected the complex organization which had evolved from Vannevar Bush's efforts to coordinate postwar research in the military services. Intended to be a temporary organization until Congress established the National Science Foundation, the board had no authority over the internal affairs of the War or Navy departments, but it was intended to assist in allocating responsibilities on matters of joint interest. Atomic energy was clearly one of these, and Bush had recognized the importance of the atomic energy committee by appointing his old friend and colleague James B. Conant as chairman and Robert Oppenheimer as a member. Because they were also members of the Commission's General Advisory Committee, their opinions were likely to have overwhelming weight in determining the future of nuclear power in the Navy. At the moment both of them were concerned about the unwarranted optimism within the public at large and even among some nuclear scientists over the prospects for nuclear power. They were more than wary of ambitious but premature proposals.

The question under discussion on July 25, 1947, was the future of General Electric's efforts to develop nuclear power for the Navy. Early in June, a few weeks before the company's original study contract expired, LaPierre and his associates in the general engineering and consulting laboratory had submitted a report of their findings and recommendations for the future. Following the suggestions of Rickover's group, the company laboratory had concluded that the best approach would be to develop a nuclear power plant for a destroyer escort as a first step toward submarine propulsion (although for security reasons the ultimate application was not specified). For the future the company recommended two projects: one to develop a reactor for the destroyer escort; the other to explore all aspects of a heat-transfer system using liquid metal.[46]

This division of the work into two projects was an attempt to adjust to the realities of the situation. Splitting the work would permit General Electric to proceed with at least the nonnuclear portion under a Navy contract. The Navy and General Electric would then have to convince the Commission to finance only the research on the reactor itself. The distinction also had the advantage of assuring full Navy backing for the heat-transfer project, because Captain Mumma and the Bureau of Ships were willing to support it. By the time of the meeting on July 25, the Navy had already sent General Electric a letter of intent providing more than $2 million to continue the heat-transfer work over a period of two years. The Navy called it "Project Genie."

On July 25 Admiral Mills urged that the committee endorse both the Genie contract and the proposed study of the reactor, to be financed by the Commission; but he did not invoke Rickover's strong arguments for the project. The Commission, which was genuinely interested in the speedy development of nuclear power, had reservations about giving the effort a military cast by supporting a joint effort with the Navy. Even more important to the Commission was avoiding any action which might further divert General Electric from the critical task of rebuilding the plutonium production facilities at Hanford.

Under the circumstances, the Navy was fortunate to get as much as it did. The Commission was willing to support the General Electric reactor study up to $30,000 in the current year, on the condition that the number of personnel involved would be cut in half. This limitation would mean keeping only two engineers on the project, but that would be better than none at all. Conant's committee favored continuing the General Electric study, which involved no major experiments, until engineering progress and economics warranted construction of an experimental reactor. The Bureau of Ships could continue its contract with General Electric for heat-transfer systems, provided the work did not interfere with the Commission's own research and development plans.[47]

It would have been too much to say that Mills and the Navy opposed the nuclear submarine, but they were not yet willing to give it the highest priority for development. Mills himself had apparently not decided how to proceed, and he was almost certain that he did not want Rickover to head the Navy's nuclear project. In August he discussed the subject with Walter J. Williams, now a civilian and the Commission's director of field operations at Oak Ridge. Williams admitted Rickover was not an easy person to work with, but he thought Mills should keep Rickover and his group on the nuclear submarine project.[48]

A Call for Action

By this time Rickover and his group were nearing the end of their tour of the Commission's installations. At each site, beginning with the Ames Laboratory at Iowa State College in mid-July, they had sought out every scientist and engineer who had any knowledge or opinions on power reactor technology. They wanted to hear the arguments for and against developing power reactors, how such an effort might be organized, and whether to stress study projects or actual construction of a reactor. They explored the details of materials specifications and discussed the type of reactor to be built first.

The replies were as varied as the backgrounds of those interviewed, but three of the interviews seemed to make a special impression. Walter H. Zinn was director of the Commission's Argonne National Laboratory near Chicago and perhaps the nation's foremost authority on reactors. Zinn told Rickover and his assistants that he favored a reactor using slow neutrons with water or helium as the heat-transfer medium. The question of shielding could be studied independently, but the choice of a heat-transfer medium would be an essential decision in designing the reactor. Zinn favored building a land-based prototype of the reactor just as soon as the chances for success were reasonably good.

At the University of California in Berkeley the Rickover group found a truly enthusiastic supporter in Ernest O. Lawrence, the director of the Radiation Laboratory, who for more than a decade had impressed scientists with his energy and imagination. Lawrence warned the naval officers that to be successful in building a submarine reactor the Navy would have to want it badly enough to spend "real cash." The $2.5 million which the Bureau of Ships was spending on heat-transfer studies was just a beginning. Lawrence thought the Navy should be willing to spend $100 million on the project. With that kind of effort, he guessed, the Navy could have the reactor in three years. Lawrence stressed the practical and psychological importance of a large project. To be credible, the project would have to be big, and if it were big it would attract good people. A big project would also make it possible for the Navy to get one of the large industrial companies as a contractor. Lawrence agreed that building a reactor would be more useful than study projects, and he urged that the Navy aim first for a land-based prototype.

Edward Teller, who was spending the summer at Los Alamos, proved even more stimulating. Like Lawrence, Teller tended to be enthusiastic about new ideas and was willing to evaluate them intuitively, at least in informal discussion. Teller told Rickover that a power reactor could be built soon, within two years if someone put the effort on it. He urged that the first reactor be simple in design to exclude extraneous matters. It would not be economical, but he thought the Navy needed such a reactor and that building it would be a big step toward nuclear power. Teller agreed that the project would involve more engineering than science, but he feared the education of most engineers was not adaptable to new methods and ideas. Scientists, on the other hand, were apt to wander from the main goal. On the whole, Teller was optimistic. He believed most people still had open minds on the subject, and he knew that Lawrence R. Hafstad, executive secretary of the Joint Research and Development Board, favored the idea of building a reactor at once. The Rickover group found Teller's ideas exhilarating, and the feeling was mutual. A few days later Teller wrote Hafstad that he was very much impressed with the Rickover group, and he thought the Navy should not lose them.[49]

Rickover himself summed up the trip in a long memorandum to Admiral Mills on August 20, 1947. Teller's glowing remarks notwithstanding, Rickover wrote: "It is significant that during our entire tour, of the many scientists contacted, not one was found who had a definite interest in and was working on the problem of furthering nuclear power." Only the Navy and the Air Force had the incentive for developing power reactors, and the problems facing the aircraft reactor seemed overwhelming, at least for the time being. Most of those interviewed agreed that the quickest way of getting nuclear power would be to build a reactor, not to study the problem at leisure on paper. Rickover urged that the Navy assign more young men to nuclear power projects at the Commission's laboratories, that the most promising basic reactor designs be selected for detailed study and experimental work, and that his own group be established in the Bureau of Ships to direct the Navy's project.[50]

Receiving no answer from Mills, Rickover wrote a second letter a week later. If Mills's silence implied his disapproval of the first suggestion, Rickover was prepared with a second which he considered less desirable but still workable. He suggested that members of his group be assigned part-time within the bureau and part-time with the Commission. Eventually Mills acted on this suggestion, but not exactly in the manner Rickover had hoped. Roddis and Dick were assigned in the bureau but in different offices and not as a group. Libbey went to the staff of the Military Liaison Committee between the Commission and the military establishment, and Dunford joined the Commission's division of military application. Rickover's own fate was the last to be decided. At one point the group heard that orders had been cut sending Rickover to Oak Ridge as a classification officer, but the orders were never received. In time Rickover found himself doing staff work on nuclear propulsion as an assistant to Mills. It was now clear that Mills had decided not to establish a nuclear submarine project under Rickover's direction, and with that decision Mills set aside any plan for priority development of nuclear propulsion.

The Challenge

It was now almost two years since Admiral Nimitz as the prospective Chief of Naval Operations had begun to think about the needs of the postwar Navy. In the fall of 1945 the idea of nuclear propulsion was little more than a subject for sensational newspaper articles, but during succeeding months the idea had taken on substance. Early in 1946 the possibility of nuclear propulsion had come to the attention of the General Board, and in January 1947 Nimitz himself had approved a recommendation supporting development of a nuclear submarine. The threat of new antisubmarine warfare methods had provided the first note of challenge.

Two years of planning and discussion had proved, however, that no idea, no matter how sound or obvious, would be realized if the need did not outweigh the obstacles to attaining the goal. The competition for scarce resources, the vast requirements for maintaining the balanced fleet, and the uncertainty engendered by readjustments and reorganization in the postwar years had all but stifled the idea that had seemed so promising in the bright light of victory in 1945. No one in a responsible position in the Navy really opposed the idea of nuclear propulsion, but few officers except those in Rickover's Oak Ridge group yet saw it as something on which the immediate future of the Navy depended.

In a larger sense the issue was not whether nuclear propulsion should be developed on a high priority but, rather, whether the potential impact of nuclear power on the Navy warranted more than routine development. Only the future could answer that question.

The Question of Leadership

During the last six months of 1945, Admiral Bowen and even some officers in the Bureau of Ships entertained the idea of an independent approach to nuclear power by the Navy.[1] Nothing would have seemed more natural to the Navy in 1945 than the creation of an organization parallel to the Army's Manhattan project, but the Atomic Energy Act of 1946 had obliterated such hopes. The act, popularly regarded as a victory for "civilian control" of atomic energy, created an independent agency with broad and sweeping authority. One of the lessons Mills and his staff had learned during the first few months of the Commission's existence was that the Navy could neither bypass nor ignore the Commission in its efforts to develop nuclear propulsion for the fleet. By the summer of 1947 the Navy had accepted the fact that it would have to live with the new Commission.

Even if Mills and his officers had been enthusiastic about the prospects for a cooperative venture with the Commission, the creation of a joint enterprise would still have been a long and painful process. Six months after it had taken control of the nation's atomic energy activities in January 1947, the Commission was still scarcely organized. A bitter controversy over the confirmation of President Truman's appointments to the new Commission had disheartened its leadership. The flood of perplexing policy questions, ranging from the international control of atomic energy to the support of basic research by the government, almost overwhelmed the small staff of the new agency. The Commission itself represented a new departure in government organization, and some of the proposed innovations both in organization and management philosophy had been difficult to put into operation.[2] The Navy had waited a year for the new legislation and now six more months for the Commission to find itself. In September 1947 the Navy was still looking for a way to establish a working relationship with the Commission to develop nuclear propulsion.

As the following pages will show, the Navy's efforts to launch a partnership resulted in a proposal for a dual organization, one which would represent both the Navy and the Commission in developing nuclear propulsion. Other concerns prevented the Commission from taking the first step in that direction until the summer of 1948. During these same months the Navy had begun to grapple with the main issues which hampered the formation of a joint enterprise. The first was whether the dual organization was to be an assemblage of engineers like Mills's group in the Bureau of Ships or a staff of scientists much like those who dominated the Commission's research and development activities. The second question was whether the first Navy reactor was to be largely the creation of the Commission's own scientific laboratories or the product of one or more industrial contractors. Not until these issues had been resolved late in 1948 could the dual organization be fairly established.

The Navy's Partner

As Mills and his officers came to know the Commissioners and their staff better, they saw little reason to be optimistic about the future. The background and experience of the new agency's leaders did not promise any easy relationship. David E. Lilienthal, the Commission's chairman, was an energetic lawyer and courageous public servant who had built a national reputation as chairman of the Tennessee Valley Authority. Lilienthal had intellectual capacity, imagination, and a sense of purpose, but he had little understanding of nuclear technology and was much closer to being a philosopher than an engineer. Of the other four Commissioners only one, Robert F. Bacher, had any technical knowledge of atomic energy, and as a physicist Bacher was more interested in basic scientific research than in reactor engineering.[3]

Because the Commissioners themselves had so little background either in the technical or administrative aspects of the atomic energy project, they relied on the General Advisory Committee established by the Atomic Energy Act. The committee of nine members was composed almost exclusively of physicists and chemists, including two Nobel laureates, and was dominated by two of the most influential scientists in the government at that time, Oppenheimer and Conant. Although the General Advisory Committee could not be said to harbor any hostility toward the idea of nuclear propulsion for the Navy, it did not consider the Navy's interest one of the really vital concerns in the nation's atomic energy program in 1947. And the committee as a whole certainly could not view the idea of a nuclear ship through the eyes of a practical engineer like Mills or Rickover.

If the Commission and the General Advisory Committee lacked engineering experience, so did the Commission's staff. Carroll L. Wilson, the general manager, was an engineer, but most of his experience had been as administrative assistant to Vannevar Bush at the Massachusetts Institute of Technology and at the Office of Scientific Research and Development during the war. Wilson had distinguished himself on the State Department staff which had produced the Acheson-Lilienthal report and had helped to organize the

initial Commission staff in the fall of 1946. However, his excellent performance in these assignments did not disguise his lack of practical experience
either as an administrator or as an engineer.[4]

For the time being, at least, the Commission's reactor development efforts
were to be directed by the division of research under the leadership of
James B. Fisk, a close friend of Wilson's and one of the most promising
young physicists in the country. Independent in his thinking, Fisk was not
moved by emotional appeals for nuclear submarines or anything else. He
was determined to see that the Commission adopt a research program that
was responsive to Commission policy rather than to outside pressures.[5] As
a result neither the Commissioners nor the staff did much to push reactor
development in 1947.

One possibly mitigating factor was that the Commission had not intended
to build strong centralized controls at headquarters but rather expected to
look to its laboratories to devise their own reactor plans. But the laboratories
were no better off than the Commission's staff in Washington. The Knolls
Atomic Power Laboratory at Schenectady was still housed in an old factory
building. At the Clinton National Laboratory in Oak Ridge, which had at
that time the largest concentration of physicists and engineers interested in
atomic energy, a series of difficulties had shattered morale. The bright hopes
for the Daniels reactor in the spring of 1946 had faded under a cloud of
technical obstacles. The Monsanto Chemical Company, which had taken
over operation of the laboratory after World War II, had decided to give up
the contract, and the Commission was having trouble finding a new contractor. Uncertainty about the future of the laboratory sapped the energy of
the Clinton staff. The laboratory desperately needed strong technical direction and firm administrative support from the Commission if the nucleus of
talented scientists and engineers was to accomplish anything.[6]

Almost as crucial as Clinton in the Commission's reactor development
plans was the Argonne National Laboratory near Chicago. Under the direction of Walter H. Zinn, one of Enrico Fermi's principal assistants in developing the world's first reactor, Argonne had succeeded the renowned Metallurgical Laboratory which had been established at the University of Chicago
during World War II. Like Clinton, Argonne could still boast of a roster of
outstanding nuclear scientists, but even more than Clinton it had the academic atmosphere of a university laboratory. Although Zinn himself was
more hardheaded and practical than most scientists, he was still a physicist
first, a man more interested at that time in reactor experiments than in nu-

clear power plants. Zinn had ambitious plans for building a breeder reactor, but the unit was to be clearly experimental with only a symbolic capacity for generating nuclear power. It was not likely that Argonne on its own initiative would undertake anything so practical as developing a submarine reactor.

Reconstruction

Rickover was determined not to accept this attitude toward nuclear power in September 1947. He was convinced that nuclear power would revolutionize naval warfare, and he was certain that the United States had the capacity to build a nuclear submarine within a few years. Yet without any organization of his own Rickover was severely handicapped in pursuing his goal. His assignment to Mills's staff helped him keep in touch with one of the few officers in the Navy who had the power to establish a nuclear project, but by staying in Washington with no one to represent him in the Commission's laboratories, Rickover would have soon lost touch with the vital technology on which his hopes rested. He continued to visit the laboratories at every opportunity. There he found some signs of life in the Commission's reactor development planning.

In October the Commission assembled a group of reactor physicists from several laboratories at Clinton to discuss the future. After each of the leaders had described reactor plans at his laboratory, Rickover took the floor. Some of those present, including Oppenheimer, scarcely knew who he was, but Rickover did not hesitate to speak his mind. He charged that the Commission was making little progress because too many physicists were involved in decisions. He wanted to see more engineers and fewer committees working on reactors. Showing his growing impatience with the Commission, Rickover asked Oppenheimer if he had waited until he had all the facts before he built the atomic bomb. Perhaps to Rickover's surprise, Oppenheimer replied that he had indeed had the facts, but he admitted that it would probably not be possible to reach that point before building anything as complicated as a power reactor.[7]

By constant badgering Rickover made certain that naval reactors were a topic on the agenda for all such reactor planning meetings. Because reactors were only of secondary concern to Fisk, it was hard to concentrate high-level attention on the subject. At the suggestion of the General Advisory Committee, the Commission in November approved the formation of a reactor development group, which consisted of reactor experts from the several laboratories. The purpose, which both Fisk and the advisory committee accepted, was to bring more engineering than scientific talent into the work.[8]

Rickover alerted Mills and obtained a spot for him on the agenda for the group's first meeting. Mills explained to the group the significance of the "true" submarine, which only nuclear power could provide. Mills admitted that nuclear power hardly seemed practical when fissionable material was extremely scarce, but he was convinced that development of the submarine reactor would solve 90 percent of the design problems facing other power reactors. Mills hastened to point out, however, that the Navy reactor was a specialized application; it would require close cooperation between the Navy and the Commission's laboratories. Mills reiterated Rickover's conviction that the submarine reactor was technically feasible and that its availability depended almost entirely on the effort expended.

Mills's presentation impressed the reactor group. The members were beginning to understand the Navy's interest in nuclear propulsion. Furthermore, they thought the Commission had at Oak Ridge a team of physicists and engineers who could start work on the Navy's request. Without making any formal recommendations to the Commission, the group concluded that the power reactor division at Clinton could probably begin such a study soon.[9]

Clinton's qualification for the naval propulsion study rested to a large extent on Rickover's tireless efforts at Oak Ridge. Early in the fall of 1947, after the tour of the Commission's facilities, Rickover had discussed the future of the Daniels reactor with some of the men in the power reactor division. Although Daniels and his associates still had hopes that the Commission would support the project, Rickover predicted that the accumulation of technical difficulties would doom the reactor. Why, Rickover asked, should the Clinton engineers continue to work on a project without a future? Would it not make sense to devote their efforts to a reactor that might be useful for naval propulsion? In 1946 Alvin M. Weinberg, the young leader of the Clinton physicists, had suggested the possibility of using pressurized water as both the moderator and heat-transfer medium in a power reactor.[10] Rickover, after his intensive study of many reactor designs, now believed Weinberg's suggestion offered real promise for the Navy project. The implication of Rickover's remarks was that at least some of the Clinton group might begin informally to shift their attention from the Daniels reactor to the pressurized-water design.

Harold Etherington, the leader of the Clinton reactor division, liked Rickover's suggestion. An experienced engineer from Allis-Chalmers, Etherington had come to Clinton with the intention of learning the elements of nuclear technology; and, like many of the industrial engineers at Oak Ridge, he was seeking the shortest route to a practical power reactor. If the Daniels reactor did not offer such a path, he was willing to consider another. During the fall of 1947 Etherington's group began quietly to study the pressurized-water reactor. By the time the reactor development group met in November, Etherington was in a position to begin formal studies.

In Washington Rickover had no success in reassembling his Oak Ridge group, but occasionally he was able to borrow the services of Dunford, Roddis, or Dick for technical meetings at Oak Ridge or Argonne. Roddis, who was following General Electric's work on the liquid-metal system for Captain Mumma, had an office just a corridor from Rickover's in the Main Navy Building. With Mumma's knowledge Roddis kept Rickover up to date on developments at Schenectady, and Rickover could conveniently discuss with Roddis his plans for keeping alive his hopes for the submarine. Dick also worked in the bureau just a few offices down the hall, and he had frequent opportunities to consult with Rickover. Dunford kept him abreast of the Commission's activities, and Libbey followed matters of interest in the armed forces from his post on the staff of the Military Liaison Committee.

For a holding action the dispersed group functioned reasonably well, but Rickover hoped sooner or later to acquire some official status. The best possible endorsement would be one from the Secretary of the Navy and the Chief of Naval Operations. Under the circumstances Rickover could hardly expect the Bureau of Ships to take the initiative, but at least he could count on support from Mills and some of the officers in the bureau. Early in October Rickover and Dick carefully drafted an exchange of letters between Admiral Nimitz, still the Chief of Naval Operations, and Secretary John L. Sullivan.

Obtaining the large number of endorsements required for correspondence at that level in the Navy was a task involving weeks of patient negotiation and painstaking revision. Rickover himself was a master of this technique, but he received help from Roddis and from two officers who were strategically placed in the office of the Chief of Naval Operations. Captain Elton W. Grenfell and Commander Edward L. Beach were both veteran submariners who had won the Navy Cross for their exploits during World War II. From their experience they could draw persuasive arguments for the extraordinary advantages of a nuclear-powered submarine. They could also guess that

Admiral Nimitz, himself a submariner, would share their views. By the end of November Rickover and his associates had the concurrences they needed before presenting the memorandums to Admiral Nimitz.[11]

The first document, which Nimitz sent to Secretary Sullivan on December 5, pointed out the Navy's established need for a ship with unlimited submerged endurance at high speed. Only nuclear power could meet that requirement. With sufficient effort, an atomic submarine could be completed by the middle 1950s. By that time, the memorandum predicted, it would be possible for a submarine to launch a guided missile carrying a nuclear warhead with a range of about 500 miles. In signing the memorandum Nimitz urged the secretary to bring "the great strategic and tactical importance of a nuclear powered submarine" to the attention of the Secretary of Defense and the Research and Development Board. Rickover also had on hand appropriate memorandums for Sullivan's signature to Defense Secretary James V. Forrestal, to Vannevar Bush, chairman of the research board, and to Mills. Much to the elation of the Rickover group, Sullivan signed the memorandums promptly. Among other things they requested the Bureau of Ships and the Commission to work out a mutually acceptable procedure for designing, developing, and constructing the submarine.

Hope and Despair at Clinton

To assure a positive response from the Commission, Rickover began stirring up interest at Clinton. On December 8, 1947, he used a meeting of industrial representatives studying nuclear technology at Oak Ridge to stress the importance of training men from industry. He urged Clinton Laboratories to establish a training program which would permit private companies to send engineers to the laboratory, where they would gain experience by working on actual design problems. Until a new operating contractor had replaced Monsanto, the laboratory could make no commitment on Rickover's proposal, but the idea seemed sound if industry was to have a real part in reactor development.

During the following week Rickover, Roddis, and Dick spent several days with Etherington's power reactor division. Etherington had already completed a very preliminary design study of a pressurized-water reactor. The next step would be to fix some of the basic specifications of the steam propulsion equipment and to begin some study of water corrosion of metals. Rickover agreed to take these matters back to the Bureau of Ships. Meanwhile,

Etherington would increase his efforts on the pressurized-water design. He already had all of his design group, or about one-third of the division, working on the project.[12]

The most interesting information Rickover picked up during his stay at Oak Ridge was new data on the rare element zirconium. One of the problems in building a pressurized-water reactor was to find a corrosion-resistant material with a low affinity for neutrons which could be used to support the reactor core and to clad the uranium fuel elements against corrosion by hot water. Relatively common materials like aluminum and stainless steel, and even beryllium, had disadvantages. Samuel Untermyer, an engineer at Clinton, had suggested using zirconium because it appeared to have a high resistance to corrosion, good mechanical strength, and good metallurgical characteristics. One apparent disadvantage—a high affinity for neutrons—now seemed spurious. Herbert Pomerance, a physicist at Clinton, had just completed experiments which indicated that impurities, principally hafnium, accounted for most of the neutron capture in earlier tests. If the hafnium could be extracted, zirconium might be an excellent core material.

Although Pomerance's work suggested a new possibility for the pressurized-water reactor, the potential difficulties were impressive. Because hafnium was chemically very similar to zirconium, it was hard to separate the two elements. Even if an economical separation process could be developed, there was always the possibility that removal of the hafnium might rob the zirconium of its desirable qualities as a metal. Another obvious difficulty was that zirconium was at that time available only in laboratory quantities at astronomical prices. Despite these obstacles, Rickover intuitively found zirconium attractive. As a starting point for design, he was willing to commit himself to using zirconium in a water-cooled reactor.

Although the work at Clinton was generally encouraging, it rested more on the Commission's sufferance than on positive support. Unless the Commission was enthusiastically behind the Navy project, the chances for success were poor. Rickover planned to take at least one small step toward formalizing the existing situation through the reactor development group, which held its second meeting later that week. With Dunford attending as a member of the group, Etherington presented some of the design considerations he had discussed with Rickover earlier in the week. The group again concurred in what Etherington's division was doing at Clinton, but again decided to make no formal recommendation to the Commission.[13]

Under the circumstances Rickover would have to pin his hopes on the

Clinton study, at least until he could organize some direct approach to the Commissioners in Washington. Suddenly a last-minute decision by the Commission threw even that modest effort in doubt. For weeks Monsanto had been planning to turn over operation of the Clinton National Laboratories to the University of Chicago on January 1, 1948. Then, a few days before the transfer was to take place, the Commission decided to replace Monsanto with the Carbide and Carbon Chemicals Corporation, which operated the production plants at Oak Ridge. As part of the decision the Commission also ordered that all work on reactor development should be concentrated at the Argonne laboratory. The sudden decision threw the scientists and engineers at Clinton into a turmoil of despair and confusion.[14]

The Navy Proposal

If anything the Commission's abrupt decision to centralize reactor development at Argonne strengthened Rickover's hand in the Bureau of Ships. In the summer of 1947 Mills's greatest concern had been that the Navy might move too quickly and thus imprudently if he gave Rickover free rein. Now the danger seemed to be just the opposite. Unless the Navy pressed the idea vigorously the Commission might delay any action on the submarine reactor indefinitely. Rickover's insistence on an aggressive effort no longer seemed unreasonable. Mills also would have had to admit that Rickover had made the best of a bad situation by stimulating interest in the Navy project at Clinton and by initiating the memorandums which Secretary Sullivan and Admiral Nimitz had signed early in December. The directives not only authorized, but called upon, the bureau to devise a workable agreement with the Commission. Early in January, Bush informed the Commission that the Research and Development Board endorsed the recommendation in the directives.[15]

By this time Rickover and Roddis were drafting a proposal which Mills could send to the Commission. The two officers started with the premise that "the problems to be solved are so intimately connected with both the Atomic Energy Commission and the Navy that neither activity can make separate engineering decisions regarding them." The aim should be to build an experimental submarine nuclear power plant. A single working level organization acting for both the Commission and the bureau would direct the project. For this purpose the Navy proposed that the Commission establish the bureau as its agent for the project and that the bureau unit directing the project would have dual status as both a Commission and Navy organization.

The proposal also included specific suggestions for research and development. Both General Electric and the Clinton Laboratories—now called the Oak Ridge National Laboratory—would undertake full-scale studies of the feasibility of liquid-metal–cooled and water-cooled reactors. A separate project would make a thorough study of the very complicated subject of shielding. Other specialized groups would begin research on structural materials, fuel assemblies, and heat-transfer systems. A permanent on-the-job training program would be established for Commission, Navy, and technical personnel. All these ideas had been key points in Rickover's recommendation to Mills in June 1947. Mills signed the proposal and sent it to the Commission on January 20, 1948.[16]

The point of contention, as Mills and Rickover expected, would be the proposed organization, not the technical proposals. The idea of a dual project certainly seemed worth fighting for. The Atomic Energy Act clearly forbade an independent Navy project, and the Commission's record during its first year of activity offered no reason to believe that the new agency could build a submarine reactor without the Navy's help. Neither did a joint effort seem promising unless there were a single organization clearly responsible for the work. A dual organization, responsible to both the Commission and the Navy, appeared to be the only solution. Roddis had pointed out that a dual organization had worked amazingly well in conducting the nuclear weapon tests at Bikini in 1946 and in preparing for the forthcoming *Sandstone* tests. Rickover, who was already adept in using correspondence between administrative units to advance the Navy's causes, saw obvious advantages in the dual organization: if either the Navy or the Commission failed to give him the support he needed, he would then be able to bring pressure through the other agency.

The Commission Demurs

Even as late as January 1948 the principal architects of atomic energy policy were the three wartime leaders Bush, Conant, and Oppenheimer rather than the Commissioners. After a year in office Lilienthal and his associates still found it difficult to master the intricacies of the technical enterprise they had inherited from the Army, and they continued to rely on the three men who between them dominated such important policy bodies as the Commission's General Advisory Committee and the Research and Devolpment Board's committee on atomic energy. Under the circumstances it was not surprising that the first response to Mills's letter came from these groups.

Conant's committee on atomic energy was the first to consider Mills's letter. Mills himself was present on February 5 to defend his proposal, and
Admiral Solberg, a member of the committee, could be relied upon to support the Navy's interests. After listening to Mills and Solberg the committee
concluded that a nuclear power plant for a submarine was feasible, but on
a "moderately long range on a time basis." Apparently the two admirals did
not yet fully share Rickover's sense of urgency about the project, or at least
not enough of it to prevail against the persuasive arguments of Conant and
of Oppenheimer, who was also a member of the committee.[17]

For months the two scientists had been deeply involved in the Commission's efforts to organize its reactor development program. For them, this
broader concern had to take precedence over the Navy project. Until the
Commission had come to some firm decisions, Conant and Oppenheimer
were skeptical of the kind of dual organization which Mills and Rickover
had proposed. For one thing, an aggressive Navy effort, particularly under
Rickover's direction, might quickly expand to occupy the vacuum left by
the Commission's lack of planning. For another, Rickover was intending to
build part of his effort around Etherington's group at Oak Ridge, and how
could that idea be reconciled with the Commission's recent decision to move
all reactor development to the Argonne laboratory?

The committee recommended a form of organization which would keep
the Navy project firmly under the Commission's control. The Navy would
assign a group of officers to work with the Commission's staff, the location
and activities of the group to be determined by the Commission's decisions
about reactor development. From reports submitted by the Navy group, the
Bureau of Ships would decide when to begin engineering studies for the submarine reactor. In the meantime Navy personnel and engineers from industry could be trained in the Commission's laboratories.

Oppenheimer had a draft of this report when he and Conant met with
their fellow members of the General Advisory Committee on February 6,
1948. The outcome was predictable. The committee concluded that the submarine reactor was feasible, that the interest and enthusiasm of Etherington's
group should be preserved by making it a new division at Argonne, and that
the Navy group should be assigned to the Commission rather than be established under dual control by the Commission and the Navy. A week later,
with both letters in hand, the Commission decided not to respond at once to
Admiral Mills's proposal but instead to wait until the staff had discussed the
committee's organizational recommendations with the Navy.[18]

Mills and his staff could not have looked forward to these discussions with optimism. Legally they could do nothing to start development of the submarine reactor without the Commission's approval and cooperation. Through Commander Dunford, who was still a member of the Commission's staff, they probably knew of the cautious position the Commission and the General Advisory Committee had taken on the Navy proposal. Zinn and Fisk would represent the Commission in discussions with the Navy. Both men were physicists rather than engineers; both had strong convictions about the course the Commission should follow in research and development; and both were unlikely to accept a division of authority in their areas of responsibility. From Rickover's perspective the prospects seemed even more dismal because Mills and Solberg were still firmly in control of negotiations on the Navy side. Rickover feared that the two admirals would be inclined to compromise if Zinn and Fisk held their ground, and in Rickover's opinion the Navy had no room for compromise.

The meeting with Zinn and Fisk occurred in Mills's office on March 4, 1948. Following the Commission's New Year's Eve decision to centralize all reactor development at Argonne, Zinn was prepared to take on preliminary studies of a power reactor that would be useful to the Navy. Zinn's plan was eventually to establish three separate groups at Argonne, one for each of the three types of heat-transfer systems then considered practicable: pressurized water, gas, and liquid metal. Research on each of these types would permit Argonne to select the most feasible design for further study. Argonne was also prepared to accept technical personnel from the Bureau of Ships and American industry to participate in these studies. In this way the Navy would be assured trained personnel when it came time to start engineering on the submarine reactor. Zinn was saying that Argonne and not the Navy would control the work at the laboratory.[19]

Fisk also made clear that the Navy's proposal for a dual organization was out of the question. Confronted with this hard fact, Mills presented a compromise plan which Solberg had drafted.[20] Abandoning the position he had taken in his letter on January 20, Mills accepted an arrangement close to that which Conant and Oppenheimer had recommended. The Commission would give the nuclear submarine the status of a formal project, and Zinn agreed to accept full responsibility for research on a design at Argonne. The Navy would participate in work at Argonne and would be responsible for engineering development necessary for actual construction of the submarine.

To Mills and Solberg the arrangement seemed perhaps a workable compromise, but from his intimate knowledge of the Commission's laboratories and personnel, Rickover found the proposal anything but acceptable. First, it placed the future of the submarine project in the hands of Argonne, not the Navy. Argonne would determine the direction of research, allocate resources, and decide when it was time to select a design for engineering development. The Navy personnel at Argonne would participate only as trainees and observers. Second, the arrangement placed the project in the control of a scientific laboratory which was oriented toward academic research. Rickover was still convinced that the Navy needed an organization experienced in practical industrial engineering, not an academic faculty of scientists. Third, the undefined period of studies before engineering would begin suggested more of the indecision which had frustrated the Navy's hopes since 1945. For these reasons Rickover was determined to keep the project out of Argonne or, if necessary, accept an Argonne study only as a temporary expedient until the Navy could bring in an industrial corporation like General Electric or Westinghouse.

Because Admiral Mills himself had negotiated the agreement with the Commission's staff, Rickover could not oppose it directly. Instead he chose what for him was an uncharacteristic strategy: inaction. He elected not to follow up the agreement to work out the details of the arrangement with Zinn.[21] As the weeks rolled by with no word from the Commission, Mills's impatience grew. With each passing day it seemed ever more likely that Rickover's contention had been correct: nothing would happen on the submarine project as long as the Commission rather than the Navy called the tune. Once Mills began to think this way, it was not hard for Rickover to stimulate his impatience.

All Rickover needed now was an occasion for Mills to express his discontent. That opportunity arrived with an invitation for the Bureau of Ships to provide speakers for the Undersea Warfare Symposium held annually in Washington by a group of scientists, engineers, and Navy personnel. Under the pressure of time Captain Mumma had asked Roddis to write three speeches—one for Mills, one for Mumma, and one for Rickover. When Rickover read the drafts he realized that it would be much more effective to have Mills, with all his prestige as chief of the bureau, deliver a hard-hitting speech staking the Navy's claim for the nuclear submarine. Mills, now eager for a chance to express his frustration, agreed and asked Rickover to draft the speech. Rickover did not bother with any elaborate statement or even with anything very original. All he did was set down the facts as he understood them, and most of those came directly from summaries he and Roddis had prepared for other purposes several weeks earlier.[22]

Rickover had correctly assessed Mills's frame of mind. When the Admiral stepped before the group of 700 people at the symposium on April 2, he was exercised enough to speak almost extemporaneously, departing frequently from his prepared text. With Commissioner Lewis L. Strauss serving as chairman and with most of the other Atomic Energy Commissioners in the audience, Mills reviewed the Navy's interest in nuclear propulsion from the tentative beginnings in 1939 and from his own introduction to the subject of atomic energy as a member of the Tolman committee in 1944. The study projects at Schenectady and Oak Ridge were useful explorations, but they represented nothing practical. Mills ventured the opinion that less than 1 percent of the work necessary to design a nuclear submarine had yet been accomplished. Furthermore, the Commission had never recognized the submarine reactor as an official project or given it any priority. Neither had the bureau and the Commission settled the organizational question. Mills reminded Strauss and his colleagues that completion of the submarine reactor would depend entirely on the initiative and energy expended. He urged the Commission to establish the submarine reactor as a formal project with a high priority.

Mills had succeeded in dramatizing the Navy's impatience with the Commission. Even Strauss, who was a master of suavity and aplomb, found it difficult to conceal his surprise at Mills's outspoken remarks. Regaining the rostrum, Strauss passed off the speech with a facetious remark: "I never thought an old friend would do that to me."

Despite the drama of the occasion, the speech could hardly have produced in the Commission the fundamental change in attitude which Mills was seeking. If during the previous months Strauss and his associates had merely neglected the Navy's request, the speech might have provoked action, but the failure to act stemmed from solid reservations. Fisk and Zinn were trying to establish a workable program for reactor development at Argonne. They were hoping to replace the haphazard pattern of individual projects at several laboratories with a single, balanced effort which would make possible an orderly study of the three basic conceptions of a power reactor: the pressurized-water system called the high-flux reactor using slow neutrons at Argonne, the sodium-cooled power-breeder reactor using neutrons of an intermediate energy at Knolls, and the sodium-cooled breeder reactor using fast neutrons at Argonne (see chart 1). Until Fisk had an opportunity to test

Chart 1. The chart shows the relationship of the three early reactor concepts to the later submarine propulsion plants.

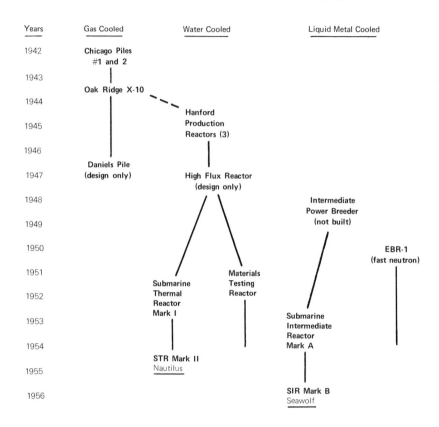

EVOLUTION OF SOME EARLY REACTOR CONCEPTS

the new organization, the Commission was not likely to make an open-ended commitment to the Navy.

In a letter to Mills on April 27, 1948, the Commission promised that the submarine reactor would be given the status of a formal project and would "be prosecuted with a high priority commensurate with the importance of this project."[23] The ambivalence of that statement suggested what little effect Mills's speech had had on the Commission. Its greatest impact was on Mills himself and on some of his fellow officers in the Bureau of Ships. They were now determined to bring the Commission to terms on nuclear propulsion for the Navy.

Industrial Participation

With this new-found conviction the Bureau of Ships could be expected to follow familiar paths. Decades of experience in building fighting ships had convinced the Navy that it had to rely on American industry for the engineering talent and industrial knowledge required to build modern warships. In the course of two world wars the Navy had built close relationships with shipbuilders and manufacturers of propulsion systems and electrical equipment. These experienced companies, spurred by a system of competition which the Navy carefully fostered, had proved their effectiveness, and the Bureau of Ships was prepared to call on them again.

Just as it was natural for the Navy to rely on experienced contractors, it was easy to understand why Admiral Mills and the Bureau of Ships were not prepared to entrust development of the nuclear submarine to the scientists at Argonne National Laboratory. True, Zinn and his colleagues knew as much about nuclear reactors as any group in the world, but they had no experience in designing and building power plants for naval ships. Argonne could help the Navy by training engineers in nuclear technology and by providing the general design for the submarine reactor; but for actual engineering design and construction the Navy would rely only on established industrial contractors, preferably at least two companies working in parallel in order to provide the incentive of competition and to assure an alternate approach should one fail.

Soon after Mills's speech had galvanized opinion in the Bureau of Ships in favor of immediate construction of a nuclear submarine, the Navy carried its demands for industrial participation and the parallel approach to the Commission. The first contact was through the Military Liaison Committee.

Rickover asked Libbey to draft a letter to the Commissioners. The letter, delivered on May 5, 1948, stated the conviction that "the most rapid progress, at this time, can be made by utilizing the parallel efforts of industrial organizations, as well as of laboratories, simultaneously to the fullest extent."[24]

Just how this might be done was the question which Solberg, Rickover, and Mumma raised with Zinn and his staff at Argonne that same week. Zinn began by announcing his plans to transfer Etherington's group from Oak Ridge to Argonne as part of the centralization of reactor development: Zinn said he was aware of the priority for the submarine, and he intended to do everything possible to develop a reactor design quickly. The naval officers were more interested in industrial participation and the implications of centralization. Did it mean that General Electric could not proceed with its studies of a sodium-cooled, power-breeder reactor? Rather than cutting back the work at General Electric or transferring it to Argonne, the Navy favored expanding the project to include a complete power plant design. Zinn said he had no objection to the idea, but he thought only the Commission could make that decision.[25] As director of the Commission's reactor development laboratory, he did not intend to try to direct reactor work at other Commission installations. The reply may have seemed equivocal to the naval officers, but Zinn understood the limits of his authority.

Just as important to the Navy was the place of Westinghouse in Zinn's plans. Solberg explained that the Bureau of Ships was about to sign a contract with Westinghouse for Project Wizard, a study of a heat-transfer system based on pressurized water, just as Project Genie at General Electric was concentrating on a sodium system. Zinn agreed that Wizard was appropriate for a Navy contract, providing the Navy understood that Argonne had complete responsibility for the reactor portion of the plant. He was also favorably inclined toward a proposed contract between Westinghouse and Argonne under which the company would furnish technical personnel and services to the laboratory for the submarine project. Zinn insisted on a sharp division of responsibility: Argonne would study the reactor; Westinghouse would develop the heat-transfer system.

The Navy's third concern was getting some work started on a propulsion system using a gas for heat transfer. Zinn's idea was that Argonne would study this system just as it was investigating water and sodium systems, but the Navy again was worried about practical engineering aspects. To wait until Argonne completed its study might preclude any chance of building the reactor within the time scale proposed by the Navy. As a compromise, Zinn agreed to let the Navy grant contracts for studies of blowers, valves, and heat exchangers based on the work which had been completed on the Daniels gas-cooled reactor.

The whole tenor of the Navy's position was that the project should be in the hands of experienced engineering contractors under the direct control of the Navy and the Commission. The same principle applied to the General Electric project at Schenectady. The role of General Electric in the submarine project now became the sticking point between the two agencies.

The Fight for Parallel Projects

Even before the meeting with Zinn, Mills and his associates in the Bureau of Ships had concluded that something more than the Argonne project would be necessary to guarantee the Navy a nuclear submarine by the middle of the 1950s. The Manhattan project had demonstrated the wisdom of parallel approaches in developing technology under the pressure of time. Mills, Solberg, and particularly Rickover were convinced that the rather limited design studies already undertaken in the Commission's laboratories had not demonstrated, and were not likely to demonstrate in the future, the clear superiority of any one of the three appoaches. They maintained that a demonstration of engineering feasibility, as opposed to theoretical possibility, depended upon actual construction and operation of a reactor.

All Zinn's assurances did nothing but increase the Navy's misgivings. Zinn interpreted the Commission's mandate for centralization to mean that Argonne would control the design studies on all three approaches and that no work beyond design studies would be started until Argonne had selected the most promising approach. To the Navy this was a hopeless procedure, and Zinn's seemingly ambivalent reaction to the General Electric project was even more alarming. Not only did the Navy consider a parallel approach essential to success, but Mills and his associates also believed General Electric was the most experienced and best qualified company for the job. If Zinn was not willing to back the General Electric proposal, Mills knew it would be difficult to convince the Commission to dilute the company's Hanford responsibilities with a big Navy assignment. The Argonne meeting made it seem all the more important to insist on a parallel approach.

After discussing the Genie project with General Electric officials, Mills and Rickover were convinced that the company could undertake the development of a complete submarine propulsion system, including a sodium-cooled reactor as well as the power machinery. Mills wrote the Commission on May 12 that such an effort would carry out the principle of parallel projects which the Military Liaison Committee had advocated in the previous week. To his letter Mills attached a summary of Solberg's meeting with Zinn, a document which set forth the Navy's interest in the parallel approach.[26]

Carroll L. Wilson, the Commission's general manager, had scarcely read Mills's letter before he felt the effects of Mills's discussions with General Electric. Harry A. Winne and the scientists at Knolls were intrigued, if somewhat confused, by the Navy proposal. Building a submarine reactor had long been of major interest to the company, but the Commission had never given the idea priority. Winne wondered whether Mills's visit meant that the Commission had changed its plans for reactor development. If it had, Winne thought General Electric should abandon the sodium-cooled power-breeder reactor and concentrate on the Navy project.

Wilson found the suggestion so disturbing that he called a meeting with Winne and his staff in Washington the following day. Wilson made clear that the Commission had no intention of changing its priorities in reactor development. The Navy had been acting on its own initiative in approaching General Electric. Wilson reiterated the point he had made many times—that General Electric's first responsibility was to assist the Hanford production plant and secondly to design the power-breeder reactor. Wilson agreed to take up the matter with the Commission but in the meantime he asked Winne to consider what impact the submarine project would have on the company's ability to meet its existing commitments.[27]

Troubled by the sudden shift in General Electric's interest, Wilson and Fisk asked the General Advisory Committee for its opinion. The committee's first reaction was to approve the idea if General Electric considered it desirable, but further discussion raised perplexing questions. Was it realistic to build a submarine reactor when the subject of nuclear power was virtually unexplored? Why was General Electric so quick to abandon its two-year investment in the power-breeder, which the committee had given a high priority? Conant feared the proposal was the result of military pressure, and from experience he questioned using military interest as justification for a development project when its practicality was not clear. Other members of the committee, namely Cyril S. Smith and Glenn T. Seaborg, looked upon the submarine project as a way of bringing nuclear power development into focus. But finally, bowing to Conant's and Oppenheimer's views, the committee approved a statement expressing its failure to understand either General Electric's desire to abandon the breeder reactor or "the military or practical urgency at the present time of reactors for submarine propulsion."[28]

Winne was playing his cards carefully to assure that, whatever happened, General Electric would have a place in developing the first nuclear power plant, whether it was designed to generate electricity or drive a submarine. The company was willing to place the priority on either reactor, but until the government settled that question Winne intended to keep all options open. In a letter to Wilson on June 3 he admitted that General Electric did not have the manpower for simultaneous development of both the submarine reactor and the power-breeder. Most of the company's research on the breeder would be useful in later design of the submarine reactor, and the company believed that the breeder would be a more flexible and therefore a more valuable facility. Winne also concluded that switching from the power-breeder to the submarine reactor would undermine the morale of the Knolls scientists, who saw the breeder project as a way of demonstrating the peaceful application of atomic energy. Another point which Winne admitted but did not advance formally was the company's fear that the submarine reactor might be too novel to build at the nearby West Milton site selected for the power-breeder, a difficulty which might leave the Knolls laboratory without an experimental reactor.[29]

When he received a copy of Winne's letter on June 11, Mills could see how far he was from his goal. The Commission obviously did not grasp the urgency of the Navy's requirement. The production of materials for weapons and the creation of a balanced research program clearly took precedence over nuclear submarines. Such weighty considerations were more than enough to quench the interest which Mills had kindled in Winne and the Knolls scientists. If Mills wanted a nuclear submarine project at General Electric, he would have to win over the Commission as well as the company.

Meeting the Soviet Threat

In some respects the Commission could appreciate the Navy's growing interest in a nuclear submarine as part of the nation's response to the rising Soviet threat in Europe as the shape of the Cold War became more apparent late in 1947. A modern, effective Navy was surely consistent with the president's support of unprecedented economic aid for western Europe and plans for a 70-group Air Force. Furthermore, the Commission had every reason fully to accept the reality of the Soviet threat. The hardening of the Soviet position in the United Nations on the international control of atomic energy, the fall of Czechoslovakian democracy to communist dictatorship, and the alarming reports from General Lucius D. Clay in Berlin had proved but the first steps on a dangerous course which the Soviet Union seemed determined to pursue. In late March 1948 the Russians had begun to cut Berlin's land links with the West, and the threat of war reached crisis proportions. So critical was the outlook that the Commission had ordered a check of procedures for the emergency transfer of atomic weapons to the Air Force and considered postponing the long-planned nuclear weapon tests at Eniwetok in April.[30]

It was one thing, however, for the Commission to acknowledge the Navy's concern and something else again to see it as more than a diffuse response to a complex set of events. The Commission obviously had not been privy to the many discussions and reports which since the beginning of 1948 had been pointing up the importance of improved submarines. Because the various bureaus and commands were continually proposing new ideas, it would be difficult to fix the origin of the growing concern about submarines. Certainly an effective catalyst had been the comprehensive study which Captain Arleigh A. Burke had undertaken for the General Board early in 1948.[31]

The purpose of Burke's study was to investigate the probable nature of warfare during the next decade and to determine the most effective contributions the Navy could make to the national defense. In beginning the study Burke's group assumed that the Cold War would continue and intensify and that the United States and the Soviet Union would be the chief protagonists in any future conflict. The committee drew up a comprehensive agenda covering not only the military aspects of any future war but also the political and economic factors involved. The General Board then sent its agenda and preliminary findings to the principal bureaus and commands for comment.

As the Burke committee expected, the comments spanned every activity and interest of the Navy, but the growing importance of submarines was a topic frequently mentioned. Much of this interest stemmed from the increasing danger of war with the Soviet Union. As a report from Nimitz's office to the Joint Chiefs of Staff stated in April 1948: "The seriousness of the Russian submarine menace is emphasized by the fact that they now have over five times the number of undersea craft that Germany had at the outbreak of World War II."[32]

Other reports asserted that the Russians had commandeered as many as twenty of the German Type XXI submarines and a large number of the technicians who had built them. The Soviet Union was deemed capable of producing Type XXI submarines in large numbers. Admiral Raymond A. Spruance, who had succeeded Nimitz as commander in chief of the Pacific Fleet, summarized the significance of these facts in a speech on February 11: "The new submarine with high submerged speed and great underwater endurance is probably the greatest threat that exists today to safe use of the sea. Until a solution is reached to the problem of how to destroy this submarine, and until the forces are made available for this work, we shall be in a poor position to operate our armed forces overseas against an enemy who has a large fleet of them and knows how to use them efficiently."[33] The Navy's Operational Development Force had reported to the General Board that "the tactical characteristics of the medium speed, deep diving snorkel equipped submarine have virtually nullified the effectiveness of most of our World War II ASW procedures, tactics, and doctrines."[34]

An Appeal to the Commission

With this background Mills was now determined to demand some direct and convincing action from the Commission. Although the Navy had had frequent informal contacts with the Commission at several levels, Mills had seldom been accorded an opportunity to meet with the Commissioners as a body. Fortunately for Mills, the chances for such a meeting had never been better than they were in the spring of 1948. In April Donald F. Carpenter, an experienced industrial executive, had taken the chairmanship of the Military Liaison Committee. As a civilian and as a former member of the Commission's industrial advisory committee, Carpenter was fully acceptable to the Commission. He had gained the confidence of Mills on the one hand and of Wilson and the Commission's staff on the other. When Mills decided he wanted a meeting with the Commission, Carpenter had no trouble arranging it.[35]

Mills played all his cards in presenting the Navy's arguments for the nuclear submarine to the Commission on June 16. Admiral Charles B. Momsen, a veteran submariner and Assistant Chief of Naval Operations for Undersea Warfare, explained how closed-cycle propulsion systems had greatly complicated the problem of detecting enemy submarines. Captain Grenfell reviewed the tactical advantages of submarines in antisubmarine warfare, an assignment which only a "true" submarine propelled by nuclear power could fulfill. The extraordinary and largely successful effort of the Soviet Union to build high-speed diesel submarines made the development of nuclear submarine propulsion by the United States all the more important.[36]

Getting down to the specifics, Mills made clear that he did not fully accept the disclaimer in Winne's letter of June 3. He thought General Electric could handle both the submarine and the breeder, particularly if the company sought outside help. He knew that the General Electric staff was interested in the submarine project. He also pointed out that the similarities between the submarine project and the breeder reactor would enable the company to develop both with little added effort.

The participation of General Electric provided the context for the discussion, but as Mills continued, he revealed the Navy's fundamental concern about bringing the "practical" approach of the engineer to bear on the project. Under questioning he was not willing to criticize the scientists working on the naval reactor at Argonne and Oak Ridge, but his remarks conveyed a sense of uneasiness. He seemed to be questioning whether the Commission's laboratories could concentrate their attention on the Navy project when interesting possibilities appeared in other reactor studies. The Navy, as both Mills and Rickover insisted, was aiming at a land-based prototype of a reactor that could power a submarine. They were not, like the Commission, concerned with broad advances in nuclear science and technology.

The Commissioners seemed to appreciate Mill's effort and acknowledged that they now had a better understanding of the Navy's interest. They intended to give the matter further study and would give Mills and Winne their answer in a few weeks. Mills and Rickover had no reason to suppose that the meeting had hurt their cause.

A New Organization

At the meeting with Zinn and Fisk early in March 1948 Mills and Solberg had abandoned their efforts to establish an independent organization in which both the Navy and the Commission would share authority. Backed by the General Advisory Committee and the Research and Development Board, the Commission had insisted upon undivided responsibility for any project involving nuclear power. The Navy was free to participate in research on power reactor designs at the Commission's laboratories and had been invited to establish some form of liaison between the Commission staff and the Bureau of Ships. Rickover had chosen not to pursue this offer as long as the Navy had hopes of creating at Knolls a submarine project independent of the Commission's other laboratories. For Mills the main reason for delay may have been his inability to decide on an officer to head the project.

If the job were to be mostly one of liaison, it would make sense to appoint someone who had gained experience in dealing with the Commission at a high level and who could command the confidence of those officials. In both respects, Admiral Solberg seemed well qualified. He had served on the Military Liaison Committee since 1946. His long association with the atomic energy project, going back before the Tolman Committee, had made his views helpful to both the Commission and the Navy. He also had the ability to be forceful without being offensive. When Carpenter had suggested early in May that the Navy appoint a liaison officer to work with the Commission's staff on the submarine reactor, Lilienthal had accepted the idea on the assumption that Solberg would get the assignment. In fact, Lilienthal went so far as to indicate that the liaison officer would in effect be accepted as a member of the Commission's staff.[37]

These expectations collapsed about the time of Mills's meeting with the Commission in June, when the Navy appointed Solberg director of the Office of Naval Research. The new assignment required Solberg to resign from the Military Liaison Committee and to sever all his ties with the nuclear project. Then, as the weeks slipped by with no response from the Commission on the General Electric proposal, the need for the liaison capabilities which Solberg would have provided seemed to decline. By the middle of July one of Fisk's assistants had started drafting a paper which would explain to the Commission why Mills's proposal should not be accepted. Perhaps Mills learned from Dunford what was happening in the Commission; perhaps the delay was indication enough. In any case, Mills decided on July 16 to give Rickover the assignment.[38]

There were good reasons for appointing Rickover. For more than a year he had sparked the Navy's effort to get work started on the nuclear submarine. Subsequent events had borne out Rickover's contention that the Navy would have to make an extraordinary effort to reach that goal. The Commission's failure to respond favorably to Mills's January proposal and its hesitation over accepting the parallel project at General Electric had convinced the admiral that the task needed the kind of hard-headed, even ruthless, direction which he knew Rickover would give it. But the decision was not an easy one for Mills. Some of the qualities which Rickover would bring to the job troubled Mills and many of his fellow officers in the bureau. Rickover flouted Navy tradition and ridiculed a system that seemed to him to give more weight to an officer's social accomplishments and willingness to conform than to his practical ability and industry. Mills could guess that once he gave Rickover a free hand, he would out-work, out-maneuver, and out-fight the Commission, its laboratories, and the Navy. He would threaten, cajole, and even insult those who stood in his way. In the process he would no doubt embarrass Mills and the Navy, but Mills was ready to do what the situation demanded. He wrote Lilienthal that Rickover would be his liaison with the Commission's headquarters.

The Rickover appointment provoked a long-overdue reorganization of the nuclear power project in the Bureau of Ships. Technically the bureau was still operating under Mills's directive of January 2, 1947, which created an organization for nuclear matters under Captains Morgan and Mumma. This group had avoided any aggressive actions on nuclear power, with the result that Mills had come to rely more and more on Rickover in his struggle with the Commission. The new organization, which Mills announced on August 4, 1948, recognized the realities of the situation. The new order established a nuclear power branch, as Code 390, within the bureau's research division.[39] This was an ideal arrangement for Rickover because the director of the division was an easy-going officer who would not try to supervise Rickover's activities. Within the nuclear power branch, Rickover began to assemble the officers of his original Oak Ridge group.

Just how the Rickover group would fit into the Commission's organization was not yet clear. Technically, reactors were still Fisk's responsibility in the division of research, but the Commission's staff was already in the throes of a reorganization which would create a separate division of reactor development. Since the spring of 1948, Wilson and his staff had been seeking a new organizational structure which would meet the criticisms of Oppenheimer and the General Advisory Committee and of Carpenter and the Military Liaison Committee. Until the new organization became effective, Rickover's best contacts were in the division of military application, where Dunford was working and whose director, General James McCormack, was more sympathetic to the Navy's needs than was Fisk. It would be another six months before the Rickover group was officially established in the new division of reactor development.[40]

The Navy Offensive

During the last weeks of July the Commission moved with measured deliberation toward a decision on the General Electric proposal. In a careful analysis for the general manager, Fisk argued that there was no reason for changing the company's assignment. Reactors using neutrons in the intermediate energy range were worthy of study, and the scientists at Knolls were the Commission's only source of such information. Only General Electric's responsibilities at Hanford took priority over the intermediate power-breeder. In Fisk's opinion, General Electric would do well to approach the naval reactor through its work on the power breeder, which would provide greater "flexibility" than the submarine prototype. This statement, in the Navy's opinion, was another way of saying that the breeder would be a more useful tool for general research than a reactor designed as a submarine prototype. For the moment at least, research, not engineering, was the focus of the Commission's concern.[41]

In accepting Fisks's recommendations on July 23, 1948, Wilson foreclosed the immediate possibility that General Electric might undertake development and construction of a prototype submarine reactor. This alone would be a severe blow to the Navy, but Wilson went even further. In his instructions to Winne he urged "that as great a portion as possible of the effort on the 'Navy reactor' at Schenectady be redirected towards the early completion of the intermediate-energy power-breeder reactor."[42] The action firmly rejected the Navy's proposal to work directly with an industrial contractor and to pursue the work as an engineering development rather than as a scientific experiment. Now Rickover would have his chance.

Just as Mills expected, Rickover seized the initiative from the day of his designation as head of the nuclear power branch. He did not wait for the formalities of organization before drafting for Mills's signature a letter to Lilienthal denouncing the Commission's decision. The letter, which Mills signed on August 2, 1948, saw "no reasonable hope" that the Commission's methods would produce a nuclear submarine in the minimum time warranted by defense requirements. Mills concluded that if the Navy was to have the propulsion plant in a reasonable time, the bureau would have to establish another project in addition to that at Argonne. If the Commission refused to act, the Navy would go it alone by negotiating contracts directly with industry.[43]

To show the Commission that the Navy and not the Bureau of Ships alone was speaking, Rickover drafted a note reporting the action to the Secretary of the Navy, and he also prepared a letter which Secretary Sullivan could use in forwarding Mills's memorandum to Secretary of Defense Forrestal. Rickover's success in getting these documents signed within forty-eight hours indicated the solidarity of the Navy's objection. As official communications, they

Chart 2. The Nuclear Power Branch was established in the Bureau of Ships, but the counterpart organization in the Commission had not yet been formed.

THE NAVY NUCLEAR PROPULSION PROJECT IN AUGUST 1948

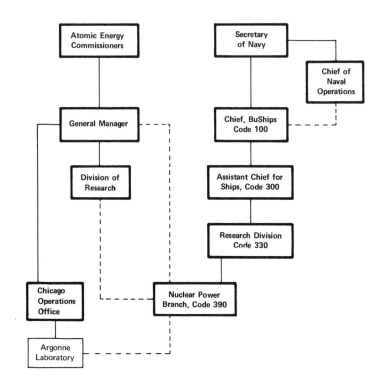

could not begin to convey the anger and disappointment which the Commission's action had engendered in the Bureau of Ships.[44]

The letter to Secretary Forrestal automatically brought Carpenter back into the dispute. In addition to being chairman of the Military Liaison Committee, Carpenter was Forrestal's special assistant for atomic energy affairs. Carpenter saw his job as primarily that of a conciliator, one who could heal the dangerous suspicions and animosities that had plagued the Commission's relations with the military services since late 1946. Here was another opportunity for Carpenter to work out a practical compromise between the Navy and the Commission. He could begin with his experiences as a member of the Commission's industrial advisory committee. He had been among the first to see the need for reorganizing the Commission's staff in order to speed action in the general manager's office. He had also advocated a stronger role for industrial engineering in the Commission's activities.

Isaac Harter, chairman of the board of the Babcock & Wilcox Company and one of Carpenter's colleagues on the industrial committee, spoke to the latter point in a meeting with three of the commissioners on August 3, 1948. Harter complained that the Commission's reactor development planning was in the hands of physicists rather than engineers. Physicists, Harter said, were needed to draw valid inferences from fundamental theory, but they were "not apt to be in full possession of the subject matter of engineering or sufficiently sensitive to the time scale of this workaday world and other material limitations which education and especially experience have taught first class engineers." He could understand why physicists had been in control of reactor development in the early years. The Navy project, however, indicated that it was time for a change. Under the Commission's system, the physicists at Argonne would not only select a contractor for construction but would also direct the work of the contractor until the reactor was completed. This procedure, in Harter's opinion, would not make the best use of either the physicist's or the engineer's talents. He thought the two functions should be separated, with Argonne concentrating on physics and an industrial contractor simultaneously on engineering.[45]

Carpenter took a similar position when he discussed the problem with Zinn the following week. Zinn told Carpenter and other members of the Military Liaison Committee that he did not like the idea of turning over construction of the Navy reactor to industry. He thought both Argonne and Knolls should do much more research on water-cooled and sodium-cooled reactors before the engineers took over. Zinn feared that at this early stage industrial companies would send only mediocre engineers to the laboratories. Carpenter agreed this might happen if the purpose were only to train technicians, but he thought industry would do its best if there were real prospects for production contracts. Carpenter thought the Navy and the Commission should jointly negotiate industrial contracts for constructing a propulsion reactor and that the Commission's new director of reactor development should administer the contract. Zinn seemed amenable to the idea and confessed that he had not understood the great urgency which Carpenter and his committee obviously attached to the Navy project.[46]

Two days later Carpenter presented his compromise to Mills and Rickover in Washington. Mills, obviously still angry over the Commission's action, placed all the blame on the Commissioners. But after much heated discussion Carpenter succeeded in convincing Mills and Rickover that they should meet with high-ranking Commission officials and members of the Military Liaison Committee. For its part, the Navy would delay any direct negotiations with contractors. The Commission would be asked to join the Navy in selecting a contractor to begin work immediately, with the understanding that the company chosen would eventually receive the entire contract for building the reactor. The Navy would take the position that General Electric was the best company for the job but that both General Electric and Westinghouse should be considered for the assignment. The Navy had no interest in interfering with the Commission's reorganization and would accept administration of the contract by the new director of reactor development. Finally, the Navy would provide liaison personnel and cooperate fully with the Commission.[47]

Because Wilson was out of town, the meeting had to be postponed for several days. Rickover used this time to prepare his case for the parallel approach. First he wanted to nail down General Electric's position on the submarine project. To protect himself against the Commission's argument that General Electric did not wish to take on the assignment, Rickover obtained from the Navy representative in Schenectady a written statement, later endorsed by Winne, that the company was "willing and anxious to design and build a reactor suitable for use in a naval vessel. This project would be accomplished with a distinct understanding that it would not significantly interfere with the progress of the intermediate pile."[48]

Rickover likewise made the best of Westinghouse's interest in the Navy reactor. For more than two years he had been cultivating this interest, and he knew that the imminence of a Navy project at Schenectady would be a powerful inducement for Westinghouse. On August 25, the day of the meeting with Wilson and the Commission staff, Rickover called George H. Bucher, the former Westinghouse president and now chairman of the company's planning and development committee. Bucher assured Rickover that the company was ready to accept his earlier suggestion that Westinghouse establish a new department in the company to handle the Navy project. Bucher said the company was also prepared to send six of its best engineers to General Electric for a one-year training course in nuclear engineering.[49] Rickover thought these commitments would be impressive. Armed with the statements from the two companies and a plan for setting up the parallel approach to the submarine reactor, Mills and Rickover set off for the meeting with Wilson.

The meeting was long and arduous, the kind which Rickover, with all his impatience for action, found difficult to endure. Both sides considered it necessary to restate the arguments they had expressed many times before. Most of the discussion centered on General Electric's ability to undertake the Navy project. Trying to avoid a deadlock, Carpenter turned the discussion to the Argonne-Westinghouse alternative.[50] Rickover and Carpenter reiterated the arguments which had softened Zinn's opposition to bringing an industrial contractor into the early phases of the project. Wilson and Fisk seemed to have less trouble with this idea than with the proposal for General Electric's participation, and Rickover assured the Commission officials that the Navy would give full cooperation to a joint Argonne-Westinghouse project.

Now Carpenter saw the basis for an agreement. Before Rickover could introduce his own proposal which might have reopened all the issues, Carpenter proposed that the Commission and Navy officials sit down together to discuss their differences with Winne and his staff. The Navy and Commission officials would explore with Zinn and the new director of reactor development how greater participation by Westinghouse could be assured. One of the first tasks of the new director would be to take control of the Argonne project and to begin discussions with Westinghouse. If there were any delay in selecting the new director, the Commission would assign this responsibility to Carleton Shugg, manager of the Commission's Hanford office, who was coming to Washington as Wilson's deputy. A Naval Academy graduate, Shugg had built an excellent reputation as an effective manager of large construction projects both in the wartime shipbuilding industry and in directing the reconstruction and expansion of production plants at Hanford. Carpenter ended by repeating the Navy's assurances that it would support the Commission's reorganization plan, the selection of the director of reactor development, and the new Argonne-Westinghouse project. Within a week both sides had accepted Carpenter's proposal.[51]

Relations with the Contractors

Four months of struggle had at last given the Navy what it wanted: a chance to bring two industrial contractors into the submarine project and permission to approach General Electric for the initial assignment. In the week following the meeting with Carpenter and Wilson, Rickover went to Schenectady to find out what the company's intentions really were. As on previous occasions, he discovered a strong interest in the submarine project, but Winne and his associates introduced a new idea which gave Rickover reason to hesitate.[52]

Winne declared that General Electric wanted to build a submarine reactor, but one using neutrons of intermediate rather than thermal energy. All of the company's experience had been on the intermediate reactor, and it would take a year to develop a comparable competence on thermal reactors. Much more important to Rickover was the fact that the intermediate reactor would use far more fissionable material than a thermal plant. At a time when uranium 235 was still extremely scarce, it did not seem reasonable to build one or two intermediate reactors when the same amount of material might power as many as six thermal reactors.

Rickover grew more apprehensive over the company's attachment to the intermediate design. He concluded that General Electric's principal interest was the chance to gain Navy support for the intermediate power-breeder reactor, which the company intended to build at West Milton, New York. A panel on long-range military objectives, which Carpenter had recently appointed, had learned from such experts as Enrico Fermi that breeding would have no practical applications for decades. Why would Winne, who was a member of the panel, continue to advocate the intermediate reactor, which had more advantages for breeding than for power generation? Perhaps, Rickover suggested to Mills, General Electric realized that its experimental reactor would not be a good breeder and wanted to recoup its investment by converting the project to naval propulsion. Rickover warned Mills that by supporting the General Electric proposal, the Navy might be assuming a partnership in a "white elephant." The Navy, Rickover advised, should insist on building a thermal reactor first, even if that meant sacrificing the advantages of a contract with General Electric.[53]

The Navy's sudden disaffection ended General Electric's hopes for a quick decision on the submarine project. It was only at the Navy's insistence that Wilson and Fisk had agreed to approach the company, and they were not likely to take the initiative in negotiations. When Mills and Rickover met with a large group of General Electric officials on September 24, only the Commission's local representative was present. Winne and his associates tried in vain to sell the idea of an intermediate reactor.[54] Sensing that an agreement with General Electric would take months of negotiation, Rickover turned his attention to Westinghouse.

Bucher and his associates at Westinghouse soon learned that they were expected to follow up quickly on the commitments they had made to Rickover for the meeting on August 25. When Rickover learned that the company could send to General Electric only two men of the five now available for training in nuclear technology, he reminded Bucher that he had used the company's promise to sell the Commissioners on the idea of a Westinghouse contract. Bucher agreed that the company could not back out now. Two weeks later Westinghouse followed through on its second commitment, to establish a separate division for its nuclear work. On October 5 the company released an internal memorandum establishing the atomic power division, which would be separate and independent from all other departments and divisions of the company. The new division would be headed by Charles H. Weaver, a young engineer who had known Rickover during the war when he was manager of the company's marine department.[55]

These decisions by Westinghouse gave Rickover a solid position for a meeting with Wilson and his staff on October 8, 1948. Mills wanted to discuss how the Commission intended to carry out the agreement of August 25, now that Shugg had reported as deputy general manager. For the moment the Navy was interested only in starting work on the water-cooled and gas-cooled reactors. Presumably action on the sodium reactor would have to wait for further negotiations with General Electric.

Wilson had already discussed with Zinn the delicate question of the division of responsibility between Argonne and Westinghouse. Zinn had suggested using the arrangement which du Pont and the Metallurgical Laboratory had followed during World War II in developing the Hanford production reactors. Using a parallel arrangement as Zinn understood it, Argonne would then be responsible for fundamental design, certain design criteria, and for approval of certain significant steps in the detailed design of the reactor; Westinghouse, as a Commission contractor, would be responsible for engineering design and construction. Wilson thought Zinn's idea might be a good starting point for a four-way discussion involving the representatives of the Commission, the Navy, Argonne, and Westinghouse.

Wilson hoped that Westinghouse would concentrate during the rest of 1948 on acquiring a basic understanding of nuclear technology. By the first of the year, Zinn expected to be ready to recommend the type of reactor to be developed. It seemed likely that pressurized water would be the choice, but Argonne wanted the three remaining months in 1948 to explore both water-cooled and gas-cooled designs. Depending on the outcome of these preliminary surveys, Wilson expected all three approaches to be in the hands of industrial contractors in 1949; Westinghouse on pressurized water, General Electric on sodium, and a third contractor, probably Allis-Chalmers, on the gas-cooled design.[56]

Mills and Rickover thought Wilson's proposal was acceptable as far as it went. Rickover doubted that Argonne would be able to develop very much solid information on both the water-cooled and gas-cooled designs by January 1949, but he liked the sense of purpose and urgency in Wilson's plan. Zinn's proposal was more difficult for the Navy to evaluate. Obviously Westinghouse would be responsible for engineering design and construction, but what exactly did Zinn mean when he proposed that Argonne control "fundamental design" and establish "certain criteria?" Was this idea an example of what Harter had called the inefficient procedure of physicists trying to do work that engineers could do better?

Both the Navy and Westinghouse officials had a better understanding of what Zinn meant after the four-way meeting at Argonne on October 26. Zinn explained that he did not intend to involve the laboratory in the purely engineering aspects, but he did remind the group that the Commission had given him responsibility for designing the propulsion plant. This meant to him that he would have to maintain full control over what he called "fundamental research," "basic research," and "development." Westinghouse would be responsible for "engineering" and "detailed engineering design." As for Argonne's control of "certain criteria," Zinn apparently meant that the laboratory, as the Commission's design contractor for the project, would review all engineering drawings and specifications prepared by Westinghouse. Zinn agreed that this division of responsibility would apply only to the first reactor, which presumably would be a land-based prototype. All later reactors would be entirely the company's responsibility.[57]

Rickover and Weaver could accept most of Zinn's proposal. Their main concern was that Argonne, lacking practical experience in industrial engineering, might incorporate in the basic design of the reactor certain features that would not meet required standards for reliability. In the abstract it was not hard to understand the difference between laboratory equipment and a reactor operating in a submarine. And yet it was extremely difficult for physicists and laboratory scientists to keep the fundamental distinction alive in their everyday work. Most of them had spent their lives designing equipment that would demonstrate a physical principle, a goal that put a premium on precision and ease of measurements, flexibility of controls, and economy. These were appropriate criteria for the laboratory, but they were less important in a shipboard propulsion plant than such matters as reliability during extended operation, simplicity of design, and accessibility for repair. It was true, as the Argonne scientists suggested, that the Westinghouse engineers could take these factors into account in the detailed engineering of components, but the engineers saw the whole system as a collection of components, all intricately related. How could the laboratory be sure that in establishing the general design criteria it was not imposing on the engineers a design not adaptable to practical engineering?

Under the circumstances existing in the autumn of 1948, Westinghouse and the Navy could express these considerations only as concerns, not as solid objections to the proposed arrangement. It probably would have been futile to attempt to define the division of responsibility more precisely on paper for a project as complex and unprecedented as the four parties were undertaking. As all the parties recognized, Westinghouse engineers would be working at Argonne and Argonne scientists would be helping in the facilities which Westinghouse expected to acquire. The limits of responsibilities would best emerge as the scientists and engineers tried to work together in a spirit of cooperation.[58] Even if a more precise definition had been possible, it probably would not have been to the Navy's advantage. Rickover realized that Zinn possessed the clear advantage of authority and experience. Argonne had designed at least six operating reactors; Westinghouse had built none. Zinn obviously enjoyed the full confidence of Wilson and the Commissioners. The best the Navy could do was to accept the arrangement and be prepared to warn the Commission if some of these misgivings threatened to become realities.

The letter contract which Westinghouse signed with the Commission on December 10, 1948, embodied the arrangement Zinn had proposed. The purpose was to build a propulsion plant for a submarine "within the shortest practicable time." Westinghouse would be required to "do all detail engineering, produce the working drawings, procure the necessary materials, and construct the Mark I plant," which would be a land-based prototype. Argonne would provide design and engineering data and, as the Commission's agent, would approve the working drawings prepared by Westinghouse. In all phases of the work Westinghouse would cooperate in every respect with Argonne in exchanging personnel and information. The contract also recognized that the aim was to design Mark I so that it would meet Navy specifications for a submarine propulsion plant. Therefore, concurrently with its work on Mark I, Westinghouse would undertake research and development for subsequent models of Mark I which could be installed in submarines. The Commission authorized Westinghouse to obtain suitable office space, laboratories, and shop facilities at government expense and specified interim financial arrangements until a definitive contract could be signed.[59]

A Place in the Commission

Rickover now had an industrial contractor and a working relationship with the Commission. All that remained in building an organization was to find a place for his group within the Commission staff and the Bureau of Ships.

Admiral Mills's directive of August 4, 1948, creating Code 390 had defined Rickover's role in the Navy at least on paper. Physically the directive had not made much difference. Rickover still occupied the small office in the temporary structure on the fourth floor at the rear of the Main Navy building. He had gradually reassembled most of his original Oak Ridge group and added a few naval officers who had expressed an interest in the assignment. But the crowded temporary space in the Bureau of Ships did not help to create the image and atmosphere of a major Navy project.

The Commission as yet had done nothing to give the Rickover group a home in its own organization. In July Rickover had discussed with Wilson his plans for eventually establishing his group as a branch in the new division of reactor development.[60] The Commission would have limited office space for the new division in its headquarters on Constitution Avenue, but most of Rickover's group had offices in the Main Navy Building, just a block away.

Actual creation of the naval reactors branch had to await the selection of a director of reactor development and organization of the new division. When Wilson's first choice for the job in September declined the appointment, the Commission began a new and protracted search.[61] By December 1948 Wilson had exhausted most of his possibilities for the directorship. Having failed to convince anyone from industry to take the job, he was ready to broaden his criteria.

One person who now became a leading candidate was Lawrence R. Hafstad, who as a professor of physics at Johns Hopkins University had helped develop the proximity fuse during World War II. After the war he had served as executive secretary of the Research and Development Board. When Mills learned that Hafstad was looking for a new job, he suggested his name to Wilson. Hafstad understood technical development, was an expert in nuclear physics, and knew something about the Commission's activities. The fact that Hafstad was acceptable to Mills must have counted as a point in his favor. On January 16, 1949, Wilson announced the Hafstad appointment. Now, at last, the Commission would have a reactor division and a director. And Rickover's group would find a permanent home as a branch within the new division.[62]

A Year of Achievement

In little over one year the Navy under Rickover's prodding had created an organization which would make it possible to build a nuclear submarine. As a member of Mills's staff without operational responsibility or authority, Rickover had managed to stimulate an interest in the Navy project within the Commission's laboratories. At his urging, the Navy had established the nuclear submarine as a formal requirement, and Rickover had convinced Mills and others in the Bureau of Ships to give it a high priority. With Rickover's help Mills had forced the Commission to recognize the project. Mills was also successful in placing Rickover in charge of the effort in both the Navy and the Commission.

Mills, Rickover, and Carpenter had helped to focus the Commission's concern on the industrial and engineering aspects of the work as opposed to the academic and scientific. Rickover had succeeded in creating a working relationship among the Navy, the Commission, Argonne, and Westinghouse, and he had convinced General Electric that the company should have a part in developing the nuclear submarine.

Almost a decade had passed since the discovery of nuclear fission had sparked the Navy's dream of nuclear propulsion. It had taken the Navy that long to create the organization and find the leadership necessary to realize that dream. Now the task was Rickover's.

4

The Structure of Responsibility

The result of the Navy's efforts by the end of 1948 was an organization involving two federal agencies (the Navy Department and the Atomic Energy Commission), two relatively autonomous groups within those agencies (the Bureau of Ships and the Commission's division of reactor development), and three research organizations (Argonne National Laboratory, the Westinghouse Electric Corporation, and the General Electric Company). Had such a diversification of effort been proposed to Admiral Bowen in 1945, he would undoubtedly have rejected it as absurdly impracticable.

Surely none of the leaders in 1949—neither Mills, Rickover, Wilson, nor Hafstad—would have chosen the complex and ill-defined pattern of organization which federal statute and practicalities had dictated. But all those involved must have been convinced after two years of bargaining that no simpler pattern was possible. They would have to learn to work together if the United States was to have a nuclear submarine.

It was also clear in early 1949 that it would not be possible to set down in the terms of a contract or in an interagency agreement the exact delineation of responsibilities between the parties. No one could prescribe precisely what tasks would be necessary to develop the submarine reactor. None of the government officials in either agency had ever designed a nuclear reactor of any kind. The organizations directly responsible for the work in each agency were new and untried, created in large part for the very task the agencies were undertaking.

Defining the relationships between the organizations and fixing their responsibilities was more a task for administrators, engineers, and scientists than for lawyers. Furthermore, this structure of responsibility could not be built in advance. It would have to emerge from the frustrating process of trying to create a new technology by committee. The success of this joint effort would depend ultimately on all the participating organizations, but the structure of responsibility would be largely Rickover's work. He had created this strange alliance, and he alone could make it function.

The Government Base

Whatever Rickover was to accomplish, he would have to start from his position as a government official—as an officer in the United States Navy and as a branch chief in the Atomic Energy Commission. By the end of 1948 his position within the Bureau of Ships and the Navy was clearer than it had ever been since 1946. Code 390 was formally a section within the bureau,

3. Lawrence R. Hafstad, director of the Commission's division of reactor development from 1949 until 1955, observes shop procedures at Argonne National Laboratory on December 9, 1954.

3

and Rickover had both authority and responsibility. The uncertainties lay not in the formal definition of authority, but in the customary vicissitudes of the Navy. Rickover could now confidently expect Admiral Mills to support him as long as he stayed within the terms of the general agreement hammered out with the Commission in 1948. But Mills was in his third year as chief of the bureau and was weary of the burdens he had carried since the beginning of World War II. When Mills retired, another admiral would replace him. Rickover could hardly expect his new superior to be particularly sympathetic or as interested in the nuclear submarine as Mills had been. Certainly no other senior officer in the Navy knew as much about nuclear technology as Mills did. And Rickover had not forgotten the firm opposition he had encountered within the bureau in his efforts to create a nuclear power section. Officers like Morgan and Mumma, who were responsible for ship design in the bureau, still gave improved diesels and closed-cycle systems priority over nuclear propulsion for new ships in the submarine fleet. Even though Rickover's position in the bureau had been formally acknowledged by early 1949, he knew that he would have to fight to hold the advantage.

Rickover's position within the Commission was less secure. In his capacity as liaison officer with the Commission he had not had much opportunity to establish reliable contacts. He knew neither the Commissioners nor the General Manager personally. Rickover's drive to establish the submarine project had impressed Williams and Shugg, but his blunt and impatient efforts to force Commission action in 1948 had irritated some of the headquarters staff and laboratory personnel.

The new division of reactor development would provide a home for Rickover in the Commission (see chart 3), but he would have to establish a working relationship with Lawrence R. Hafstad, the new director. The two men had known each other during the years of Hafstad's service with the Joint Research and Development Board. An engineer with practical experience in both private industry and large government research and development projects, Hafstad could appreciate Rickover's technical and administrative ability. He had backed Admiral Mills's efforts to gain support for the Navy project as a practical first step toward developing nuclear power plants. But Hafstad was also a nuclear physicist with many connections with the scientific community. His appointment rested in part on his ability to balance competing demands and to work out compromises. Suspicious of all scientists, Rickover did not expect from Hafstad automatic support for his hard-driving and uncompromising approach.

Chart 3. The dual organization in its initial stage. The Pittsburgh Area Office and the Bettis Laboratory had just been established.

THE NAVY NUCLEAR PROPULSION PROJECT IN JANUARY 1949

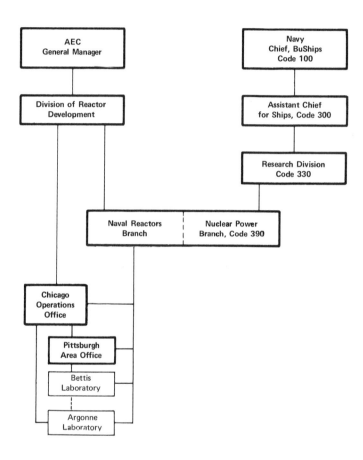

Rickover did not find a congenial home in the division of reactor development. The zeal with which Rickover pursued his goal did not make Hafstad's job any easier and posed a potential threat to the director's authority. Because reactor materials and skilled personnel were scarce, Hafstad faced a real danger that Rickover's demands would upset his effort to achieve a balance in distributing these limited resources among the reactor projects the Commission was already supporting. The rest of the division in early 1949 consisted mainly of George L. Weil's small staff, which had followed reactor development for Fisk in the division of research. Weil, a physicist and student of Enrico Fermi, thought the Commission had a responsibility to support studies of a variety of reactors. Until scientists in the Commission's laboratories had an opportunity to examine some of these reactor schemes, Weil questioned the wisdom of a heavy investment in a submarine reactor.

The Dual Organization

Under the circumstances, Rickover would have done well to survive in either the Navy or the Commission; fortunately the dual organization gave him maneuvering room that neither agency alone would have provided. As an experienced practitioner in the Navy bureaucracy he knew how to take advantage of a complex structure. In the first place, his two roles gave him immediate and direct access to both organizations. It was almost impossible for either agency to deprive Rickover of information he needed or to act without his knowledge. Second, his dual role helped to set him apart from others in each organization. Because he could represent the Navy in the Commission's offices and the civilian agency in the Bureau of Ships, he could sometimes avoid administrative red tape which would delay or blunt his actions. Third, the dual organization permitted Rickover to assemble a complement of personnel and resources which neither organization would have been willing to provide alone. Fourth, and perhaps most important, Rickover's dual role permitted him to take the initiative when neither agency was willing or able to act. He could write a letter for Hafstad's signature requesting Navy action and then draft an approval of the request for Mills's signature. This procedure did help sometimes to move issues off dead center, but it also had its difficulties. Obtaining the necessary signatures was never automatic and often required weeks of patient persuasion.

The Field Offices

As a member of the new reactor division, Rickover had direct and effective access to the laboratories and contractors which would build the nuclear submarine. Under the 1948 reorganization, the Commission's Chicago Operations Office reported directly to Hafstad and in turn held administrative controls over both Argonne and Westinghouse. Furthermore, the function of the Chicago office meshed nicely with Rickover's method of operation. The office had been established in 1947 primarily to handle contracts with a number of universities and research organizations stretching from Ohio to California. The most important of these from Rickover's perspective was the contract with the University of Chicago for operation of the Argonne National Laboratory.

So vast were the administrative responsibilities of the Chicago office that the field staff could not hope to take part in technical decisions involving the contractors. Such matters were reserved for direct discussion between the reactor division in Washington and the individual laboratories. The distinction between technical and administrative matters was not always sharp, but the system worked because it had been developed and followed by the one man who had served as manager of the Chicago office from the time of its creation. Alfonso Tammaro had been a civil engineer with the Army Corps of Engineers and the Manhattan District before joining the Commission's staff in the autumn of 1946 as a contracting officer. Because the Chicago office was to be mostly concerned with contract administration, Carroll Wilson, the general manager, had sent Tammaro to Chicago as acting manager in the summer of 1947. Since at that time he had no background in nuclear technology, Tammaro rejected the idea that his office should second-guess Washington or Argonne on technical matters. Rather, as he viewed it, his job was to see that Washington decisions were carried out as effectively as possible. Rickover was pleased to let Tammaro handle routine administrative matters as long as he followed instructions on policy and procedure.[1]

To help manage his vast network of contracts, Tammaro had established a number of field offices which placed Commission officials at the sites (usually within the offices) of the principal contractors. Immediately after signing the letter contract with Westinghouse in December 1948 Tammaro had created the Pittsburgh Area Office and appointed Lawton D. Geiger as area manager. Geiger, like Tammaro, had been with the Manhattan District during the war. An engineer with construction experience, he had been serving as area

manager at the Ames Laboratory at Iowa State College. His background
would be valuable as Westinghouse set about building a new laboratory for
the submarine project. Quiet and methodical, Geiger was a conscientious and
effective administrator, particularly on procurement and contracts. His loyalty
to Rickover and the project was unwavering, and Rickover quickly came to
rely on Geiger for difficult assignments.

Such was the government organization of which Rickover was a part. It
was complicated, full of nuances and pitfalls, and never fully predictable.
Rickover would always complain about it, but he had studied it carefully and
would use it to his best advantage.

Argonne

From the time of his first visit to the Chicago laboratory in the summer of
1947 Rickover had been taking the measure of Argonne and of its director,
Walter Zinn. For all intents and purposes Argonne was Zinn. The Canadian-
born physicist never let others forget that he and he alone ran Argonne. The
laboratory had been created to serve as a regional center for nuclear research
in the Midwest, but Zinn had never done much more than pay lip service to
the board of governors representing the other participating universities. His
first allegiance was to the University of Chicago, and he strove to maintain
the laboratory's identity as a contractor facility rather than as a government
installation. The university's contract with the government required the laboratory to carry out certain missions, primarily in reactor development for the
Commission, and he insisted upon the right to determine how he would allocate resources and personnel to fulfill the contract.

Zinn had not only a strong personality but also great prestige as a scientist.
As one of Enrico Fermi's most famous protégés, Zinn had established himself as a national authority on nuclear reactors. He had drafted the Commission's first reactor program in 1947 and had reluctantly accepted the Commission's request to make Argonne the center of the Commission's reactor
effort early in 1948. Although Zinn preferred to leave to Fisk, and later to
Hafstad, all decisions concerning reactor development at other Commission
laboratories, he claimed absolute control over what happened at Argonne. As
Rickover had already discovered in some sharp exchanges with Zinn, the
Argonne director did not intend to let Rickover or anyone else tell him how
to run his submarine project.[2]

Zinn officially established the naval reactor division on December 3, 1948, but most of the personnel had been working at Argonne since September. Many of the engineers had come from the Daniels project at Oak Ridge and included men from Westinghouse and the Allis-Chalmers Manufacturing Company as well as from Oak Ridge National Laboratory.[3] The director of the new division was Harold Etherington, a British-born engineer whose background in atomic energy went back to the Manhattan project. He had been in charge of testing and production research on compressors which Allis-Chalmers had built for the gaseous-diffusion plant at Oak Ridge. At the war's end, Etherington took a leave of absence from Allis-Chalmers and stayed on at Oak Ridge, where he experienced the uncertainty that afflicted the nation's reactor planning in the postwar years. When the Commission decided to centralize reactor development at Argonne, Etherington had agreed to go to Chicago knowing that he was to head the Navy project.

The new division also included a dozen Navy personnel, most of whom had come from the Bureau of Ships to gain first-hand experience in nuclear technology. Four lieutenant commanders, all engineering duty officers, would later have an important role in the Navy nuclear program: Eli B. Roth, Sherman Naymark, Jonathan A. Barker, and Marshall E. Turnbaugh. One unrestricted line officer, Lieutenant Commander Eugene P. Wilkinson, had agreed to become Geiger's assistant at Pittsburgh and had come to Argonne for technical background. He already had hopes of becoming the first commander of a nuclear submarine. The civilians from the bureau were mostly specialists in components of steam propulsion systems. Included in the group was Alvin Radkowsky, a bureau employee who had just received a doctorate in physics from the Catholic University in Washington. Radkowsky would later become the senior physicist on Rickover's Washington staff. Although the naval personnel (except for Wilkinson) had been sent to Argonne at Mumma's instigation, Rickover had them reporting directly to him on their training assignments by 1949.

Etherington and his Oak Ridge group had just completed a study of the feasibility of the pressurized-water reactor for the nuclear submarine. Although he and his associates discovered some difficult technical problems in the design, Etherington was convinced that it offered the most promising approach. The study had been so preliminary, however, that it still seemed wise to investigate other types of reactors. Some members of his division were eager to evaluate other designs, and Etherington saw some merit in allowing enough analysis of other types to confirm his tentative choice of the water-cooled reactor. In December 1948 Etherington proposed what he called a

4. Walter H. Zinn (left), the director of the Commission's Argonne National Laboratory, explains research apparatus to Sir Edwin N. Plowden, chairman, United Kingdom Atomic Energy Authority, and Chairman Lewis L. Strauss at the laboratory in May 1956.

5. The Bettis Atomic Power Laboratory at West Mifflin, Pennsylvania, as it appeared in 1956. Built for the Commission by Westinghouse, Bettis developed the pressurized-water reactor design for the Navy.

4

5

"Phase I" study of each basic type in order to establish design variables, prepare rough reference designs, and summarize the main problems. From these studies Etherington thought it would be possible to select one reactor type for intensive development. He thought he could complete the Phase I studies by September 1949 and start the design of Mark I sometime in late 1950.[4]

Zinn, Etherington, and the Argonne staff had a head start on the submarine reactor, and they were pursuing an independent course. In time Westinghouse would be prepared to start engineering work on the reactor, but that moment seemed a long way off. If Rickover expected to control the Argonne project in the meantime, he would have to come to terms with Zinn.

Westinghouse

Since 1946 Rickover had established many contacts with Westinghouse, but by far the most useful was with Gwilym A. Price, the Westinghouse president. During a conversation in May 1946, Rickover had convinced Price that Westinghouse could no longer afford to ignore atomic energy. Price feared that General Electric, with the Hanford project and the Knolls laboratory, already had the competitive edge. Recognizing the importance of the new venture and deeply impressed by Rickover, Price was determined that the company should enter the project wholeheartedly and assign its best people to it. To concentrate men and resources, he had established the atomic power division as a separate department of the company with no other responsibilities than for the submarine reactor. The new division would be tied to Westinghouse management through the company's senior operating vice-president, but Rickover knew that as a last resort he could always appeal directly to Price.[5]

After seeing Weaver work for several months, Price was convinced he had the right man to head the new division. Weaver had proved to be aggressive and interested in atomic energy. He had already expressed to Price his concern that Westinghouse might not be moving fast enough to take advantage of its opportunities. Weaver was young—only thirty-four in 1948. Twelve years earlier he had joined Westinghouse and had acquired a background more in sales than in engineering. His experience in working with Rickover during the war would be an added advantage.[6]

Price was running a risk in setting up a new division and giving Weaver authority to recruit men throughout the company. The individuals Weaver demanded were top-flight. In some cases the men were reluctant to move. In other instances their superiors feared a disruption of efficient operations. This reaction was natural in those who argued that Westinghouse made its profits on conventional products and that the company should give them first priority. But Price realized that such an attitude would never give the company competence in radically new technology like atomic energy. Price and others at Westinghouse saw that Rickover was offering an opportunity that the company dare not miss. Westinghouse not only needed Navy contracts but also had to be in a position to enter a future civilian market for power reactors. Price acted to overcome the inertia which often impeded acceptance of a new technical opportunity. He accepted Rickover's offer to send some of his staff to Pittsburgh to deliver a series of lectures on atomic energy for senior management. Moreover, Price himself attended, and many senior executives followed his example.[7]

One of the first tasks Weaver faced was that of finding a site for the new project. He wanted a large facility, one capable of housing about 600 employees and providing 150,000 square feet of floor space, preferably near the Westinghouse research laboratories in East Pittsburgh. After investigating several sites, Weaver recommended Bettis Field, which for some years had been the main Pittsburgh airport. Thirteen miles southeast of downtown Pittsburgh in West Mifflin, the site consisted of 160 acres, most of which were flat. Two hangars and an administration building on the site would provide temporary office space and shops until new structures could be built on the open space of the airport, with the runways serving as construction roads. Six days after signing the letter contract with the Commission, Weaver requested approval of the Bettis site. In January 1949 the Commission agreed to acquisition of the site by Westinghouse under an arrangement which would permit later government purchase.[8] By that time Weaver and Geiger were already moving into the old airport buildings.

The letter contract which Weaver and Tammaro had signed on December 10, 1948, described only in the most general terms the task the company was to undertake. Presumably, actual working relationships would develop as Weaver organized his staff and began to consult Argonne on design of the Mark I reactor. Some of these relationships could be expected to find their way into the formal contract which Weaver began discussing with Geiger early in 1949.

The negotiations, which continued intermittently for six months, resulted in a contract, and hence a form of relationship, which was unprecedented. The chief architect of the Westinghouse contract was James T. Ramey, a young attorney on Tammaro's staff in Chicago. Ramey had joined the Commission's headquarters staff in 1947 after serving for several years in the legal division of the Tennessee Valley Authority in Knoxville. With a strong interest in administrative law and management, Ramey had seen in the unique relationships between TVA and other regional agencies the opportunity to develop new contract forms to replace the conventional government instruments, with their pages of fine print and legal technicalities.

Ramey's TVA experience was particularly valuable in negotiating contracts like the one with Westinghouse. It was impossible to define exactly what Westinghouse would do. It would take the company the better part of a year to train staff and to build new facilities at Bettis. No one could guess what would result from the Argonne studies by that time and what specific tasks Westinghouse would perform in designing the reactor. Ramey's suggestion was that the two parties give up the idea of trying to define precisely the obligations of each party against all the contingencies which might develop in the course of the contract. Instead he urged that the contract incorporate broad, general language which would reflect the willingness of the Commission and the company to enter into a cooperative venture in a spirit of mutual trust and goodwill. As Ramey drafted the provision, Westinghouse and the Commission would declare their intent "that this agreement shall be carried out in a spirit of partnership and friendly cooperation with maximum of effort and common sense in achieving their common objective." These phrases expressed the spirit of the arrangement which Ramey called the "administrative contract." It proved to be a common instrument in the Commission's contracting procedures.[9]

In other respects the contract followed patterns already firmly established in Commission policy. Because the scope and nature of the work to be performed by Westinghouse were indefinite at best, Tammaro proposed to use the cost-plus-fixed-fee contract form, which he had employed in contract negotiations since early Manhattan project days. Under this form Westinghouse and the Commission would estimate from financial plans the operating costs which Westinghouse would incur during the coming fiscal year. The Commission would determine the fee based on this estimate from a fee schedule, which was not part of the contract. The schedule specified a fee of 5 percent on the first $5 million of cost, plus 4 percent on the second $5 million, plus 3 percent on the next $10 million. On an estimated adjusted operating cost (exclusive of fee) of $2,431,430 for fiscal year 1950, the Commission agreed to pay a fee of $121,570. To avoid any suggestion that this was a cost-plus-a-percentage-of-cost contract (which was illegal), the draft stipulated that the amount of fixed fee would not be adjusted even if actual costs differed from the estimate. The draft also provided for a general and administrative overhead rate of 5.1 percent in a manner similar to that established in Navy contracts with Westinghouse.[10]

In contract negotiations during the spring of 1949 Geiger kept in close touch with Rickover and Dick. Ramey made sure that the draft followed the terms of the letter contract as closely as possible in order to minimize the risk of reviving old issues between Argonne and Westinghouse. The principal items for negotiation were the definition of costs that would be reimbursable under the contract and the determination of costs to be included in the fee base. Other items for discussion were accounting and auditing systems, employee compensation and incentive payments, insurance, personnel policy, and industrial relations. Tammaro, Ramey, and Geiger were able to complete most of the negotiations with Weaver through correspondence and occasional informal meetings. The definitive contract, which was fully acceptable to both parties, was signed on July 15, 1949.[11]

Asserting Authority

The complex organization of the Navy project and the vaguely defined relationships between the government agencies and contractors could in the wrong hands have been a source of trouble. But Rickover welcomed the arrangement with Westinghouse as one which would provide necessary flexibility, particularly for his own organization. During the hectic war years in the electrical section he had relied on day-to-day technical direction rather than a written contract to produce the equipment he was developing. The very heart of his disagreement with the bureau system was his contention that a purchase order or contract guaranteed nothing. How the contract was enforced was all that mattered.

The contract could be general in its terms, but it clearly established Rickover's authority in the important technical decisions, and here he approached his task with the attitude of a suspicious housewife making sure that the butcher kept his thumb off the scale. He did not see himself as a casual agent of a faceless bureaucracy but as a personification of the government itself. He believed that he was, in a very personal sense, the "customer," and he was determined that he would get full value for "his" money. His use of the word "customer" did not suggest an individual who stood at a shelf glancing over prepackaged merchandise. To him the word involved an imperious demand which the seller had to satisfy. From the very beginning Rickover insisted on full value from every contractor, and he expected them to spare no effort.

Early in 1949 Rickover realized that he would have to assert his authority quickly at both Argonne and Westinghouse if the work at both sites was not to flounder. The key issue, as he saw it, was the type of reactor to be developed for the Navy. At Argonne Etherington was already embarked on his Phase I studies which would evaluate the various reactor types which might be used. Although Westinghouse was not yet prepared to start work on any reactor, settlement of this question would enable the company to concentrate its effort. Delay might lead Westinghouse to pursue a study as diverse as Argonne's. Weaver was already drawing up an operating plan which called for the Bettis laboratory to engage in "scientific trouble shooting with a first-rate scientific staff."[12] Without a firm goal, Argonne might well drift off into years of speculative research on all kinds of reactor designs. Without a specific design to pursue, Westinghouse might take a similar course. But if Argonne could be forced to concentrate its efforts, Westinghouse would follow.

Rickover was convinced that the water-cooled reactor was the proper assignment for Argonne and Westinghouse. From his year at Oak Ridge and his continued close study of reactor development he believed that this approach was the most promising. Etherington's preliminary study of the water-cooled reactor at Oak Ridge was the most detailed analysis yet made of any reactor for submarine propulsion, and the conclusion was that the obstacles did not appear insurmountable. Nor would the focusing of Argonne and Westinghouse on the water-cooled reactor mean that other possibilities were being eliminated. To investigate the gas-cooled approach, Rickover already had Allis-Chalmers studying the heat-transfer characteristics of helium. On the liquid-metal approach, both Argonne and Knolls were busy. Project Genie at the Schenectady laboratory would help in providing data on sodium systems.[13]

Assignment of the water-cooled reactor to Argonne and Westinghouse did, however, have its risks. Although Etherington admitted that the water-cooled approach looked most promising, he was not ready to make a commitment without further study. Working in a new technology, Rickover would be taking an exceptional responsibility in making a decision on a technical matter when the experts hired to make the evaluation considered a choice premature. Although other approaches were being investigated, a decision to proceed with the water-cooled reactor early in 1949 would be irrevocable for all practical purposes. Rickover had no reason to believe at that time that he could induce General Electric to accept a submarine reactor project on a high priority. No other company in the United States except Westinghouse had the capacity to take on such an assignment with the kind of schedule Rickover was contemplating. If the choice of the water-cooled reactor did prove premature, the impact on the Navy project could be severe.

Rickover, however, was now prepared to force Argonne to make the decision he wanted. He asked Argonne to determine which approach would be the best if the choice were to be made at that time. The reply came back on March 21, 1949, over Zinn's signature. On the basis of existing knowledge, the water-cooled approach was the most promising. Rickover could have expected no other answer.[14]

Rickover's action was crucial. His purpose was to make certain that Argonne and Westinghouse would do engineering—not research. It was a point he was to hammer at many times. In an area as new as reactor technology, the unknowns were so great and the possibilities so intriguing that the lure of research was irresistible to many scientists. In Rickover's mind, research meant investigation and exploration. Engineering meant creating something new to reach a fixed goal. Research was vital, but in his program it had to be controlled. Forcing a decision on Argonne was the act of an administrator who had a firm grasp of the technical issues and knew intimately the people involved.

Management Appraisal

Rickover from the start insisted upon continually appraising contractor performance so that he could intervene as soon as he saw weaknesses that threatened progress. Through his own representatives he learned—daily if necessary—what was happening at each of the laboratories. The stream of reports and correspondence made him aware of every operational detail. From scanning this material he could detect potential trouble spots. As these began to form a pattern, Rickover would send one of his Washington staff to investigate. If the situation appeared serious, he would make the trip himself.

Rickover's first management inspection trip to Bettis came in the early summer of 1949. At the laboratory he confirmed many signs of weakness. Reactor physics and engineering had no firm direction. Weaver had not yet found a technical director to assist him. Under the pressure of time, Weaver had authorized work on technical problems before the essential data were available from Argonne. Rickover believed that Westinghouse was not doing enough to recruit and train new personnel. Occasional lectures did not constitute an educational program which would strengthen the laboratory.

Weaver thought most of his difficulties came from poor liaison with Argonne. That failure explained why he had authorized some activities which seemed premature. Bettis was already pursuing studies that Argonne had requested, such as building and testing pumps and fabricating samples of zirconium metal. The division had already begun work on twenty-five projects which Etherington had drawn up and sent to Bettis in advance of Zinn's approval. Still, the fact remained that Bettis did not have a firm and detailed grasp of Argonne activities.[15]

The Argonne visit in July revealed a different problem and perhaps a reason why Weaver did not have all the data he wanted. For months Etherington and his staff had been trying to fix the size of the core, the central portion of the reactor which would contain the uranium fuel elements, the control rods, and the channels through which water would remove energy from the reactor. Rickover wanted to reach agreement on core specifications because the design of hundreds of other components rested on this decision. Etherington understood this point, but he was not willing to commit himself until he had resolved some of the conflicting considerations. Rickover could see the difficulties, but thought that perhaps Argonne was attempting to be too self-sufficient. Etherington might consider calling in such experts as Fermi or Eugene P. Wigner as consultants. Rickover also believed Argonne should make more use of the engineering capabilities of Westinghouse. In all aspects of the project Rickover saw the need for more personal contact, not just between Argonne and Bettis but also with the Bureau of Ships in Washington.[16]

Whether at Argonne or Bettis, Rickover's methods of appraisal were much the same. He inspected facilities and saw the work that was being done. He and his staff had followed the contractors' efforts closely and knew the key personnel. Rickover and his men could question the scientists and engineers in detail about their work. Occasionally the process was bruising. Weaver might think criticisms of Bettis were unfair when his organization was not getting the information it needed. Etherington and his physicists might believe that Rickover was underestimating difficulties in getting fundamental nuclear data and that he was overlooking real accomplishments. To Rickover the purpose of the meetings was not to bask in the glow of achievements but to ferret out technical obstacles and management weaknesses.

The impact of the conferences did not end when the participants adjourned. Members of Rickover's Washington staff and representatives of the local

Commission office followed the discussions closely and knew what points to watch for in the weeks ahead. The Washington group took extensive notes at each meeting and consolidated them into a formal report. Later Rickover discussed the report with the individuals involved to make sure they understood the problems and what they had agreed to do about them. The conferences and the report soon became an effective and distinctive tool of the naval reactors branch.

Report to Management

Rickover knew from his earlier experiences in the Navy and with industry that management surveillance consisted of doing more than telling operating personnel their shortcomings. Sometimes it was necessary to go directly to senior management with examples. Rickover had discovered that he could not rely on the internal communications within a large corporation to reveal problems.

In September 1949 Rickover decided it was time to discuss Weaver's work with his Westinghouse superior—in this case, with Latham E. Osborne, the senior operating vice-president. Rickover told Osborne he was worried about Weaver's plans for the coming year. The funds Weaver was requesting for research were about double the amount for engineering, a proportion Rickover found difficult to reconcile with the goals of the project. He guessed that much of the research would duplicate work at Argonne. Even more important, Rickover found few signs of a comprehensive, well-thought-out plan, and no evidence of a real schedule. Furthermore, Rickover complained that the company had done little to correct the deficiencies he had pointed out in June. Weaver still did not have a technical director, and there seemed to be little zeal for the project within Weaver's division. Rickover said he had noticed that many individuals were reluctant to attend evening or weekend meetings. Apparently, Rickover observed, many of the Westinghouse people considered that building the Mark I was just another job.

Attempts at Coordination

It would be many months before Etherington and Weaver could respond effectively to Rickover's basic complaint. Competent scientists and engineers could not be hired and trained in a day, and it would take much longer to build them into a team with the kind of dedication Rickover was seeking.

Under the lash of Rickover's questions and demands, the two leaders began to move in that direction.

Both Etherington and Weaver followed some of Rickover's suggestions for improving the technical competence of their divisions. In September 1949 Weaver, with Rickover's approval, appointed Charles M. Slack as his technical director. Slack knew little about nuclear technology, but he was a physicist with experience in designing complex equipment such as X-ray tubes. Weaver hoped that Slack would help him maintain closer ties with activities at Argonne. Both Westinghouse and Argonne sent men to the Bureau of Ships in Washington to learn about such matters as shielding requirements for reactors, pumps for the heat-transfer system, and the layout of machinery in a submarine. As an additional guide for the laboratories, Rickover's staff drew up a tentative set of requirements for a submarine power plant. Although the specifications were similar to those which Rickover's group had discussed on several occasions, it was helpful to have them in writing. As an aid to communication within the laboratory, Etherington had started a biweekly newsletter reporting on all current work, even when the results were only preliminary. This device met Rickover's idea that technical reports should reflect the situation as it existed and not be used to gloss over failures or problems.[17]

By September 1949 Argonne and Westinghouse had each recruited about seventy scientists and engineers and had established an initial organization. Etherington had divided his division into four groups along functional lines. The nuclear engineering section dealt primarily with reactor physics, an area which included control systems, instrumentation, and shielding. The engineering analysis group was primarily interested in the heat-transfer system for the reactor. Under mechanical design came the structure of components and the general layout of Mark I. Development of the fuel elements, studies of corrosion, and the effects of radiation on materials came under materials engineering.

Although Weaver could see the advantages of organizing his division on the Argonne pattern, he found it easier to follow the company's customary structure for an operating division. Weaver divided his organization into two departments: research, which included instrumentation and controls, chemistry, and physics; and engineering, which involved plant and component design for all equipment from the reactor to the turbine, reduction gear, and condenser.[18]

Once the two contractors had established their organizations, it was easier to build lines of communication directly between units with similar responsibilities. At Rickover's request, Etherington and Weaver created what they called the naval reactor coordinating committee. To provide balance the two leaders served as co-chairmen. Meetings were to alternate between Argonne and Westinghouse, with the host of each session drawing up the minutes. As its name implied, the committee was to coordinate technical work and promote effective liaison between the two groups. Inevitably the first meeting dealt with procedures, but there were also reports on pumps and materials as well as efforts to frame a schedule.[19]

The committee seemed to be a step in the right direction, but Rickover thought it looked weak. He was suspicious because none of his representatives at Argonne or Bettis had attended the first meeting. His Washington staff concluded that the committee could not be effective, balanced as it was between two organizations. All it could do was exchange information and act when there was mutual agreement. Despite good intentions on both sides, the coordinating committee was not the management instrument that Rickover wanted. There was no way of resolving disputes and no way of acting in the absence of agreement. The basic flaw, to use Rickover's words, was that the "customer" was not present.[20]

Rickover in fact was fully aware of the situation and had no intention of accepting it. Almost a whole year had passed since Argonne and Westinghouse had agreed to undertake their joint venture, and they had not yet succeeded in devising a satisfactory structure of responsibility. This fact was deplorable in itself, but there were new reasons for concern during the autumn of 1949. On September 23 President Truman announced that the Soviet Union had successfully detonated a nuclear device. More significant than the test itself was the obvious implication that the Soviet Union had now mastered the essential elements of nuclear technology. American scientists and engineers, including Zinn, Etherington, and Weaver, could no longer proceed on the comfortable assumption of an American monopoly. No longer were they simply meeting a Navy requirement; now they could well be in direct competition with the Russians.

On October 4, 1949, Hafstad met with his staff to see what could be done to speed up the Navy project. He had concluded that Argonne work was weak and showed little signs of improving. Part of the difficulty appeared to be the priorities which Zinn assigned his reactor projects. Hafstad understood that the Navy reactor was at the bottom of the list. He explored the alternatives. One was to strengthen the Argonne division and put the submarine reactor project at the top of the list. The second was to assign the whole task to Westinghouse. The difficulty, Hafstad admitted, was that there were no signs that Westinghouse could do the job.

Rickover proposed a broader solution, more far-reaching and less drastic than reassigning the work at Argonne. Argonne should give the Navy project top priority and the Westinghouse division should be improved. These were obvious steps, but Rickover went further. Schenectady should also be brought more closely into the effort. Knolls should have a Navy reactor project which would be a longer-range effort than the Argonne-Westinghouse reactor. As far as the workload at Knolls was concerned, the new project would rank, in Rickover's opinion, second only to the power-breeder.[21]

When pressed, Zinn agreed that the Navy unit should come first on the Commission's priority list for reactors. His opinion rested solely on his understanding that the Navy considered the reactor of vital importance if war should break out in the next five or ten years. Somewhat cautiously Zinn pointed out to Hafstad that he had not seen any reasoned explanation of how the Navy would use the submarine nor had he heard any qualified military expert give advice on the importance of the project. Because the Argonne-Westinghouse reactor would be an inefficient user of fissionable material, Zinn would not have put the project into first place if the Commission had not stressed military applications of atomic energy.[22]

Zinn agreed with Rickover on October 15 that progress had not been satisfactory. He could cite many reasons, including the distractions created by uncertainties in the Commission's reactor planning, budget matters, and the shortage of personnel; but the main trouble was still the lack of coordination with Westinghouse. Rickover's suggestion was immediate and typical: he called a meeting with Zinn and Weaver to thresh out the problem.[23]

The meeting on October 31 was no haphazard occasion. Rickover's staff had prepared a detailed agenda setting forth the issues to be discussed, and both Zinn and Weaver had accepted it. In the meeting Rickover kept the discussion close to the agenda. The purpose was to induce both sides to acknowledge their failure to build an effective team. Once Rickover had accomplished this, he could propose his solution—a policy board consisting of himself, Weaver, and Zinn. The board's chief function would be to draft a schedule for the project. To do the leg-work, Rickover proposed a scheduling committee composed of the leading technical representatives of the three parent organizations. The logical choices were Etherington, Slack, and Lieutenant Dick.[24]

In any endeavor Rickover insisted that one person be in charge, and under the circumstances it made sense to select Zinn. Zinn, however, was reluctant to accept. For one thing, he had many responsibilities, while Rickover and Weaver had only the Navy reactor to consider. Furthermore, Zinn knew from his experience that, although the responsibility might be his, authority usually rested with someone in Washington. He agreed that the Soviet detonation made the Navy reactor an urgent project for national defense, but the circumstances were hardly the same as those under which General Groves had worked. Then the nation had been at war and Groves could command support at the highest government levels. To himself, Rickover probably would have admitted that Zinn's appointment as chairman was inconsistent with his philosophy that the customer and not the contractor had to be in control, but he accepted the realities of the situation. Zinn was an acknowledged expert and was director of the Commission's reactor development center. He commanded the respect and confidence of the Commission. Rickover could hope that by persuading Zinn to take an active part, the policy board might be effective.[25]

In appointing Dick as executive secretary of the board, Rickover could be certain that his views were represented. Dick was energetic, tenacious, and intensely loyal. As Rickover's project officer for the reactor, Dick thoroughly understood the technical problems, the facilities, and the people. Dick wrote the board's minutes, a task that inevitably enabled him to point up decisions and see that they were implemented. Even more important, Dick's presence gave the naval reactors branch strong and constant leverage on the project as the work on Mark I progressed from design to engineering and primary responsibility moved from Argonne to Westinghouse.

As it turned out, the name "policy board" was a misnomer. It neither set policy, which was a prerogative Rickover jealously reserved for himself, nor did it make decisions, which usually could not be delayed until the time for a meeting. Between November 1949 and April 1950 the board met only seven times. Usually the meetings began with discussion of the scheduling committee's report, in some cases a brief document, in others more than a hundred pages. The main questions usually concerned who was to do what to make up for slippages in the schedule. The board helped to point up trouble spots and bottlenecks. Once these had been acknowledged, those responsible would feel the full weight of Rickover's pressure for action. The board was mainly a device which Rickover used when he wanted to deal with Weaver and Zinn at the same time.

General Electric—
A New Possibility

Building the Westinghouse-Argonne team had consumed much of Rickover's attention during the first half of 1949, but he still wanted a second contractor to pursue an alternate approach to nuclear propulsion. In fact, the difficulties the two laboratories were experiencing supported his arguments for a second project. Should Bettis and Argonne fail to make headway on the water-cooled reactor, the Navy might have to accept General Electric's terms for developing the liquid-metal–cooled design.

The trouble was that General Electric was refusing to abandon its civilian power project. In the summer of 1948 the company had insisted that any work it did on a naval reactor be based on the technology it had developed for the power breeder. In itself this position was not unreasonable, because much of the technology for the civilian project was applicable to the approach that Mills and Rickover had envisaged in 1947 for the submarine reactor. It also made sense both from the company's and the Navy's point of view to use capabilities already available. There remained the same danger, however, that faced Rickover in 1948: by accepting the General Electric proposal, the Navy might find its own interests subordinated to the company's civilian project. Rickover had convinced Mills in 1948 that the company's offer was simply an effort to find support for a reactor which seemed ever less likely of fulfilling its original purpose. Early in 1949 Rickover was avoiding that trap as carefully as he had in 1948.

General Electric's leaders were well aware that their reactor had little chance of succeeding as a breeder of fissionable material. The fault lay not in engineering but in the laws of nature; assumptions based on fragmentary data had proved wrong. To Winne and others at General Electric this discovery was disappointing but did not seem necessarily fatal. The project would explore one of the most interesting reactor types. It could be useful in developing nuclear plants for generating electric power or for driving a submarine. After more than two years of study, the project had gathered a good staff of scientists and engineers, and the Commission had approved a site for the reactor at West Milton, New York, a few miles from the Knolls laboratory east of Schenectady. In February 1949 Kenneth H. Kingdon, the technical director of the laboratory, had confidently sent the Commission a preliminary feasibility report, which he hoped would convince the Commission to authorize construction.[26]

Contrary to Kingdon's expectations, the feasibility report did more harm than good. After reading it, Weil had strong doubts about proceeding with the power-breeder. If it could not attain its original purpose as a breeder, it might not be worth the large investment in fissionable material, technical manpower, and money. Surely the submarine reactor made more sense than a civilian power reactor of doubtful technical merit. Weil concluded it might be better to delay construction until there was more conclusive data on breeding. A canvass of the Commission's reactor experts resulted in a somewhat more optimistic view. The reactor might be of some value even if it did not breed. There was also some concern that cancellation might destroy General Electric's strong team of experienced reactor engineers. Adding up these mixed reactions, the Commission did not have a very convincing case for the power-breeder.[27]

Although General Electric had placed most of its hopes on the power-breeder, the company had not forgotten the Navy project. In April 1949 a small group of physicists and engineers at the Knolls laboratory had completed a comparative study of several approaches to a nuclear propulsion plant for a submarine. One idea was to combine data from Genie and the power-breeder and to build an experimental propulsion plant aboard a surface ship. Only after operating the power generation system with a conventional boiler would the reactor be installed. At an estimated cost of $54 million and almost a decade of development, the Knolls group thought, it would be possible with some confidence to build a submarine reactor.[28]

Rickover read this study with dismay. As he had suspected, the company was still determined to build a civilian power reactor using the power-breeder design. Even more discouraging was the evidence that the Knolls staff had not begun to comprehend the kind of effort and commitment required to build a submarine propulsion system. In comparison with the plan and time schedule Westinghouse was drafting, the idea of building several prototypes over the period of a decade seemed preposterous.

But Rickover did not underestimate the enormous pressure which General Electric could bring to bear on the Commission by reason of the inescapable tie between Knolls and Hanford. As long as the Commission was dependent upon General Electric for continued operation and expansion of the plutonium production facilities at Hanford, it was impossible to dismiss the power-breeder even if its original purpose was evaporating. In addition, there was a strong feeling among the Commission's most trusted advisers that the reactor program had been buffeted so often by indecision that cancellation of the Knolls project could have a disastrous impact on the Commission's effort generally. With no enthusiasm the Commission had reaffirmed the choice of West Milton in August 1949 and had authorized additional funds for site development. Major construction, however, would not be approved until the company had completed a new feasibility report, one including firm estimates of construction and operating costs. The report was to be submitted by February 15, 1950.[29]

Rickover appraised the situation warily. If the power-breeder were ready for construction, the engineers and scientists who had been designing it could be reassigned to a Navy project. But what kind of an effort should that be? Rickover had rejected all the approaches the Knolls group had proposed in the comparative study; they were too complicated, expensive, and would take too much time. He could, however, support construction of the power-breeder at West Milton if in fact the personnel were reassigned and if the reactor would provide enough data to be an adequate prototype for a shipboard plant.

On August 11, 1949, during one of his frequent trips to Schenectady, Rickover tried to convince Winne that the time had come for General Electric to begin work on a Navy reactor. Rickover maintained that the company needed a new project, not only to keep its skilled manpower occupied, but also to enter the new and promising field of nuclear propulsion—an area in which Westinghouse was already working.[30]

Rickover's arguments fell on fertile ground. In a letter to Carleton Shugg on August 22 Winne repeated the familiar story of the Navy's efforts to draw General Electric into the submarine project. He admitted that the company had been unwilling to follow that course because it would interfere with work on the power-breeder. Now that situation was changing. Very soon manpower would be available, but Winne warned that the company would need more money to get started.[31]

The Definition of Responsibility

Winne's concession that General Electric might now have room for a submarine project was encouraging, but many uncertainties remained. A critical question was where the Navy project would be placed in the company's sprawling and decentralized organization. Neither the Navy's nor the Commission's work was concentrated in any one of the company's departments, and no one high official in the company was responsible for it. To an outsider, the lines by which Kingdon at Knolls reported to his superiors were vague. Others besides Rickover in the Commission's staff wanted a clearer delineation of responsibilities.

The issue of organization came to a head in a meeting in Schenectady on August 30, 1949. The Commission representatives—Rickover among them —sought simplification of the company's administration of atomic energy projects. James C. Stewart, the local Commission manager, wanted to be able to deal with one individual in the company. C. Guy Suits, the General Electric vice-president in charge of research, reacted sharply to this criticism. He maintained that the proof of any organization was in its results, and he claimed that Knolls was the best laboratory the Commission had.

Rickover could accept Suits's criteria for measuring excellence if not his evaluation of Knolls, but there was another factor underlying the arguments on both sides: the Commission and General Electric had never been completely satisfied with their relationship. The Commission thought the company sometimes put its own interests ahead of its atomic energy projects, and the company often complained that the Commission was indecisive. In more general terms, the structures of responsibility within the Commission and the company as well as between them had never been clear.

The heated debate on August 30 did have some results which Rickover could appreciate. Winne promised that the company would re-examine its organization. The company also agreed to transfer the control of Project Genie from the general engineering and consulting laboratory to Knolls, which was also working on sodium systems. Rickover already had three engineers at Knolls doing paper studies on submarine reactors. Now he was to be responsible for Genie as well. From this small nucleus Rickover might be able to build a submarine project, especially since the transfer of Genie would place it under Commission contract and hence out of the reach of the design group in the Bureau of Ships.[32]

Yet the larger issues remained unsettled. It was not clear where the Navy project would find a home. Knolls was a multipurpose scientific laboratory and would be reluctant to give the Navy work the concentrated attention Rickover demanded. The atmosphere at Knolls was clearly more favorable to research than to engineering. Kingdon, the technical director, was a scholarly physicist whose temperament clashed with Rickover's and, more importantly, whose interest was research and civilian power. He could never be

—nor did he wish to be—a hard-headed manager who would drive a project through on schedule. If, in Rickover's sense of the term, Kingdon had established no real authority at Knolls, General Electric had no effective authority over the laboratory.

As the Commission deliberated over the priority of its assignments to General Electric, the company's officials considered how they might organize the Navy project. Suits favored putting it at Knolls, where he thought it would fit logically with the work on the power-breeder. At the same time he did not overlook the possibility of creating a new organization which might eventually move into the production of commercial atomic power equipment. Winne thought the project could be put under the general engineering and consulting laboratory, the nucleonics department, the apparatus department, or Knolls. Apparently he did not consider Suits's suggestion of an independent department reporting directly to the company's president. In any case, no decision was likely until the company had completed its feasibility report on the power-breeder and the fate of that reactor had been determined.[33]

During the autumn of 1949 the power-breeder was still the main obstacle in the path of the Navy project. From Rickover's perspective that difficulty could be resolved in either of two ways. If the power-breeder were ready for construction, he could press for an early decision to clear the way for work on the submarine. If it were not, he could attempt to terminate all work on the power-breeder at once. Getting a decision in either case would not be easy. The Commission had delayed answering Winne's letter of August 22 while it debated over the most appropriate response to the Soviet detonation. Not until November did the Commission inform Winne that the company was to give its first priority to the Hanford production plant, then, in descending order, to the power-breeder, the Navy project, and finally to research.[34]

Even that listing of priorities depended heavily on the feasibility study of the power-breeder. Unable to gain any extension of the deadline, General Electric had no choice on February 15, 1950, but to submit a summary report lacking the details necessary for a reliable evaluation. Even worse, the study seemed to admit that the original conception of the power-breeder was no longer valid. When Kingdon could not assure Rickover that the reactor could be used as a prototype for a ship propulsion plant, the last reason for the project drained away. After listening to a final appeal by Winne and his associates on March 17, 1950, Wilson decided that the Commission would not authorize construction of the power-breeder at that time.[35]

New Ties with General Electric

Deferral—or more realistically, cancellation—of the power-breeder did not in itself provide the basis for a full-fledged Navy project at General Electric. It was still necessary to define the goals and the structure of responsibility. This task was complicated by the multiple role which General Electric had in the Commission's activities. The threat to the Commission's production efforts, and not to the submarine project, had been the principal reason for dropping the power-breeder. The cancellation was also unwelcome news to certain leaders of the American power industry and members of the Joint Committee on Atomic Energy, who looked upon the power-breeder as the last best hope for developing a power reactor in the United States during the 1950s.[36]

Rickover's first concern was to allay the fears of those who saw the submarine project, on one hand, as a threat to the production of nuclear weapons for national defense and, on the other, as a blow to the promise of civilian nuclear power. To meet the first concern, Rickover welcomed a meeting with Senator Brien McMahon, chairman of the Joint Committee, on March 21. McMahon, an ambitious and effective young senator, had made his reputation as a sponsor of the Atomic Energy Act of 1946 and had served briefly as chairman of the committee during the closing months of that year. With the return of a Democratic majority to Congress in 1949, McMahon had again become chairman and, together with William L. Borden, the new executive director, had launched an intensive campaign to increase the nation's stockpile of nuclear weapons.

During the bitter debate over the thermonuclear weapon during the fall of 1949 and the winter of 1950, McMahon had led the Joint Committee in its successful drive to override the opposition of most members of the Commission and the General Advisory Committee. On the crest of that triumph, McMahon had become the most ardent spokesman of those in the Congress, the Department of Defense, and the Commission who favored a new emphasis on the military uses of atomic energy. Once Rickover had assured McMahon that the cancellation of the power-breeder would not harm the production effort, the senator proved fully receptive to Rickover's arguments for the nuclear submarine. Such a ship, like the hydrogen bomb, could become a key to the nation's defense. This meeting marked the beginning of a close and active alliance between the naval reactors branch and the Joint Committee, which would outlive McMahon by decades.[37]

In response to the civilian power advocates, Rickover concentrated his attention on the reactor subcommittee of the Joint Committee. This group, including Congressmen Carl T. Durham and Carl Hinshaw, the only engineer on the committee, had recently visited the Commission's reactor facilities at Argonne, Oak Ridge, and Schenectady. Enthusiastic presentations by Zinn, Weinberg, and Kingdon had impressed them with the potential advantages of power and breeder reactors. In the course of a hearing before the subcommittee on April 3, Hafstad and Rickover succeeded in convincing the members that the Commission had not expected General Electric to do the impossible in building the power-breeder, but merely that further study had indicated the advisability of postponing construction. Rickover could also assure the Congressmen that postponing the power-breeder had not damaged the Navy project. Quite the opposite, it would enable General Electric to put more effort on the nuclear submarine.[38]

Once these potential sources of opposition had been removed, Rickover was prepared to act quickly. The next day, April 4, Rear Admiral David H. Clark, who had succeeded Mills as chief of the Bureau of Ships, had discussed with Wilson and Shugg the Navy's hopes for building a submarine reactor prototype at West Milton. This time the Commission and the Navy were in complete agreement on their priorities and goals. Only research for the production plants at Hanford took precedence over the Navy project.

The following day Clark joined Rickover, who had already gone to Schenectady for discussions with General Electric. They found Winne and Kingdon reconciled to switching the focus of Knolls from the power-breeder to the Navy reactor. Rickover had smoothed the path to that conclusion a week earlier by inviting Harry E. Stevens, an old acquaintance from Oak Ridge days, and several other General Electric officials to the Bureau of Ships in Washington, where they received a full briefing on technical aspects of the Navy project. After the April 4 meeting Winne sent the Commission a formal letter proposing that about one hundred employees, or about half the technical manpower at Knolls, be committed to the Hanford project. The remainder would be shifted from the power-breeder to the submarine project. The aim would be to build a land-based sodium-cooled prototype at West Milton as soon as possible. Rickover later informed the Commission that it would be possible to begin construction of the prototype in 1951 and to have it in operation in 1953.[39]

As soon as the Commission accepted Winne's proposal on April 12, Rickover set about organizing the project and establishing controls. His first concern was a definition of his responsibility within the division of reactor development. This point was particularly important because the Knolls laboratory would continue to have major responsibilities not involving the Navy project. A discussion with Hafstad that same day resulted in an agreement that Rickover would continue to have technical responsibility for the submarine project while general program direction of the Knolls laboratory would be assigned to the stationary reactors branch. On the allocation of personnel, the two branches would attempt to reach agreement among themselves, with any dispute to be settled by Hafstad.

Within the General Electric organization itself, Rickover had some success in establishing a distinctive structure for the submarine project and clear lines of authority. In June the company announced that Knolls had been established as an organization completely independent of the General Electric Research Laboratory. For general manager of Knolls the company had selected William H. Milton, an experienced electrical engineer who had recently been commercial vice-president in charge of customer relations in Washington. Milton was experienced in government contracts, particularly with the Navy, and would bring a new sense of administration to Knolls. Lines of authority were clarified when the company announced that Milton would report to Suits and that Kingdon would serve as technical director of the laboratory.[40]

Milton's aggressive and practical approach to his new assignment impressed Rickover during the summer of 1950. Milton was quick to strengthen his staff and to investigate weaknesses which Stewart or the Navy representatives found in the Knolls operation. These first steps were encouraging, but Rickover was still concerned about the future. Milton did not have control of all General Electric work on the naval project. The division of responsibility between Milton and Kingdon was not clear, and there was reason to doubt whether on major issues Suits could speak for the company as a whole. There was a new spirit within the Knolls staff, but Rickover's group still thought Knolls showed a lack of concern with scheduling. In Rickover's opinion the Knolls staff was too heavily loaded with scientists; he had more confidence in engineers in other parts of the company. When Rickover suggested more use of General Electric personnel in the Navy project, Milton mentioned that the company had to make a profit, and that the submarine reactor was a non-profit venture. Knolls under Milton was still far from being the independent department which Bettis was under Weaver.[41]

In Rickover's mind, the source of difficulty at Knolls was his lack of effective control. Unlike Bettis, Knolls had vital functions to perform for the Commission's production effort and also claimed a role in general research and development on power reactors. These added functions were the responsibility of others in the division of reactor development and always appeared to Rickover as a potential source of competition. Furthermore, Stewart was less willing than Tammaro or Geiger to confine himself to administrative matters. From the very beginning, General Electric had taken an almost stubbornly independent course toward the Navy project, and it was to Rickover's credit that after four years of trying he had succeeded in establishing the rudiments of a workable relationship with the company. But the arrangement was far from perfect in Rickover's estimation, and he was prepared for trouble in the future.

Idaho: An Organizational Puzzle

In addition to the projects at Bettis and Knolls, Rickover also found it necessary early in 1950 to establish a working relationship with the National Reactor Testing Station, which the Commission was building in the Idaho desert, 40 miles west of Idaho Falls. The idea for the station had grown out of the need for a remote facility where experimental reactor designs could be tested without endangering population centers. There in the remoteness of southeastern Idaho, the Commission would build a dozen experimental reactors over the next decade.

Establishing the national testing center presented the Commission with an unusual organizational problem. Scientists and engineers at several laboratories would at different times be proposing to build experimental reactors at Idaho. The groups which designed these reactors would have to bear the responsibility for building and operating them. At the same time, the Commission needed a local office to manage the site and coordinate activities. The functions of the Idaho office could hardly follow those prescribed for Chicago or Schenectady.[42]

Recognizing these facts, the Commission had established the Idaho operations office in the spring of 1949 under Leonard E. Johnston, who had made a reputation as a field administrator at Schenectady, where he was Stewart's predecessor. Johnston's duties, stripped down to essentials, were to manage the station and administer contracts for building and operating the reactors which Hafstad's division assigned to the testing station.[43] Johnston had gone to Idaho with all the conviction and determination of a strong administrator.

Finding himself in a remote area, where labor and materials were scarce, Johnston believed that sound management required him to exercise broad authority over construction activities, the procurement of materials, and the hiring of labor. Specifically he hoped to negotiate area-wide agreements with local contractors, suppliers, and labor unions. In this way he hoped to make the testing station an integral part of the community, a goal he could reach only if he had full control over all facilities on the site.

Johnston's conception of his role as manager ran directly against Rickover's ideas on management. Rickover fully intended—and the contract so specified—that Westinghouse would build the Mark I. He had created an administrative structure to achieve that end. The company not only would control all activities at Bettis but would also hire construction subcontractors for the Mark I facilities at Idaho. Rickover would not be able to hold Westinghouse responsible if the construction contracts were negotiated and administered by Johnston and his staff. Rickover was determined not to let Johnston disperse the responsibility he had carefully concentrated.

The deadlock lasted for two months, but Rickover knew from the beginning that he would never have to accept Johnston's demands. In the heat of the argument Rickover pointed out that in accepting Johnston's proposal the division of reactor development would be assuming the responsibility which the Commission had given to Westinghouse for Mark I. Hafstad recoiled from this prospect. Rickover—speaking as a Navy officer rather than a member of the division—also threatened to find another site for the Mark I. This was not a practical idea if Westinghouse was to meet the schedule for Mark I, but the threat was effective. By acting in his Navy capacity, Rickover was suggesting that he would take the matter to the Commissioners if Hafstad and Johnston did not resolve the issue, and none of the Commission staff wanted that.[44]

In the end Rickover relented to the extent of assuring Johnston that he would keep in mind the broader interest of the Idaho office, but he could not yield on the central point of Westinghouse responsibility. The solution was an artful compromise which made a gesture in Johnston's direction but firmly backed the position Hafstad and Rickover had taken. Westinghouse would be responsible for constructing Mark I. Johnston would be the Commission's authorized representative on the construction of Mark I, but Geiger in turn would represent Johnston on the project. Thus Johnston was given theoretical authority while actual control remained in the original structure of responsibility from Hafstad to Rickover, Geiger, and Weaver. To Rickover the titles and organization charts meant nothing; only the realities of responsibility mattered.[45]

The Emerging Structure

When Rickover succeeded in establishing the second Navy project at Knolls, the basic structure of his organization was complete. In outline this structure was deceptively simple. On the technical side, he and his staff in Washington set the goals and specifications which guided the laboratory studies and development projects at Argonne, Bettis, and Knolls. On the administrative side, the lines of authority extended from his naval reactors branch through the division of reactor development to the Commission's field offices at Chicago, Pittsburgh, Schenectady, and Idaho Falls.

In terms of its operation the organizational pattern was much less precise. The basis of Rickover's authority was his dual role which tied him to both the Commission and the Navy. Because he quickly sensed the possibilities of this arrangement, he was able to turn it to his advantage. Instead of a double infringement on his authority, the dual organization became a vehicle for unusual independence. Rickover achieved this independence, however, by avoiding routine procedures that would fix organizational patterns. In one instance he would act as a naval officer, in another as a Commission official. This unpredictable and pragmatic approach gave him the freedom he sought. The dual organization itself simply provided the opportunity for independence.

Another source of imprecision in organizational practice was the great variety in the groups which made up the Navy project. The organizational structure and style of Westinghouse contrasted sharply with that of General Electric. Although Argonne and Knolls were both Commission laboratories, they had few other similarities. The Commission field offices all had distinctive characteristics. Some of these differences were the result of varied responsibilities; others were the consequence of conditions existing at the time the Navy project was established.

This diversity prevented Rickover from following any uniform or fixed pattern of organization. Even if he could have done so, he would not have established a rigid system. He passionately believed that success in building a submarine reactor lay in flexibility. He wanted to be able to meet each problem as it arose in the way he thought best. He would not be committed to a fixed structure because every new situation involved a unique combination of personalities, talents, and technical considerations.

Behind this flexible response to each situation was an unyielding authority based on supremely confident determination and the most rigorous form of self-discipline. Rickover himself made the decisions in the sense that he acted on all the evidence available and took personal responsibility. He did not simply ratify as administratively acceptable the proposals submitted by the contractors. To make such decisions he had to have detailed technical information. For this reason he and his staff had to be insatiable consumers of technical data and probing inspectors who cross-examined those responsible for technical assignments.

This philosophy of management had a direct impact on the structure of responsibility. It meant that Rickover and his staff had to have direct, frequent, and uninhibited contact with the contractors—Rickover himself with senior management, his staff with specialists in the laboratories and technicians in the shops. No aspect of the contractor's operation could be immune to inspection or criticism. No member of the contractor's organization could escape personal scrutiny and evaluation. Rickover personally questioned technical staff at all levels in the contractor's organization. In a very real sense, the lines of communication were direct from Rickover's office in Washington to the manager's desk.

Within the contractor's organization the Rickover approach to management was bound to have far-reaching effects. The new atomic power division in Westinghouse bore little resemblance to other company divisions. In many respects, its ties to the naval reactors branch in Washington were much closer than to the company headquarters in Pittsburgh. The impact was much more frustrating in General Electric, which had a long tradition of maintaining its independence from customer influence. In accepting the Navy assignment, General Electric found itself yielding control of a segment of its organization. In both instances, the Navy project moved in the direction of becoming isolated and independent of the parent company. This new entity, theoretically a part of the contractor's organization but in many respects an integral part of Rickover's project, offered new and unexplored possibilities for managaing engineering enterprises.

Building a nuclear submarine required nothing less than all the resources, talents, and energies of those involved. The organization that Rickover created reflected this commitment. How well that structure of responsibility operated in designing and building the first reactor prototypes is the subject of the next chapter.

Emerging Patterns of Technical Management

The structure of responsibility described in chapter 4 became the initial administrative framework for the naval nuclear propulsion project. It defined in a general way the relationships between the various institutions involved —the Commission, the Navy, the Commission's laboratories and research contractors, the Bureau of Ships, and some of its contractors. The structure of responsibility also prescribed the limits of Rickover's authority and predetermined to some extent the administrative system which Code 390 would use.

Important as the structure of responsibility was, however, it did not begin to explain how the naval reactors branch succeeded in building the world's first nuclear propulsion plant. To discover what Rickover's organization actually did in directing technology requires a much deeper examination of the organization than a view of the general structure of responsibility provides. It is necessary to probe the composition and character of the naval reactors branch itself, to see the intimate relationships between Rickover and his staff and to understand how responsibility was designated, how technical decisions were made, and what the nature of those decisions was.

Recruiting and Training

When the Bureau of Ships established Code 390 in the summer of 1948, Rickover could reassemble only a portion of the Oak Ridge team. Only Roddis, Dick, and Emerson were still in the bureau. Libbey was serving with the Military Liaison Committee, and Dunford would not be returning from the Commission staff until Hafstad organized the division of reactor development in January 1949. Among the civilians, Blizard had stayed with the reactor physics group at Oak Ridge and Amorosi had gone with Etherington to Argonne. In the field Rickover could list only Geiger and Wilkinson, who were organizing the new Pittsburgh office. The engineering duty officers and civilian engineers whom Mumma had sent to Argonne from the bureau were not responsible to Code 390.

By any standards the personnel of Code 390 were inadequate for the task at hand, but in terms of Rickover's conception of his assignment the organization scarcely offered a place to begin. Rickover had every intention of building the kind of organization he had created in the electrical section during World War II. That is, he wanted Code 390 to exercise the kind of control over design which had existed in the bureau before the war. Building such an organization would be difficult, not just in the ordinary sense of recruiting competent engineers but also in that it would run counter to the course the bureau as a whole was following in ship design and construction. The wartime experience had convinced many senior officers in the bureau—men like Mills, Morgan, and Mumma—that the only practical way to design and build the variety of highly complex ships which the modern Navy needed was to move most of the design and procurement functions to the shipyards and field installations. The bureau codes in Washington would perform only such broad management functions as issuing directives, approving general plans and specifications, and supervising field activities. As a reflection of this system, Mumma saw the engineering duty officer spending his years in the junior grades gaining practical experience in technical assignments in naval shipyards and laboratories to prepare himself for broader administrative and management responsibilities in the bureau during his later career in the senior grades. The Commission, influenced largely by Chairman Lilienthal's experience in the Tennessee Valley Authority, had gone even further and had permitted its laboratories to define to a large extent their own research and development programs. Rickover accepted the decentralization of design and procurement functions as essential in modern military technology, but he insisted upon retaining very tight controls over field activities.[1]

To build this kind of competence into Code 390, Rickover was careful to select only those who could demonstrate some practical knowledge and skill in engineering. He did not care whether they were officers or civilians if they could meet the high standards required in a design group. In fact, one of the first men he recruited was a civilian, Jack A. Kyger, who had earned a doctorate in chemistry at the Massachusetts Institute of Technology in 1940. Kyger had helped develop uranium processing techniques at the Mallinckrodt Chemical Works in St. Louis and had served as a chemist at Oak Ridge during the war. When Rickover met him at Oak Ridge in 1946, Kyger was chief of the engineering materials section in the laboratory. Because Rickover saw materials as one of the most critical problems in reactor development, he considered Kyger a valuable asset.

The first military additions to the original Oak Ridge group were three engineering duty officers, all lieutenant commanders who had completed graduate work in naval construction and engineering at the Massachusetts Institute of Technology. Robert V. Laney had been a classmate of Roddis and Dunford at MIT and had heard about the project when the original group visited the west coast during the summer of 1946. Excited by the prospects of working on the development of nuclear propulsion, he had applied for duty in Code 390. Archie P. Kelley and Jack A. LaSpada had both taken introductory courses in nuclear physics and had volunteered for the project. All three officers appeared to have the incentive and basic engineering ability Rickover was seeking.

As he had done at Oak Ridge, Rickover personally supervised the training of the new officers. He gave each of them the same kind of assignments the original group had tackled at Oak Ridge in preparing surveys of special materials, reviewing and summarizing technical papers, and drafting critiques of other reactor development projects. Roddis, Dick, and Dunford were responsible for detailed supervision of training, and the new officers very soon became involved in every aspect of the work in Code 390. During the last half of 1948 and most of 1949 the staff was still small enough so that everyone could have a part in virtually every activity.

Rickover also required Laney, Kelley, and LaSpada to undertake an intensive course of self-education in various phases of reactor technology. The course of study, carefully outlined for Rickover's approval, included the mastery of advanced textbooks in physics and engineering, special study assignments, and field trips to Commission installations. As outlined, the course would require "a total of 854 hours study or 16 hours per week, excluding time spent on field trips or special duty assignments."[2]

The "self-education" method was rigorous enough to provide adequate training for a few officers in the early phases of the project, but it could not meet the long-term requirements for Code 390. One solution was to make use of existing naval training programs. Early in 1949 Rickover investigated the possibility of adding courses in nuclear engineering to the naval architecture and marine engineering curriculum at the Massachusetts Institute of Technology. When Roddis and Dick visited the institute in January, they found interest among both the naval officers and the engineering faculty. The group suggested the addition of a survey course in nuclear physics to the existing graduate program. In addition selected graduates of that program would be assigned at MIT for another year for advanced study in nuclear physics and engineering. Rickover approved the proposal and arranged through the bureau to start the advanced course in June 1949.[3]

Most of the officers sent to MIT during the first two years were later to have positions of responsibility in the nuclear propulsion project. Lieutenant Commanders John W. Crawford, Jr., and Edwin E. Kintner in the 1949–50 class were both to have major technical assignments in Code 390 and in the field. Lieutenant Commanders John J. Hinchey and Arthur E. Francis in the 1950–51 class would make their greatest contributions as Rickover's representatives in shipyards where nuclear ships were to be constructed. Captain Robert L. Moore, Jr., whom Rickover initially considered as a possible deputy and his eventual successor as head of Code 390, later became supervisor of shipbuilding for the bureau at the Electric Boat yard during the construction of the *Nautilus*. While these officers were in training at MIT, Rickover maintained close touch with them, both to see that they were making satisfactory progress and to check on thesis work which he expected to have a practical application to the activities of Code 390. Rickover also used reports from these officers in proposing changes in the curriculum.[4]

At the same time Rickover was attempting to recruit qualified civilian engineers for the project. Because very few engineers had any knowledge of nuclear technology, he proposed organizing a new school at Oak Ridge somewhat along the lines of the one he and his original group had attended. Rickover insisted, however, that the new school place more stress on reactor technology and less on nuclear physics. With the help of Alvin Weinberg, the director of research at the Oak Ridge laboratory, Code 390 laid plans to start the first class at Oak Ridge in March 1950 and the second in September.

Rickover himself picked most of the students in the first class. I. Harry Mandil, an electrical engineer who had served as a reserve officer in Rickover's section during World War II, agreed to come back to the Navy as a civilian and enrolled in the one-year course at Oak Ridge. After completing the course Mandil would return to Washington and would be in charge of developing all new reactor systems in Code 390 for more than a decade. Howard K. Marks, who left the Puget Sound Naval Shipyard to take the Oak Ridge course, would spend most of his professional career as a senior engineer in the nuclear project. Another recruit from the wartime electrical section, Joseph C. Condon, would be responsible for component development in Code 390 for several years. The first Oak Ridge class also included several engineers Rickover had met in his work with manufacturers of electrical equipment during the war and five engineers from various bureau codes. Finally, Rickover persuaded the Electric Boat Company to send two young engineers to Oak Ridge on the grounds that the company would probably become involved in building one of the first nuclear submarines. Again Rickover took an intense personal interest in both the school and the students. In his opinion, nothing was more important for the future of the project than sound technical training.[5]

The Oak Ridge School of Reactor Technology, as it came to be called, became a training center not only for engineers in Code 390 but also for those working in the laboratories and in the offices of many contractors. Starting from the first class of twenty students, enrollment soon increased to 120. Within six years thirty-four scientists and engineers had completed the course under the sponsorship of Code 390, another twenty-two had come from other bureau codes and naval shipyards, and sixty contractor employees had completed the course. Because Rickover had taken the trouble to establish the school in a Commission laboratory as a project of the division of reactor development and not just the naval reactors branch, Oak Ridge was able eventually to provide a large number of trained engineers for the emerging nuclear industry.[6]

Another potential source of engineering manpower were officers and civilians in the Bureau of Ships. In the spring of 1950, when the engineers whom Mumma had sent to Argonne completed their training, most were available for assignments directly or indirectly related to the nuclear project. Most of the civilian specialists returned to their original codes in the bureau, where they could be expected to contribute some expert knowledge of the technical requirements of nuclear plants. Radkowsky, the only physicist in the group, joined Code 390 in October 1950. Most of the officers at Argonne were assigned to Commission field offices and laboratories—Turnbaugh to Pittsburgh to replace Wilkinson, who had returned to sea duty; Naymark to Argonne and then to Schenectady; and Roth to the Commission's Chicago office, where he could assist Tammaro on Navy matters. Only Barker came to Code 390 in Washington, where he replaced Libbey and became the group's expert on the test irradiation of reactor components.

During the summer of 1950 Rickover had also recruited Robert Panoff from the submarine propulsion section in the bureau. Panoff had been on Rickover's staff during the war as a civilian and had specialized in submarine propulsion systems in the bureau during the postwar period. He had never forgotten the high standards of technical excellence which Rickover had enforced both in the Navy and in industry. Panoff's stubborn insistence on quality and his seasoned knowledge of the bureau and its ways would be a major asset to Code 390. As the code's specialist in ship applications and relations with the bureau, Panoff did not take time out for the Oak Ridge course but moved directly into the problems of designing shipboard equipment for the first nuclear propulsion plants.

Other civilian engineers followed Kyger from Oak Ridge to Code 390 during 1949. Theodore Rockwell III had attracted Rickover's attention in conferences on reactor shielding at Oak Ridge. A Princeton graduate, Rockwell had worked as an engineer at Oak Ridge during the war. In Code 390 he would concentrate on a wide variety of technical problems in chemistry, radiation, and coolant technology. In time he would become one of Rickover's principal staff assistants. Others who had come earlier from Oak Ridge were Frank Kerze, Jr., a metallurgist, and William H. Wilson, an engineer, both of whom had gained experience in research on materials for nuclear application during the war.

As the nuclear project grew in the early 1950s, Rickover and his staff tried to provide the Bureau of Ships with the kind of engineering talent required in building nuclear ships. Engineers from Code 390 conducted training courses for officers and civilians in other codes, ranging all the way from the chief of the bureau to technicians in specialty codes. As the volume of work required Code 390 to bring in additional personnel without preliminary training, Rickover ordered the staff to set up a series of technical courses in the office. Junior engineers were constantly being prodded into classes taught by the senior staff on reactor theory, shielding, elementary physics, or mathematics. Rickover also arranged to have professors from universities in the Washington area give lectures for the staff, and he urged young engineers to enroll in night courses in nuclear engineering, naval architecture, management, administrative law, or public speaking. There was even an office course to teach clerks and secretaries the rudiments and terminology of nuclear engineering.[7]

In all these endeavors the central purpose was to build a staff which could take an active and effective part in designing and building nuclear propulsion plants for the Navy. First, that required people with the incentive and talent for creative engineering. It also demanded rigorous, practical training in the special skills and knowledge. As head of the project, Rickover considered teaching one of his most important responsibilities. He took training seriously, gave it a large amount of his time, and constantly strove to improve the effectiveness of these training activities.

Internal Organization

Just as Rickover drew on earlier associations and principles in recruiting and training his staff, so did he rely on experience in organizing and directing Code 390. During the first year, when his staff consisted of little more than the original Oak Ridge group, he could depend upon the informal personal

relationships which had always characterized that group. But as new personnel continued to arrive, he began instituting some of the techniques he had
employed in the electrical section during the war.

One of these was the use of "pinks." In July 1949 Rickover ordered all
stenographers in Code 390 to submit to him a pink copy of everything they
typed, whether it was incomplete, in draft, or in final form. At the end of
each day, Rickover carefully read all the pinks and annotated them with terse
comments, exclamation marks, and epithets which called attention to gram-
matical errors, careless expressions, vague terminology, and poor administrative tactics. Occasionally Rickover found it necessary to mark up two or three
drafts of the same document, but usually one critique was enough.

The pink system was obviously a good training device, but it also had a
more important function. It was one of the many ways Rickover kept in touch
with what was actually happening. The pinks permitted him to follow in detail the work of each staff member. If he saw an unsatisfactory response to a
question from a laboratory, he could sometimes intercept it in draft form before a commitment could be made in writing. Even when he accepted the
contents of a letter, he could add marginal admonitions which would alert the
staff to future dangers. Most important of all, the pinks furnished Rickover
with a source of questions, which were for him the fountain of technical management. He had learned over the years to question everything he read, no
matter who wrote it. The abrupt questions, sometimes only one word, cut
through unexamined assumptions or opened new areas for investigation.

In fact, it would not be much of an exaggeration to say that every idea,
every policy, and every decision in Code 390 began with a question. Rickover
never saw Code 390 as a static organization of engineers, each methodically
reviewing written reports and initialing routing slips. Rather, he saw to it
that the code became a loose confederation of men harried by overwhelming
technical problems and responsibilities, all too worried about the crisis of the
moment to give any thought to rank, protocol, or organization charts. No
matter how hard any individual worked, there was always Rickover or one
of his staff or even a laboratory scientist to raise a question, and the questions
were never too large or too small for Rickover to take up personally if he
thought it necessary.

The questions were almost always technical in context, and because sound
questions could affect the future of the project, they had to be taken seriously. If the question was really important, Rickover would assemble his
senior staff for a discussion. These meetings were spontaneous, animated, and

often abrasive. No technical questions could be too embarrassing to ask, and Rickover expected everyone to express his frank opinion, regardless of age, rank, or position. Everyone, including Rickover, stood on his own feet and argued his point on technical grounds. Silence was interpreted as assent, and a silent participant was only postponing the day of reckoning if he did not really agree with what was being said.

By their very nature, these meetings with Rickover could seldom be scheduled in advance. The participants varied according to the question under discussion. Sometimes the group settled the question quickly; sometimes the meeting became a shouting match in which several participants fought passionately for what they believed to be right. Even losing an argument to Rickover did not always provide the loser with an acceptable excuse for abandoning his position. If he still believed he was right, Rickover expected him to raise the question again later, even at the peril of sustaining a tirade for resurrecting an issue which had been settled. Sessions with Rickover could be bitter, disheartening, and deflating, but they could also be challenging and inspiring. Whether he "won" or "lost," each participant had the consolation that he had been able to argue his position directly and that his views received serious attention.

The most important advantage of the Code 390 meetings was that they assured that decisions were made on a sound technical basis. It was all too easy, especially in a military organization, for juniors to defer to seniors even when they knew the decision was ill-founded. Rickover had suffered superiors who made technical decisions on matters which they did not understand and then arranged to present their opinions in such a way that no one would dare to contradict them. Rickover's refusal as a junior officer to accept such decisions when he considered them wrong accounted for some of his unpopularity in the Navy. The rough-and-tumble technical meetings in Code 390 were designed to avoid this danger.

This kind of operation precluded the customary form of Navy organization, which was based on a hierarchical arrangement of positions with fixed duties assigned on the basis of military rank or civil service grade. Instead, Rickover created an essentially flat organization without precise titles or hierarchical levels. Titles were invented only to justify civil service grades, and the only evidence of hierarchy was that some members of the staff had more frequent entrée to Rickover than did others. Those who saw Rickover often came without formal designation to be part of his senior staff. Rickover assigned each of these men specific responsibilities in accordance with individual talents and the immediate needs of the project. When new needs or problems arose, Rickover reassigned or combined responsibilities as required. He was often willing to give a man far more responsibility than he had ever exercised before; but if the man failed, Rickover did not hesitate to relieve him.

Although each individual member of the staff had a reasonably clear understanding of his responsibilities, it was almost impossible to reduce the organization to a single chart or functional statement. The organization changed from week to week as personnel shifted or as new functions developed. Never worrying about assigning consistent titles to coordinate organizational units, Rickover made new assignments as the need arose. It was not at all unusual for one individual to be in charge of one function and a subordinate in another. In fact, there was usually some overlap in responsibilities, particularly between project officers and heads of technical sections. Some of this overlap was intentional on Rickover's part to assure him that more than one of his senior staff was worrying over every important question.

From the beginning Rickover used a combination of project officers and technical groups as his organizational base. During 1949 and early 1950, when most of the work centered on feasibility studies of the most promising propulsion systems, the project officers were Roddis for liquid-metal reactor systems, Dick for pressurized-water systems, and Dunford for gas-cooled systems.[8] Each project officer provided Rickover with an extra set of eyes and ears sharply focused on each project. The three officers were in constant contact with the contractors, mostly in terms of asking technical questions and suggesting new ideas. Another important function was coordinating the activities of the various contractors working on the project to avoid duplications and oversights in exploring technical questions. The project officers at this time were also responsible for a wide range of related functions such as contract administration, contractor evaluation, security, budgets, and reporting. (See chart 4.)

At the same time, Kyger was working with the technical groups, which were expected to concentrate on a wide range of design and development affecting all projects. Initially the technical groups were involved in such matters as investigating the physics of reactor designs, selecting materials for reactor systems, developing effective shielding against radiation, and starting the preliminary design of components. Here, as in the projects, Code 390 did not do the actual technical work. There were no drawing boards or test equipment in the Washington headquarters. Rather, the task was to direct the work

Chart 4.

Code 390 became the Nuclear Power Division; General Electric had joined the project; construction had started on prototype facilities at West Milton and in Idaho; and Electric Boat had begun ship design.

THE NAVY NUCLEAR PROPULSION PROJECT IN NOVEMBER 1950

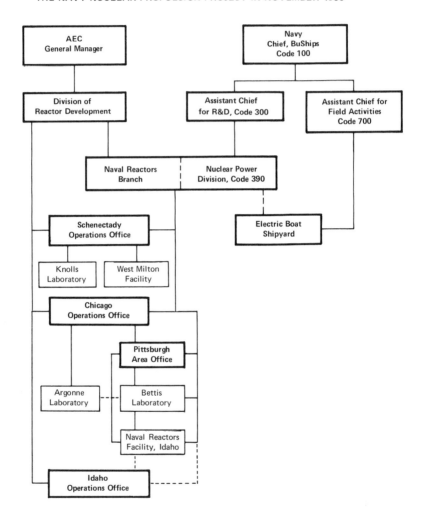

of each contractor to see that the technical data and later the equipment needed for the project were produced on time and in a useful and accurate form. The technical groups were supervising for the most part the same contractors who were working for the project officers. The overlap of responsibilities often caused friction within the organization, but it enabled Rickover to check one group against the other. If a project officer failed to detect or report a technical problem, a technical group might do so. If a technical group overlooked a vital point, the project officer could be expected to complain.

From week to week as the work evolved, Rickover changed his organization to meet new situations. By the summer of 1950 the idea of using a gas-cooled reactor in a submarine had been dropped as an immediate objective and that project had disappeared. Laney had replaced Roddis as head of the liquid-metal project and Dick continued to lead the water-reactor project. Kintner and Crawford had just completed the new course at MIT and had been assigned as assistants to Laney and Dick. Dunford was now in charge of submarine applications, which included the development of the steam propulsion system and all other problems of placing a nuclear reactor in a submarine hull. Panoff, who had just joined Code 390, would work under Dunford for a time before he took over the section himself. Roddis was now assisting Kyger and Rockwell with the growing number of problems confronting the technical groups.[9]

Although there were the usual changes in personnel and inevitable shifts in assignments, the basic structure of project sections and technical groups persisted into the early 1950s. As the work at Bettis and Knolls grew, Laney and Dick took on increasing technical responsibilities. Under Rickover's system they were completely responsible for everything related to their projects. This meant that they were answerable to Rickover for every question or criticism he might raise in their areas. They were expected to foresee needs, detect problems, and propose courses of action. Rickover usually discussed important issues with the senior staff and then made the decision himself. The senior staff during this period continued to include Dick, Roddis, Dunford, Kyger, and Laney; but as time went on, Rockwell, Panoff, and Mandil tended more and more to participate in the important meetings.

The Technical Environment

Code 390, as it emerged in the early 1950s, reflected Rickover's personal experience and his philosophy of technical management. The task, in his view, was one for engineers rather than administrators, for men who could understand the intricacies of design and manufacturing, who could take the initiative in engineering and direct the work of contractors. Thus the form of the organization at any particular moment rested as much upon the status of technical development at that time as it did upon the technical qualifications of the engineers assigned to Code 390. To move beyond the generalities of technical management, it is therefore necessary first to understand the fundamentals of the technology in which Rickover's group was involved.

In the early 1950s the task of Code 390 was to direct the design and development of two land-based prototypes: the Mark I version of the submarine thermal reactor and the Mark A version of the submarine intermediate reactor. Development studies for the Mark I at Argonne had centered around a system using pressurized water to transfer energy from the reactor to the propulsion equipment. Unless the system was pressurized, the water would boil and create bubbles, an activity which engineers at that time believed would make the reactor more difficult to control. For this reason the reactor would have to be placed in a large steel tank or pressure vessel similar to that shown in figure 1.

The pressure vessel would enclose the fuel elements containing uranium 235 in metallic form. The fuel elements would be fabricated with great care to assure high integrity against failure in an environment of high radiation and severe temperature changes. The entire core of fuel elements would be assembled with exceptional precision to guarantee satisfactory operation of the reactor. Water, which would both transfer the heat from the fuel elements and moderate the fission neutrons to thermal energies, would be pumped into the pressure vessel and forced through hundreds of channels between the fuel elements in the pressure vessel.

Figure 2 shows how the cooling water would circulate through the power equipment and back to the reactor. Heated (and now radioactive) water would leave the pressure vessel and flow to the steam generator or boiler, where energy from the water would be used to produce steam in the secondary system. The main coolant pumps would then return the water to the pressure vessel. The major purpose of the primary system was to contain all radioactivity. The entire primary system would have to be enclosed in a radiation shield and all components in the primary loop would have to be designed to operate for long periods without leaking. The steam in the secondary system would not be radioactive and thus the steam propulsion machinery would not have to be shielded.[10]

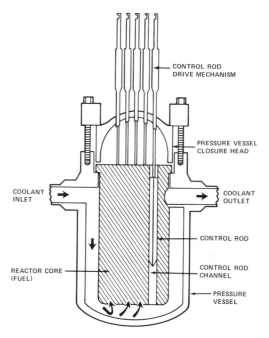

Fig. 1. A schematic diagram of a pressurized-water reactor.

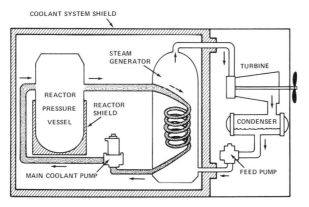

Fig. 2. Schematic diagram of a pressurized-water nuclear propulsion plant. Relative sizes not indicated.

Most of the new design features of the primary system were related only to the presence of radiation, but the steam generator or boiler had other novel requirements. Most marine boilers at that time consisted of assemblies of tubes carrying water. Surrounding the tubes were extremely hot gases derived from the combustion of fuel. Heat from the gases flowed across the walls of the boiler tubes and flashed the water to steam. Nuclear propulsion required radical changes in boiler design because the heat source was relatively cool water rather than furnace gas. Flowing through the boiler tubes, the water would give up its heat, which would move from the inside to the outside of the tube, where it would convert water into steam.[11]

Although steam propulsion systems in naval surface ships had been commonplace for almost a century, the Navy had always found steam impracticable for a submarine. One of the truly perplexing engineering problems Rickover faced was how to arrange the steam generators, piping, turbines, and condensers within the limited confines of a submarine hull. Even if this could be accomplished, there was the all-important matter of providing enough air conditioning to keep temperatures in the steam machinery areas down to habitable levels.

Some of the characteristics that distinguished the sodium-cooled Mark A from the water-cooled Mark I stemmed from the difference in neutron energies. Designers of the Mark A had chosen a higher (intermediate) neutron velocity for the reactor. Intermediate neutron velocities would be achieved by partially moderating the fast neutrons created in the process of fission. The moderator would be a series of beryllium reflectors surrounding the fuel elements. Although beryllium was toxic and difficult to shape, its outstanding nuclear properties had attracted attention in the early 1940s, and this widespread interest had produced a substantial body of information about the metal.

The second distinctive feature of the Mark A was its use of sodium as the heat-transfer material. Liquid sodium had excellent thermal properties: high thermal conductivity, relatively high specific heat, and a large volumetric heat capacity. These qualities offered the possibility of attaining higher temperatures than in the Mark I and of using more efficient steam equipment. The use of sodium also eliminated the need for the high pressures required in the system for the Mark I. A potential but then undemonstrated advantage of sodium was the possibility of using an electromagnetic pump which involved no moving parts and, hence, could be completely sealed against leakage.

The Mark A design did have sobering disadvantages. For one thing, it would require more fissionable material than the Mark I for a comparable power capacity. But the most telling drawbacks were linked to the use of sodium. Neutrons within the reactor would transmute some of the coolant into sodium 24, an isotope which has a half-life of about fifteen hours and emits gamma radiation of very high energies. As a result every component and pipe in the primary system would have to be shielded and, unlike the Mark I plant, the compartment containing the reactor and the primary system could not be entered for maintenance until many hours after the reactor had been shut down. Sodium also had the inconvenient property of being a solid at room temperatures, which meant that it would freeze in the pipes unless they were continuously heated when the reactor was not operating. In addition, sodium reacted violently with water. Although Knolls and other laboratories had learned much about minimizing the dangers of handling sodium, a leak which brought sodium into contact with water could be disastrous.[12] The Mark A was too promising to be overlooked, but it did not offer any easy shortcuts to a submarine propulsion system.

For a summary of the essentials of the two reactor systems, see table 1.

Table 1.   Submarine Reactor Systems

	Submarine Thermal Reactor	Submarine Intermediate Reactor
Prototype Designation	Mark I	Mark A
AEC Contractor	Westinghouse	General Electric
Prototype Location	National Reactor Testing Station, Idaho	West Milton, N.Y.
Fuel	uranium 235	uranium 235
Moderator	water	beryllium
Coolant	water	sodium
Neutron Energy	thermal (low)	intermediate
Ultimate Use of System	*Nautilus*	*Seawolf*

Laying the Foundations of Technology

From the beginning Rickover insisted upon focusing his attention on specific projects which would lead to a practical nuclear propulsion system. He was ruthless in eliminating research that did not contribute directly to these projects. This focus did not mean, however, that Rickover took a narrow view of what was needed to achieve success. He did not attempt to throw some hardware together and then tinker with it until it worked. Quite the contrary; Rickover was determined to build the propulsion plants on a solid technological base. Reliability, essential in submarines, depended upon a thorough understanding of the physical, chemical, and nuclear forces operating within each component of the system. In Rickover's opinion, one of the weaknesses in the Commission's reactor projects was that the designers overlooked engineering fundamentals in their impatience to build an operating reactor. Rickover had no intention of repeating such mistakes.

From his first days at Oak Ridge Rickover had understood the fact that the necessary technological base for designing propulsion reactors did not exist in the United States or anywhere else in the world. His almost instinctive reaction as an engineer was to begin to assemble available data and then to add new information in a systematic way. The initial papers prepared by the Oak Ridge group were a step in that direction, and it became a primary function of Code 390 and the contractors to lay the foundations for reactor technology.

The layman's common impression of the nuclear sciences was that they involved extremely complex and esoteric conceptions that were far beyond the understanding of ordinary men. In some areas of reactor physics this impression was correct; in other areas it was less true, although real ability in engineering was required. In general, however, the striking feature of the research initiated by Code 390 during the early 1950s was its elementary nature, its attention to the sorts of basic measurements and analyses which physics and engineering students performed in college classes. It was exactly the sort of research which many scientists and graduate engineers would disdain and yet it was precisely the kind of information Code 390 needed before the reactors could be designed.

In all of classical physics and engineering perhaps no material was more commonly used than water. Its very abundance, its convenient properties, and man's vast experience in using it made water an exceptionally attractive material as a heat-transfer medium in a reactor. These considerations had entered Rickover's decision to develop a water-cooled reactor. Yet in reviewing existing data on water technology, Rickover and his associates were surprised to discover how little was known about the properties of water itself or its effects on materials. There was little understanding of how metals, or even oxygen, became dissolved or suspended in boiling water in conventional steam plants. Water corrosion effects on stainless steel systems for reactor plants were even less known. Changes in the temperature, flow rate, or chemical composition of the water could create deposits which would carry radioactivity to external portions of the plant, foul heat transfer surfaces, or cause sticking and galling of mechanisms. Under Kyger's direction, William H. Wilson and later Mandil coordinated a variety of laboratory studies on corrosion and wear in water systems.[13]

Because research on sodium systems antedated the Mark A project, Code 390 did not have to initiate all the studies at Knolls and other laboratories. Rather the task was to focus research activities and compile results. To speed the work, Rickover suggested that the Office of Naval Research and the Commission cooperate in preparing a handbook of all available information on using liquid metals for heat transfer. This handbook, first published in June 1950, was the first of a series on reactor engineering to appear over the next decade under Code 390 sponsorship. The *Liquid-Metals Handbook,* like those which followed it, created a literature that was an essential part of the technical foundations not only for the Mark A and Mark I projects but also for reactor development in general.[14]

Equally as important as the heat-transfer medium in a practical nuclear propulsion plant was the shielding which would protect personnel from the extraordinary amounts of radioactivity generated within the reactor. During the year at Oak Ridge Rickover had realized that the massive shielding used in the land-based production reactors at Hanford provided little applicable experience in designing an effective shield for a submarine plant. In the fall of 1946, while at Oak Ridge, Rickover had asked Libbey and Blizard to compile a technical summary of information on shielding. The report explained the types of radiation and the possibility of constructing shields from different combinations of materials. Far from explaining away the problem of shielding, the report heightened, if anything, Rickover's concern over the difficulties of controlling radiation.[15]

So vital to Rickover was the question of shielding that he insisted upon examining all the fundamental assumptions involved, even the accepted standards for radiation protection. He approved the conservative radiation standards adopted for the wartime project, and he accepted the possibility that a very low level of radiation might later be found to have some effect on man. To check these ideas, Rickover invited Hermann J. Muller, the world-famous geneticist, to discuss radiation effects with the Navy group. These and other discussions gave Rickover a firm understanding of the subject and helped him to take a practical and effective approach to shielding design.[16]

Even before Code 390 was established, Rickover had arranged a Bureau of Ships contract with the Massachusetts Institute of Technology for shielding studies. At the first national shielding symposium at Oak Ridge in September 1948 he urged the nation's scientists and engineers to concentrate on the practical problems of shielding design, but despite his prodding not much progress was made in that direction until Rickover forced an agreement at the bureau in November 1949 on the materials to be used in the Mark I shield. Rockwell worked with the Oak Ridge staff in organizing a series of experiments to test the performance of these materials in the X-10 reactor at Oak Ridge. Westinghouse then had to translate the experimental results into shielding design which Rickover personally evaluated against civilian standards for radiation exposure. Conservative as these standards were, Rickover accepted the possibility that they might have to be revised later, and Rockwell's shielding group continued to compile basic data on shielding, most of which later appeared in the *Reactor Shielding Design Handbook*.[17]

Just as the development of reactor shielding brought Code 390 into fundamental studies of natural phenomena, so did the design of the reactor itself. Here Rickover relied heavily upon physicists, chemists, metallurgists, and other specialists who could provide authoritative judgments on underlying theories and conceptions. Initially Rickover depended upon the outstanding scientific resources of the Oak Ridge and Argonne laboratories, but he began almost at once to broaden the base of this resource, primarily at Bettis and Knolls. To assure that scientific research in the laboratories was properly coordinated, Rickover built real strength in reactor physics and metallurgy in Code 390. Radkowsky, after a year of practical experience at Argonne, became not only an effective overseer of physical research within the project but also a creative innovator in his own right. The originator of several new design principles of water-cooled reactors,[18] Radkowsky could hold his own with any physicist in the project and helped to assure that reactor designs were based on a sound analysis of physical theory. Robert S. Brodsky, working under Radkowsky and Rockwell, specialized in the application of high-speed digital computers in reactor physics and design. As a result, the Navy project soon became a center for information on computer codes for this purpose.[19]

Rickover's development strategy inevitably led the Navy project into studies of new materials. As we saw in chapter 3, he early recognized the potential advantages of zirconium and was not deterred by the fact that the metal was extremely expensive and not available in commercial quantities.

More important in Rickover's mind was zirconium's low affinity for neutrons, which promised a more efficient use of uranium than was possible in reactors containing aluminum members. As long as uranium was a very scarce material, Rickover was determined to conserve that resource even if it was necessary to use expensive materials like zirconium. Rickover's engineering experience also led him to believe that the price of zirconium would drop quickly once it was in commercial production.

In addition to zirconium, the Navy project also required substantial quantities of beryllium for the Mark A plant and later hafnium for the Mark I control rods. All of these materials were relatively unknown in American industry, and their use would require extensive study of their physical, chemical, metallurgical, and nuclear properties. For each metal, research included a study of ore-bearing materials, methods of extraction, processes for reducing the material to metal, and special techniques for fabricating, treating, and testing the metal.[20] In each instance the Commission's laboratories and many other research institutions were involved in these fundamental studies of materials. The task of coordinating and directing these activities fell on the technical groups in Code 390.

The demonstrated competence of Code 390 even in such sophisticated disciplines as reactor physics and metallurgy did more than assure sound direction of research in the Navy project. It also gave Code 390 the ability to exercise positive leadership in development rather than just a passive review of the work of others. By focusing scientific resources on practical problems, Code 390 produced the technical data needed for Mark I and Mark A. But far more important in the long run were the new ideas generated in all these technical areas. Furthermore, Rickover made certain that this new information would be available by stimulating the preparation of a dozen technical handbooks. In addition to those already mentioned, these included *The Metallurgy of Zirconium* (1955), *A Bibliography of Reactor Computer Codes* (1955), *The Metal Beryllium* (1955), *Corrosion and Wear Handbook* (1957), *The Metallurgy of Hafnium* (n.d.), and the three-volume *Physics Handbook* (1959–64). Summing up years of fundamental studies in many laboratories, these handbooks became important building blocks of a new technology.

Producing Materials

The use of unfamiliar materials like zirconium, hafnium, and beryllium in submarine reactors posed problems for Code 390 going far beyond the compilation of fundamental data. It was also necessary to establish commercial facilities to produce these materials or to develop new processes that were suitable for large-scale production. The difficulties faced in producing large amounts of pure zirconium illustrate the kind of stimulus which Code 390 could bring to technical development.

In 1947, when Rickover decided to use zirconium as the principal structural material in the core of the Mark I, there were two promising methods of producing the material. The oldest was the de Boer or "crystal bar" method, which produced small bars of metal by decomposing zirconium tetrachloride on a hot filament. The process was capable of producing zirconium of high purity, but the quality of the product was not always predictable and the costs were high. The Foote Mineral Company of Philadelphia—the sole producer of crystal bar in the United States—sold the product in 1948 for prices between $135 and $235 per pound, and output during that year was slightly more than eighty-six pounds.

The second process, under development at the U. S. Bureau of Mines' laboratory in Albany, Oregon, rested on work by William J. Kroll, who tried to produce pure zirconium by reducing the tetrachloride with magnesium. The product was a zirconium sponge which could be melted and consolidated into metal. Although the process was promising, it had serious weaknesses. The greatest flaw was that metal produced from the sponge was not particularly resistant to corrosion. Various ways of correcting this deficiency, however, had been suggested.[21]

If Code 390 could have negotiated contracts directly with the producers, the strong technical direction which Rickover's staff brought to other matters could have been applied to zirconium production. But the Commission had already assigned responsibility for procuring zirconium and other metals to its New York office. To obtain zirconium for the Navy project, Geiger at Pittsburgh had to work through a chain of Commission offices extending to New York and back to Washington. Unable to bring pressure directly on the production plants and laboratories, Rickover became increasingly impatient during 1949 with the slow pace of development, even though Code 390 had not yet established precise specifications for the zirconium to be used in Mark I. Finally, in March 1950, Rickover appealed to Hafstad and the Commission's general manager for authority to deal directly with the producers in filling the Navy's requirements for zirconium. Believing that Rickover was pushing too fast and that recent experiments at Oak Ridge would improve the product of the sponge process, Hafstad was reluctant to give Rickover

6. Alexander Squire, a Bettis engineer who played a major role in the first mass production of zirconium metal, examines some of the crystal bar zirconium produced for the Mark I reactor.

independent authority; but it was impossible to ignore his impassioned arguments that delay was threatening the schedule for Mark I. In May 1950 the Commission gave Rickover his own hunting license for zirconium.[22]

Rickover was ready to move. From a survey of all contractors engaged in zirconium research or production, Code 390 had gathered data on processes, requirements for feed materials, the possibilities for increased output, and cost estimates. A detailed report to Rickover on July 11, 1950, concluded that the best chance for meeting the Mark I requirements was to build a crystal-bar plant at Bettis. Geiger had already found a way to stretch the Westinghouse contract to cover zirconium production. Weaver had submitted plans to install the capacity to produce 3,000 pounds of corrosion-resistant zirconium per month. Two days later Rickover accepted the spectro-chemical criteria and corrosion standards which Argonne and Bettis had proposed for Mark I zirconium and authorized Weaver to build the plant in one of the high-bay areas at Bettis. Geiger had done a splendid job in clearing away administrative problems and helping Bettis obtain the necessary materials. A little more than twelve weeks later the plant was in operation.[23]

The zirconium incident demonstrated how quickly and effectively Code 390 could move on a technical problem affecting the Navy project. In the larger dimensions of the Commission's reactor development effort the more leisurely approach taken by the Commission staff would have led to a satisfactory production process eventually. But Rickover would not accept the delay involved and he refused to permit the success of the Navy project to rest with an organization not in his control. The lesson seemed to be that projects should control not only the central activities in technical development but also the source of vital materials. From Rickover's perspective the argument made sense, but its extension to all aspects of the project would ultimately conflict with the aims of the Commission and the Navy.

Directing the Contractors

Although Rickover and his staff became involved in a wide variety of activities as the project grew, the central task was always the direction of technical work in the laboratories, shipyards, and component fabrication plants. In a broad sense, it was a responsibility common to every development project in the government. Somehow the sponsoring agency had to monitor contractor activities and see that the product met specifications. The Navy project was nothing unique in this respect.

What did set Code 390 apart from most other government projects was the way in which Rickover and his organization went about doing that job. In oversimplified terms, a common government approach could be described as a process of passive review. The project officers in government were primarily administrators, although they often had some relevant technical training or experience. At the very least they saw their responsibilities in terms of contract administration; at the most they might presume to suggest a course of action to a contractor or question a contractor's decision. Direct intervention in technical decisions, however, was considered inadvisable. It would mean that the project manager or administrator was second-guessing the experts who had been hired to do the technical work. The assumption on which the decentralized project system existed was that the central project office could not presume to possess the expert technical knowledge and experience required in all the specialized aspects of a complex development effort.[24]

Rickover had organized Code 390 on a different but not entirely converse assumption: that the central office had to have enough technical competence to control and evaluate contractor activities as well as to administer contracts. The actual research and development activities were still to be dispersed to contractors in the field, not because of a lack of technical competence at headquarters but simply because it was impractical to perform such work in Washington.

As we have seen earlier in this chapter, the distinctive feature of Code 390 was that it was an informal organization of technical specialists rather than a hierarchical structure of administrators. Full responsibility for contractor activities rested with the project officers and technical groups in Washington. Every staff member, whether a military officer or a civilian, was answerable to Rickover for every incident or action affecting his area of responsibility. His task was to keep Rickover informed and to be sure that his response to every situation conformed to the current operational policy of Code 390.

In exercising his responsibilities, each staff member occupied a precarious position between Rickover and the contractors. He had to be meticulous about carrying out the spirit as well as the letter of Rickover's orders; yet he was expected to exercise his own initiative in questioning the contractor's activities or proposing new ideas. The staff member was Code 390's daily contact with the contractor; yet he knew that the contractor could go directly to Rickover if he found the ideas coming from Code 390 negative or off the mark.[25]

Most difficult of all for the Code 390 staff member was the proper exercise of authority. There was never any question that he was to influence the direction of contractor activities. Yet it was never permissible to do this by issuing a direct order. He could suggest and object, but he could not order.[26] The distinction rested on Rickover's conviction that the responsibility for technical work had to remain with the contractor. In this respect he accepted the general practice in the government that the contractor should be permitted to do the job he had been hired to perform. But to Rickover this did not mean that the contractor was beyond criticism. Thus the staff member in Code 390 was expected to question everything dubious he saw, to express Code 390's dissatisfaction, and to suggest ways of improving the situation. The contractor had to produce convincing evidence that the equipment would work, and this required almost endless testing.

In actual practice Code 390 followed contractor activities so closely that the distinction between a suggestion and an order was sometimes not very clear. A common difficulty was that contractors tended in time to consider every suggestion as an order to be followed blindly. Rickover found this reaction unacceptable because it permitted the contractor to shirk his responsibility and did not provide for the free exchange of honest opinions on technical matters. Sometimes, however, the source of the difficulty lay with Code 390, as when the suggestion was so sharply pointed that it amounted to an order or when Code 390 had already decided what course the contractor should follow and successively rejected solutions until the "right" one was presented. Like all principles in management, this one had its limitations, but for the most part it created the kind of environment Rickover wanted for technical development. ·

Rickover understood intuitively how his system of contractor direction was to work, but he did not set down these principles in hard and fast terms. Even had he thought of doing so, he would have found pronouncements of this kind contrary to his conviction that technical development was not a matter of rules and organization but of competent engineers applying their talents to specific technical problems in an effective way. Both the Code 390 staff and the contractors had to learn by doing, and the learning experience was not always pleasant.

Because they were the first major contractors on the Navy project, Argonne and Bettis had to bear the brunt of the learning experience. At the time the division of responsibility between the two laboratories was defined in December 1948, it was impossible to predict just how the work would evolve. Rickover, always more interested in technical questions than administrative niceties, assigned work where the capabilities lay at the moment. The result was that by 1950 Argonne was involved to some extent in what seemed purely engineering activities while Bettis had some assignments which bordered on fundamental research.

Even more disruptive to the laboratories was the constant barrage of questions, criticisms, and demands from Washington. Lieutenant Dick, the Mark I project officer, could be as outspoken and unyielding as Rickover himself, and engineers in Code 390 technical groups proved tenacious and unrelenting in their criticism both of personnel and results. The continued application of Code 390's sharp spurs did goad both laboratories into extra effort, but the results were not always impressive. By early 1951, reports from the laboratories showed that many aspects of the project were falling far behind schedule. Most ominous of all were reports on the design of such reactor internals as fuel elements and control rods.[27]

Influencing Mark I Design

Coming up with a practical design of fuel elements had proved an especially difficult task. Not until March 1950 did Argonne and Bettis decide that it would be feasible to assemble a fuel element consisting of a uranium-zirconium alloy clad with zirconium. Once that decision was made, the laboratories had struggled for months to develop a workable process for producing the hundreds of fuel elements needed for the reactor. Even the processing of zirconium metal proved extremely tricky. Bettis found that zirconium easily picked up contaminants during processing and fabrication, and these could destroy the corrosion-resistance of the material. Once pure zirconium was available, similar precautions were necessary in forming the uranium alloy and then in bonding the material to the zirconium cladding. Any hope that fuel element fabrication could be farmed out to a subcontractor disappeared during 1950.[28] Even after years of experience, Bettis did not seem to be getting much closer to a workable process in the spring of 1951 than it had been a year earlier.

The uncertainties surrounding the development of the control rod mechanisms for Mark I were even more complex. The control rods containing a neutron-absorbing material would move in grooves between the fuel elements. When the control rods were inserted into the core, they would shut down the reactor. As for all components of the reactor, there was first of all the straightforward problem of choosing the best materials. Argonne had early decided to use an alloy of silver and the neutron-absorbing material cadmium, which would be bonded by hot rolling to strips of stainless steel.

The laboratory was confident that the control rods would work, but no one could be sure until some had been fabricated and tested in a hot-water environment similar to that expected in a reactor.[29]

As Bettis began to inch forward on zirconium production in 1950, an alternative to the silver-cadmium rods appeared. The zirconium process involved the extraction of the closely related element hafnium, which was itself a good neutron absorber. Thus the production of zirconium gave Bettis a growing stock of pure hafnium, and the question arose of whether hafnium would make a better control rod material.[30] An alternative design always offered an engineering opportunity, but it could also introduce an uncertainty which could delay the project. No one could then foresee that late in 1952, when Mark I was virtually complete, it would be necessary to shift from cadmium to hafnium.

The design of the drive mechanism for the control rods was complicated because the task involved conflicting considerations. On the one hand, there was a distinct advantage in being able to control each rod individually in order to achieve optimum power distribution in the core throughout its life. On the other hand, the rods had to operate within the reactor pressure vessel. Either the entire drive mechanism and motors had to be inside the pressure vessel, or leak-tight seals would have to be developed to connect any external portions of the mechanism with the control rod. To avoid the difficulty of maintaining effective seals for the control rods, in 1950 Argonne proposed using just two drive shafts and a complicated mechanical system of racks, pinions, and gears within the pressure vessel to power two groups or "gangs" of rods. The mechanism, however, was so complex that it did not appear to be a promising engineering solution. Both Argonne and Bettis were studying a number of alternative systems, and this trend toward multiple solutions again suggested delay.[31]

The design of the fuel elements and control rods were important concerns of Code 390 in 1950 and early 1951, but they were by no means the only ones. In each case, Lieutenant Dick, as the project officer, tried to coordinate activities at Argonne and Bettis. Kyger, Roddis, Gerald H. Welsh, Mandil, Kerze, Wilson, Crawford, and others in the Code 390 technical groups tried to follow activities in the laboratories through reports, daily telephone conversations, and weekly visits. Naymark at Argonne, and Geiger and Turnbaugh at Bettis, served as Rickover's personal representatives. Through this network of knowledgeable technical staff, Rickover was able to bring an extraordinary amount of pressure on the laboratories. The truth was, however, that all the suggestions, criticisms, reorganizations, and personnel changes did not measurably improve the situation during the first six months of 1951. Everyone recognized the growing crisis, but no one seemed to be able to do much about it. To cite only the two examples used in this chapter, fuel element manufacture continued to be plagued by a swarm of minor technical troubles, all of which added up to failure; and for the control rod drive mechanism, Rickover and Code 390 were growing even more disenchanted with Argonne's "gang" system, even though there was little reason to believe that individual external drives would be practical.[32]

"Quaker Meetings" at Bettis

During the dozens of meetings with his senior staff during the spring of 1951, Rickover gradually evolved a strategy to meet the burgeoning troubles on Mark I. The scores of technical problems, he and his staff concluded, were merely symptomatic of a more fundamental deficiency that lay in the relationship between Code 390 and the contractors. Somehow the pressures of recent months had polarized relationships rather than brought the laboratories closer to Code 390. Rickover believed the underlying problems could be discovered only if a group of engineers from Code 390 and Bettis set aside their immediate concerns and concentrated on the more basic issue. He suggested a sort of "Quaker meeting," in which both sides would meet together and, if necessary, just sit in silence until they could begin to talk with each other as individuals rather than as spokesmen for their organizations. The group decided to give the idea a try, and Rickover had no trouble getting Bettis to participate. Westinghouse had used special conferences to solve difficult organizational and technical problems in the past.

The group, known as the project review board, first met at Bettis on June 5, 1951. Philip N. Ross, a senior engineer at Bettis who had worked with Rickover in the electrical section during the war, served as chairman with two other Bettis men. The representatives of Code 390 were Dick, Panoff, and Roth from the Commission's Chicago office. Assigned a conference room at Bettis where they would be isolated from project activities, the members of the review board began the arduous and sometimes painful process of getting to know each other as individuals, of stripping away the institutional allegiances that concealed true feelings on both sides.[33]

It took ten days of discussion to break down the barriers of hostility, distrust, and misunderstanding which had grown up during months of pressure and frustration on the Mark I project. Finally, when the individual members were able to talk "with" rather than "at" each other, they found they could also discuss the project objectively. A test of this new-found perspective came when Rickover asked the review board to study the impasse in the design of the control rod mechanism. Concentrating on the technical problems and ignoring organizational loyalties, the board soon concluded that there was no valid reason for continuing work on the "gang" system. The complicated mechanical arrangement would never be reliable; nor could it be maintained once the reactor was critical. For better or worse, Bettis would have to use individual drives with external motors and find some way to seal the shafts against the leakage of water and radioactivity.

Perhaps the members of the review board exaggerated the importance of the "Quaker meetings" in breaking the control rod impasse, but they did come to some more general conclusions about organizational relationships which could have an impact on the future. Exercising great care to be objective, the group analyzed some of the practices which had caused a breakdown of effective communication between Code 390 and Bettis. One of the most revealing of these was the "fire drill" syndrome. When a Code 390 staff member in Washington discovered a discrepancy or error in some laboratory activity, he usually called his counterpart at Bettis on the telephone. Fearing that any delay in response might bring additional calls from Washington, perhaps even from Rickover, Weaver often summoned an urgent meeting of the senior project staff. These meetings could divert group leaders from their regular responsibilities for several days. Sometimes Weaver appointed a special committee just to investigate one particular problem, and it was not unusual for another crisis to emerge and another committee to be appointed before the first had completed its work. Thus, instead of systematically pursuing each aspect of development according to a logical plan, Bettis came to be more and more preoccupied with "fire drills."[34]

The board found that the responsibility for the fire-drill syndrome rested on both sides. Code 390, in its zeal for action, tended to consider every problem a crisis demanding immediate attention. Bettis tended to overreact to these requests and did not stop to evaluate the priorities involved in shifting key technical staff from regular work to the crisis. Underlying the actions on both sides, the board detected a lack of confidence in the other side and even a suspicion of the other side's integrity. All too often engineers in the laboratory would not accept criticisms or instructions from Washington at face value, but rather looked for hidden motives or "political" connotations. There was also a common impression in the laboratory that Code 390 constantly forced technical decisions without adequate supporting data. An objective examination of such incidents convinced all members of the board that such impressions were in fact not valid but resulted from the way in which Code 390 brought "pressure requests" to Bettis. Such problems, in the board's opinion, could be avoided by promoting better understanding and mutual respect between the two organizations.

Taken out of the context of the crisis at Bettis and Argonne in the spring of 1951, these conclusions sound very much like platitudes, and yet they were in the long run to be more important than the specific organizational changes which the board recommended. The conclusions were perhaps only inadequate symbols of a deeper understanding which the engineers in Code 390 and Bettis had reached. This outcome tended to support, and perhaps was even the product of, Rickover's predilection to rely on individuals rather than organizations in engineering development. The fundamental flaw, the board decided, was not in the organizational deficiencies it found in both Code 390 and Bettis but in the inability of the two groups to focus their discussions on purely technical issues. That was a difficult task for all men, but Rickover saw it as absolutely essential in the process of technological development.

Relationships at Argonne

Rickover never had an occasion to use the "Quaker meeting" technique at Argonne. By the time he tried the idea at Bettis, Argonne was already moving out of the Navy project. Besides, to be successful, conferences of this type had to start with a certain unity of purpose. Bettis had been created specifically for the Navy project, and Price as president of Westinghouse was determined to make the project a success. There was never any question that Bettis would do whatever was necessary to build a nuclear submarine, even if it did not always agree with the technical strategies and procedures imposed by Code 390.

At Argonne the situation was entirely different. Argonne had existed as a laboratory long before the Navy project started, and its responsibilities to the Commission extended far beyond the nuclear submarine. Zinn, the tough-minded director of Argonne, took every precaution necessary to keep Rickover from establishing the kind of influence at Argonne that he exercised at Bettis. The Navy project would never be more than just one of Argonne's assignments, thus falling short of the total commitment Rickover demanded from his contractors. Although Rickover hoped to continue using Argonne for fundamental research, he was transferring all engineering activities on the Mark I to Bettis as quickly as possible. Once Argonne's work on the Mark I was completed, Zinn had little interest in having the laboratory play a subordinate role as a research contractor for the Navy.

Knolls: A Struggle for
Technical Control

Code 390's relationships with Knolls fell somewhere between the high degree of cooperation achieved at Bettis and the almost complete lack of Navy penetration of Argonne. General Electric's larger responsibilities to the Commission and the company's interest in developing a commercial power reactor fostered an attitude of independence which never developed at Bettis. Especially important at Knolls was the fact that the laboratory did not owe its existence to the Navy project as did Bettis. As a result, physicists rather than engineers set the tone at Knolls just as they did at Argonne. Kingdon, the technical director, was a physicist and had authority in the laboratory equal to Milton's as general manager.

Rickover had every intention of establishing at Knolls the same kind of relationship he was creating at Bettis. The task would be more difficult at Knolls, but Code 390 could draw on the Bettis experience. When Laney took over as project manager for the Mark A in September 1950, Rickover sent Dick to Schenectady to help Laney get started. Rickover himself visited Knolls in October 1950 to explain the management controls he intended to establish. These included an array of detailed reports and schedules to be developed with Code 390. Again, Rickover's intention was to develop a true engineering partnership which would permit an open and frank exchange of technical ideas. Laney as the project officer and LaSpada as Rickover's local representative were to be the principal contacts with Milton. Kyger and Roddis as senior staff in the Code 390 technical group would work through Kingdon on engineering details. Kerze, Rockwell, and Kelley also had counterparts at Knolls on various aspects of the Mark A design.[35]

Despite his efforts to establish an effective working relationship with Knolls, Laney had to admit by early 1951 that he was making little headway. Knolls, in the opinion of the Code 390 representatives, was still a research and development laboratory. Kingdon and the physicists were still in control, and there was no sense of urgency about building Mark A on a definite schedule. The Code 390 representatives contended that Knolls continued to think of Mark A as a research facility and not as a prototype of a submarine propulsion plant. Knolls, in Code 390's opinion, was not organized as a development project. There was the possibility that Code 390 might have been at fault in not making its objectives clear. There was even some feeling at Knolls that Code 390's constant intervention and criticism meant that the Navy did not really want to see the submarine built. But Code 390 still believed that the division of responsibility between Milton and Kingdon was largely responsible for the absence of a strong sense of purpose at Knolls. In a meeting with General Electric officials in March 1951 Rickover stated plainly that he considered the management of the laboratory unsatisfactory.[36]

In response, General Electric reorganized Knolls in June 1951, but the change was hardly an improvement in Rickover's estimation. Milton was given full control over engineering and reported directly to Henry V. Erben, an executive vice-president who was experienced in production matters. But Kingdon was still in charge of research and would report to the General Electric vice-president for research, thus preserving or even strengthening the duality in laboratory management. To make matters worse, General Electric planned the reorganization and informed Rickover only two days before it was to become effective. Once again the company had deliberately thwarted Rickover's attempt to establish a technical partnership. In the opinion of Ralph J. Cordiner, the president of General Electric, Rickover was just one of the company's many customers and as such he would receive no more and no less consideration than any of the others.

In January 1952 Laney was still complaining about the unsatisfactory relationships between Code 390 and Knolls. "Our relations are marked by mutual suspicion and distrust rather than by understanding and collaboration. We are not working as a team."[37] Rickover raised this problem with General Electric officials and persuaded them to participate in a "Quaker meeting" of the kind that had been successful at Bettis.[38] Dick and Panoff joined Laney as the Code 390 representatives, presumably to bring some flavor of the Bettis experience to the sessions. The three men were also careful in choosing the Knolls personnel to be on the team, but all these precautions were to no avail. The group never even began to establish rapport, and the meetings were abandoned after a few sessions. As relationships between Knolls and the Navy worsened in 1952, Milton, who had been brought to Knolls primarily to serve as a conciliator between the company and Rickover, had little reason for remaining. What Rickover demanded was not a conciliator who could slide over the issues but a hard-headed counterpart who could face the issues squarely on their technical merits. In May 1952 Milton resigned to be replaced by Karl R. Van Tassel, an electrical engineer who had been with General Electric since 1925.[39]

Lessons for the Future

Although matters improved under Van Tassel, Code 390's relationships with General Electric still fell far short of the cooperation achieved with Westinghouse. In the final analysis, the difference between Rickover and General Electric lay in the definition of "customer." Rickover believed the customer in a major development project had to function as a partner on the purely engineering aspects. He had no intention of trying to "run" the company by interfering in company finances and administration, but he believed that the building of a nuclear submarine required knowledgeable technical direction from the Navy. General Electric, however, saw Rickover as a customer much like those who ordered toasters or turbines. Looking back on the situation in 1954, Rickover wrote that the company's attitude had been: "Give us money, do not bother us, and we will do the job." That was an approach Rickover could never accept, because he had long since learned that it would not produce reliable equipment.

Rickover's experiences with Westinghouse and General Electric illustrated both the advantages and limitations of his management approach. At Westinghouse, where his definition of the "customer" prevailed, it was possible to establish a kind of joint effort that worked effectively. At General Electric, where his definition had not been fully accepted, the continuing conflict between customer and contractor probably dissipated energies that might have been used for more constructive purposes. In time Knolls became a highly effective laboratory for engineering development, but at great cost to both sides in time and effort. In any case the fundamental intent of the Rickover approach seemed unassailable. In a complex development effort involving a new technology and a tight schedule, the government could not simply place an order and expect the contractor to fill it. Unless the government officials themselves had sufficient technical competence to evaluate specifications, contractor performance, and the quality of product, the government's interests were not likely to be protected. Creating and maintaining that kind of technical competence in a government organization was a back-breaking task, but it was on this principle that Rickover had staked the future of the nuclear submarine.

6　Prototypes and Submarines

In developing nuclear propulsion systems for submarines, Rickover and his group had no choice but to work within the Commission's organization. Rickover's immediate goal, however, was not just a reactor and steam plant but an operating submarine which could serve as a combat vessel in the fleet. The design and construction of the ship itself was not Rickover's responsibility but rested with the Bureau of Ships, of which Code 390 was only a part. Rickover had the task of producing the propulsion plant, but he had to rely on the bureau chief and the other codes for the hundreds of technical decisions and approvals required in designing and building the hull and providing the thousands of items of equipment that were part of a fighting ship. He needed the bureau's support to obtain the necessary authorizations and appropriations from higher echelons in the Navy, including the Chief of Naval Operations and the Secretary.

As a professional naval officer who had spent a large part of his career in Washington, Rickover was familiar with the ways of the Navy and the Bureau of Ships. Compared with the Commission, the Navy seemed to Rickover in some ways an old-fashioned, unenlightened, and tradition-bound bureaucracy whose organization and methods were not equal to the task of exploiting the advantages of modern technology for the fleet. Building one nuclear submarine would not do much to help the Navy meet the challenge posed by technology, but Rickover hoped he could use the project to convince the Navy to accept some of the methods and approaches he was using in the nuclear project.

For an engineering officer in the middle echelons of the Bureau of Ships, Rickover's intention was surely ambitious, but he had the advantage of supreme confidence in the soundness of his position. He began his drive for the *Nautilus* early in 1949 and with it his implicit attempt to transform the bureau into a new kind of technical organization. During the next four and a half years he never ceased to challenge old ideas and prejudices or to propose new approaches and methods. Inevitably opposition grew in the bureau and the Navy as officers and civilian leaders came to realize that Rickover's bid to develop nuclear propulsion was likely to succeed. In the summer of 1953, with the successful operation of the Mark I as evidence of his success in technical development, Rickover faced the ultimate challenge: the Navy's decision to effect his retirement by neglecting to promote him to rear admiral. The outcome of that struggle would indelibly stamp the later development of the nuclear Navy.

Establishing a Requirement

The administrative procedures which transformed an idea for a new type of naval vessel into a ship operating in the fleet were time-consuming and complicated. In a formal sense line officers assigned to the staff of the Chief of Naval Operations defined the military characteristics of the ships required. Officers in the technical bureaus translated these requirements into designs and, when these were accepted, supervised construction. Actually the interaction between line officers and engineers in exploring new types of ships was far more extensive than the formal procedures suggested. Officers in the technical bureaus frequently telephoned or dropped in on their counterparts in Naval Operations, just as Rickover kept in touch with the divisions of undersea warfare and atomic energy. Often personal friendships going back to Annapolis days or previous assignments provided the base for informal ties between officers in operations and the technical bureaus. Although the bureaus often suggested ideas for ships, aircraft, or weapons, they were usually reluctant to invest very much in research and development without a formal requirement from Naval Operations.

To obtain such a requirement for a nuclear-powered submarine, Rickover early in 1949 approached Lieutenant Commander Charles B. Momsen, Jr., a young officer who had succeeded Commander Edward L. Beach in the atomic energy division in Naval Operations. In addition to being the appropriate officer for Rickover to contact, Momsen was the son of Rear Admiral Momsen, who was head of the undersea warfare division in the same office. The obvious advantages of a nuclear submarine attracted the interest of both Momsens and they were willing to assist Rickover in initiating the administrative actions leading to a requirement.

In March 1949 the Chief of Naval Operations requested the Navy's submarine conference (directed by Admiral Momsen) to make a comparative analysis of the closed-cycle and nuclear propulsion systems for submarines. To prepare a reply to the Chief of Naval Operations, Admiral Momsen appointed an ad hoc committee which included his son. The members consulted the numerous studies already available, talked with various officers in the Bureau of Ships, including Rickover, and reported to the submarine conference on May 18, 1949.[1]

The report pointed to the clear superiority which operational officers saw in nuclear power. The committee found that in terms of submerged range the two systems were not even comparable. Only by using a snorkel could the closed-cycle begin to challenge the nuclear submarine in submerged operation, and in the committee's opinion postwar radar techniques had already made this tactic unacceptable. An exposed snorkel would be detected at extreme ranges almost as easily as a submarine on the surface. The committee found the nuclear submarine superior to the closed-cycle in many respects: submerged cruising speeds, endurance at any speed, security from enemy antisubmarine warfare tactics, ability to complete missions under all weather conditions, and over-all characteristics of an all-purpose submarine.

Even more impressive were the implications of these operational advantages for the Navy. The committee declared that "the nuclear power plant is a fundamentally new means of submarine propulsion which will probably make a profound impression on submarine design and the whole art of waging undersea warfare. The advent of the true submarine, capable of unrestricted operations in a medium which covers $5/7$ of the globe, may revolutionize the entire character of naval warfare." The committee recommended that "the Navy support very strongly the early development of a nuclear propelled submarine for evaluation purposes." Work should be continued on the closed-cycle system but only as an interim measure.[2]

The report of the submarine conference provided a basis for a formal requirement from the Chief of Naval Operations to the Bureau of Ships. Rickover and his staff helped young Momsen prepare a document calling upon the bureau to develop a nuclear propulsion plant capable of driving a submarine at high speed for extended operation. The propulsion plant was to be ready for operational evaluation and installation in a submarine hull by 1955.[3]

The reasons for choosing the 1955 date are obscure, but apparently some of the planning within the Department of Defense at that time was in terms of five-year periods. A year earlier, in 1948, Robert Oppenheimer had concluded in a special report on the long-range military uses of atomic energy that a submarine reactor was feasible. He foresaw the possibility of having a test-stand reactor in five years, a shipboard reactor in ten years, and nuclear-powered combat ships in fifteen years. Not only as chairman of the long-range objectives panel in the military establishment, but also as chairman of the Commission's General Advisory Committee, Oppenheimer's views were influential.[4]

Approval of the Momsen draft by Admiral Louis E. Denfeld, the Chief of Naval Operations, on August 19, 1949, did little more than give formal status to the development of a nuclear propulsion plant.[5] Within the Bureau of Ships there were other research and development projects to improve submarines. Rickover obviously would have to compete with them for limited funds and resources.

Nuclear Propulsion in the Bureau

In addition to the nuclear propulsion project, the Bureau of Ships was carrying on three other main efforts to achieve the high-speed submarine sought by the fleet. Two of these did little more than draw upon existing technology. As a stopgap, the bureau was converting some of the fleet submarines to "Greater Underwater Propulsive Power," a project which received the acronym *Guppy*. The conversion consisted of installing a larger storage battery, providing a snorkel system which would permit charging the battery while the vessel was submerged, removing some projecting hull fittings, and streamlining others. More advanced, but still based on conventional technology was the *Tang*-class, which the bureau had designed in 1947. In addition to the *Guppy* improvements, the *Tang* submarines were intended to achieve better performance from shorter hulls with greater diameter and from a new type of diesel engine. In June 1949 the bureau approved research on a high-speed submarine which would test various hull forms for resistance, stability, and control.[6]

The two other approaches—closed-cycle and nuclear-propulsion—were far more demanding on technology. The Germans had done a great deal of work on closed-cycle design during World War II, and since 1945 the Bureau of Ships had financed several investigations into a number of ways of providing oxygen for the combustion system. Late in 1949 the Navy's engineering experiment station across the Severn from the Naval Academy would begin testing the *Kreislauf* cycle, which recirculated cooled and cleaned exhaust gas to a diesel engine with the addition of sufficient oxygen to maintain combustion. By the end of the year, the bureau would be conducting model basin tests on the hull form for a closed-cycle submarine.[7]

There was little doubt that the Navy could build a closed-cycle engine, but many officers experienced in undersea operations were not enthusiastic. They believed that some of the chemicals necessary for the system would be dangerous in a submarine. Still, there were sound reasons for continuing the effort. Even if the ultimate superiority of nuclear propulsion was already evident, the technical obstacles to its achievement were formidable and the

7. Rear Admiral David H. Clark (right) succeeds Vice Admiral Earle W. Mills in March 1949 as Chief of the Bureau of Ships. Secretary of the Navy John L. Sullivan is between the two officers. *U.S. Navy*

schedule uncertain. Another factor was that uranium was still in short supply, and it seemed unlikely that atomic energy could ever meet all the needs of the submarine force.

The task of balancing these efforts in the Bureau of Ships fell to Rear Admiral David H. Clark, now chief of the bureau. An engineering duty officer with broad experience at sea and in naval shipyards as well as in Washington, Clark recognized the importance of nuclear propulsion, and he was ready to support Rickover when his proposals seemed consistent with the over-all objectives of the bureau. But Clark gave Rickover no special consideration. He saw Rickover only as the manager of the nuclear power branch (Code 390), just one of the branches under the newly established assistant chief of the bureau for research and development (Code 300).

Contractors and Dates

The August 1949 memorandum from the Chief of Naval Operations was vaguely worded, calling for the propulsion plant to be ready for "operational evaluation and installation in a submarine hull by 1955." To a few people in his office Rickover broached the possibility of having a nuclear submarine ready for sea by January 1, 1955, a breath-taking idea considering the status of nuclear technology. Rickover's staff estimated in October that the Mark I would have to be completed as early as January 1952 if its operating experience was to have any influence on the design of the shipboard plant. Working back from the January 1955 date in terms of the shipbuilding activities, Lieutenant Dick thought that the Mark I had to be in operation no later than May 1, 1952. He reasoned that unless the weight, size, and location of all the major components of the propulsion plant had been determined by that time, it would be impossible to fix the hull design early enough to have the ship completed by January 1, 1955.[8]

Setting a goal was one thing; drawing up a schedule and assigning work was another. Although Rickover had not yet convinced General Electric to drop the power breeder for the Mark A, it seemed only a matter of time before the company would become a full-fledged partner in the submarine project. Despite the fact that Westinghouse had almost a year's lead in designing the Mark I, Rickover was hardly willing to commit himself to only one type of reactor for the first submarine. Until he had solid engineering data to support eliminating one approach, he intended to continue both, although this course probably meant building a submarine for each reactor type. He would need a shipbuilder to work with each.

In fact, the Bureau of Ships had already established such a pattern. For some time the Portsmouth Naval Shipyard had built submarines fitted with Westinghouse equipment, and the Electric Boat Company, a private yard in Groton, Connecticut, had built those using General Electric machinery. It made sense to Rickover to use the same general arrangement, so that Westinghouse and Portsmouth would comprise one partnership and General Electric and Electric Boat the other.[9]

Although General Electric was not yet fully committed to the Mark A, Rickover chose to approach Electric Boat first. At Groton on December 6, 1949, he discussed the General Electric project with O. Pomeroy Robinson, the general manager of Electric Boat, and Andrew I. McKee, the chief design engineer. Both men were veterans of the shipbuilding industry, and Rickover had known McKee as a fellow naval officer in the Bureau of Ships during World War II. After studying engineering at Cornell University, Robinson had gone to work in 1915 as a machine shop chaser at the New London Ship & Engine Company, a subsidiary of Electric Boat. By 1918 he was a draftsman and working on diesel engine development, but he left the company for travel and broader experience. He returned to Electric Boat in 1922 and stayed on through the depression years, which were particularly bleak in shipbuilding. When business began to pick up again in 1938, he had been appointed general manager. Robinson was intensely proud of his company. His office window gave him a commanding view of the yard, the Thames River, and the town of New London on the far bank. He knew many of the men in the yard by name. Having gone through one lean period and been faced with another, Robinson tended to take a hard look at expenses. He hired few people, made sure they were good, and tried to keep them.[10]

Rickover began by explaining General Electric's role in the project. Anticipating that General Electric would soon take up the Mark A design in earnest, Rickover wanted to have an experienced shipbuilding company ready. He could easily convey his enthusiastic conviction that the future of submarines rested with nuclear power. To give Electric Boat a start, Rickover promised to arrange a series of lectures on nuclear technology for Robinson's staff and to send some of his engineers to Oak Ridge for more comprehensive studies.[11]

Robinson had every reason to welcome Rickover's tentative proposal. At its peak in 1944 the company had a working force of more than 12,000 men and was launching a submarine every two weeks. During World War II Electric Boat delivered sixty-four submarines to the Navy—more than any other shipbuilder—but when the war ended, the building ways stood empty.[12] Although in 1949 the company was doing some work on the snorkel in partnership with Portsmouth, these small contracts were not enough to keep the yard in business. To keep the company alive, Robinson was building highway bridges and accepting any work he could find.

A week later, Rickover and a delegation from General Electric inspected the drafting rooms, shops, and shipway facilities at Groton. In the course of the visit Rickover explained how each company could help the other. Because General Electric knew little about submarine design, Electric Boat could assist in laying out the machinery in the reactor compartment and the steam generating system and in constructing the radiation shield. Electric Boat, in turn, would have to depend on General Electric for reactor technology. There would have to be mutual education, some of which could be done by an exchange of personnel. Electric Boat would become a subcontractor to General Electric on a cost-plus-fixed-fee basis. Rickover, anxious to get an agreement, proposed setting January 20, 1950, for having a letter of intent signed.[13]

As General Electric and Electric Boat began negotiating, Rickover approached the other two organizations he hoped to form into a team. On January 12 he and Weaver arrived at the Portsmouth Naval Shipyard. Located on the Piscataqua River in New Hampshire, Portsmouth was the naval yard with the longest experience in constructing submarines. The yard officers listened as Rickover explained the relationship between Argonne and Westinghouse and described the schedule. Weaver believed that to meet the timetable, Portsmouth would have to assign one man full-time to the project immediately, and about thirty people by the end of the year.

The yard's response was disappointing. With present commitments, the yard commander explained, such a build-up of personnel would be impossible without delaying construction of the first submarine of the new *Tang* class. If Westinghouse wanted Portsmouth's advice—say, a visit every two or three weeks—this could be handled informally, but nothing more was possible. Convinced that Portsmouth would not give the project the priority he demanded, Rickover reached across the desk for the yard commander's telephone and called Robinson. The Electric Boat official assured Rickover that he would be willing to consider building two nuclear submarines.[14]

Portsmouth's reaction to Rickover's project contrasted sharply with Electric Boat's. The clue to the difference was that Portsmouth was a government yard, an integral part of the Navy shore establishment under Navy command. The yard was less susceptible than Electric Boat to immediate economic pressures. The civilian employees at the naval yard had built many submarines, but they were bound by a network of government regulations that made it hard to shift them from job to job. The naval yard could not act quickly by hiring new men or by paying them what they might receive in private industry. While it would be easy to exaggerate these factors, the naval yards did lack flexibility. Although Portsmouth's refusal was in some ways a disappointment, Rickover knew that his management approach would be more easily applied in private establishments, where he was the customer, than in government installations where he was just another naval officer.

Rickover, Roddis, and Dunford arrived at Robinson's home that same evening. Although a contract would have to be negotiated, the purpose was clear. Electric Boat would aid in designing and building both prototypes and both submarines. It would be a heavy assignment for the yard. The submarines built during World War II were of one basic design. To cope with a nuclear submarine would demand a high degree of adaptability on the part of Robinson and McKee. For Rickover, too, it was a risk. Up to this point he had been able to establish parallel approaches in his operations so that he would not be dependent upon one organization. Because of the Portsmouth refusal, his parallel lines merged at Electric Boat.

The building organizations Rickover assembled were superficially complex. As in any large project, there was a web of contractors and layers of subcontractors. It was not Rickover's intent to build a nuclear ship just for the sake of building it quickly. He demanded detailed designs supported by engineering analysis before authorizing purchase orders and subcontracts. His own preference was for lump-sum contracts for construction work because they involved the least administrative supervision. Under the main contractual arrangements completed in 1950, Westinghouse and General Electric were responsible for the design and construction of the reactor plants. Electric Boat, under separate subcontracts, was to assist in the designs and was to construct the hull portions of the prototypes. Westinghouse had another contract with the Rust Engineering Company for the design of the supporting facilities which would be needed at Arco, Idaho. General Electric had a similar contract for the West Milton site. The arrangement reflected Rickover's determination to keep intact the responsibilities of Westinghouse and General Electric.[15]

A Shipbuilding Program

Although Rickover had lined up Electric Boat to work with Westinghouse and General Electric, he still needed approval of the January 1, 1955, date in the Bureau of Ships. In the autumn of 1949 that approval was urgent because the Navy was already preparing its shipbuilding program for the 1952 budget.

Beginning with the Bureau of Ships, Rickover and his staff prepared for Admiral Clark's signature a memorandum calling for the cooperation and assistance of the bureau codes. The language of the memorandum seemed routine, but it contained Rickover's startling proposal. Again he interpreted the requirement for operational evaluation to mean "that we should have a submarine ready to leave the building yard, complete with a nuclear power plant, on January 1, 1955." He undercut any possible objection from the bureau by noting that the Commission had "a major portion of its reactor development program scheduled to meet this date."[16] Clark's signature established the bureau's recognition of the January date and Rickover's own responsibility for meeting it.

Including the nuclear submarine in the 1952 shipbuilding program was the prerogative of the Chief of Naval Operations. In January 1950 Rickover asked his staff to draft a memorandum for Clark's signature to Admiral Forrest P. Sherman, the new Chief of Naval Operations, who would present the building program to Congress in a few months. Using the January 1, 1955, target date, the staff set down a tentative schedule for completing the land-based prototypes and the shipboard plants for the Mark I and the Mark A systems. It was not yet possible to determine which system would be ready first, but in either case the size and weight of the propulsion plant would require a new submarine hull design.[17]

Determining the size, mission, and armament of the proposed vessel rested with the ship characteristics board. Again Code 390 worked closely with Lieutenant Commander Momsen. Discussions with Rickover's staff had convinced him that nuclear propulsion would require an entirely new approach to submarine design. To prevent the bureau from simply modifying some existing plans to accommodate a nuclear propulsion plant, Momsen proposed a new torpedo arrangement and depth requirement. The ship characteristics board did not accept the new features, but it did approve the general plan in March 1950. Since the board was part of the office of the Chief of Naval Operations and included a representative of the Bureau of Ships, the action amounted to an agreement on the part of these two organizations to build a nuclear submarine. Under these terms, however, it would only be a test vehicle.[18]

Rickover was also paving the way in Congress. On February 9, 1950, he appeared as the sole witness before the subcommittee on reactor development of the powerful Joint Committee on Atomic Energy. He discussed the limitations of conventional submarines, portrayed the advantages of nuclear propulsion, and forecast the probable development of nuclear submarines by the Soviet Union. To Congressmen already worried about the recent Soviet development of an atomic bomb, Rickover's warnings were impressive.[19]

Within the Navy the General Board still had the function of reviewing the Navy's proposed shipbuilding program for the secretary. In briefing the board on March 28, Rickover was careful to speak as a representative of the Commission. He was enthusiastic about what "the Commission" was doing to develop the reactor. As he had told Clark, the Navy could not afford to be caught without a hull for a propulsion plant it had requested another agency to develop. The board not only approved the project early in April 1950 but also reversed the action of the ship characteristics board by reinstating the specification for torpedo tubes.[20]

The recommendation of the General Board meant more to Rickover than a simple modification of plans. Without torpedo tubes, the submarine would have been only a test vehicle on which other codes might have been tempted to test their own ideas and experimental equipment. Rickover feared that it might prove too expensive later to convert such a ship into an attack submarine. By avoiding the test-vehicle stage, Rickover could hope to have a combat submarine ready to leave the building yard in January 1955.

The General Board's action in April 1950 was tantamount to approval by the Secretary of the Navy. Later in the month Admiral Sherman presented the Navy's shipbuilding program for fiscal year 1952 to the House Committee on Armed Services. The proposal included the construction of two new types of submarines: one using nuclear power, the other a closed-cycle system. When President Truman signed the authorization act on August 8, Rickover had the authority he needed.[21]

Rickover's success in adding the nuclear submarine to the shipbuilding program showed his grasp of the bureaucratic machinery of the Navy and of the government at large. It subtracts nothing from the accomplishment to suggest, however, that Rickover had certain factors in his favor. One was the eagerness of some influential submarine officers to gain the advantages of nuclear power. They were captivated by the vision of a true submarine even if they did not understand all the technical difficulties. To these officers, nuclear propulsion was far more attractive than the closed cycle, and it is significant that the closed-cycle submarine already authorized was never built. Another favorable factor was the growing interest in nuclear technology within the armed forces. The nation had just embarked on the quest for the hydrogen bomb. If the search was successful, the new weapon would certainly enhance the striking power of the Air Force. In early 1950 Rickover was offering the Navy its own doorway into the nuclear age. Certainly many senior officers in the Navy—perhaps some on the General Board which voted to restore the torpedo tubes in the first nuclear submarine—saw this possibility. In any case, Rickover had in a matter of months converted a small research and development project into a plan to build a fighting ship.

Concurrent Development and the Prototypes

For anything as revolutionary as a nuclear submarine, prudence dictated a carefully planned sequence of research and development such as Oppenheimer had assumed in 1948 or as General Electric had proposed a year later. As a future Chief of the Bureau of Ships was to tell a Senate committee: "In other more orthodox engineering fields, when all the factors are better known, the Navy normally would not even ask for shipbuilding authorization before a complete laboratory demonstration of equipment in support of such programs." In this sequence, the development contractor would build a test-stand —or bread-board—reactor plant, in which the components would be dispersed so that each could be observed in operation and modified if necessary. After the test-stand reactor had proved successful, a propulsion plant would be built to propel a vessel, but this prototype would also be used for testing and evaluation. Perhaps after fifteen years of operation, the system would be ready for installation in a combat submarine.[22]

Rickover had a very different approach in mind. He planned to combine the functions of a test-stand reactor and a shipboard prototype into one facility—a land-based prototype. The reactor and the steam plant would be arranged in the prototype as they would be on an actual combat submarine. By omitting the test-stand phase Rickover would lose a certain flexibility for testing components, but he was convinced that this advantage was not only illusory but unnecessarily expensive and actually detrimental. A test stand

offered the opportunity to postpone certain decisions on the design and configuration of the actual plant. In his approach engineers had to come to grips
from the first with the physical characteristics of a plant which could fit inside a hull and be simple enough for a Navy crew to operate. Rickover's
strategy was to determine what the over-all characteristics of the plant would
be and to work toward them from the beginning rather than approach the
final plant through several evolutionary phases. This harsh note of practical
realism—designing, manufacturing, testing, and assembling the components
as they would be aboard an operational submarine—affected the preliminary
design at Bettis, Knolls, and the supplier manufacturers. By having the land-
based prototype closely resemble the shipboard plant, Rickover also saw
that he would gain construction experience which would be priceless in the
shipyard. After all, not only the reactor itself but also the use of steam in a
submarine would be novelties for Electric Boat.

To a certain extent Rickover's approach could be described as "concurrent" as opposed to "sequential" development. Not only would he combine
the test-stand reactor and the shipboard prototype in one land-based plant;
he would also begin construction of the submarine long before the Mark I
prototype was completed. Instead of taking each step in sequential order from
prototype design to completion of the submarine, Rickover intended to develop Mark I and Mark II concurrently. Rickover was to sum up the idea
in the catch phrase "Mark I equals Mark II."[23] The essential idea was that
Mark II was to be so much a copy of Mark I that a change in Mark I would
automatically appear in Mark II, and to a large extent development followed
that pattern. It was not true, however, as some contractors later learned, that
Mark II necessarily equaled Mark I. Sometimes research and testing showed
up weaknesses or made possible improvements which were incorporated in
Mark II but which were not used in Mark I. Thus Mark II would not be an
identical copy of Mark I; it would be better.

The assumption of sequential development had led the ship design division
in the Bureau of Ships to conclude it would be impossible to meet the January 1955 date. Even if Rickover could complete the Mark I by May 1952,
the division estimated that it would take four months of operation to demonstrate that nuclear propulsion was safe and reliable. That would mean the
bureau could not start contract design of the ship before September 1952.
On the basis of recent experience, the division estimated that this stage would
take twelve months and construction thirty-six months. Even on the most
optimistic schedule, it did not seem possible to complete the submarine be-

fore September 1956. Furthermore, the design division saw nothing sacred in the January 1955 date. Rickover had set that himself, and in so doing he was scheduling work over which he had no authority.[24]

Under ordinary circumstances the argument that Rickover had no authority to set the schedule by himself was valid, but he did have two points in his favor: Clark had accepted the date, and Rickover could speak for the Commission on reactor work. Although Rear Admiral Frederick E. Haeberle and the officers in the ship design division considered Rickover's schedule unreasonable, they made a real effort in the spring of 1950 to meet it. In June they came up with two proposals. The first, which would meet Rickover's completion date, would require the bureau to place a contract with the shipbuilder five months *before* the Mark I had even begun to operate. Some material for the submarine would have to be delivered and even more be fabricated seven months *before* Mark II reactor design could be fixed. Believing the risks in this schedule to be too great, the design division preferred a second timetable maintaining the sequential approach, which, through careful compression of each step, would see the vessel completed on June 1, 1955.[25]

The issue was finally settled in Haeberle's office on July 7, 1950. Again speaking as a Commission official, Rickover insisted that the Mark I would be completed by January 1952, and that the shipboard reactor would be delivered to the building yard by July 1, 1953. Haeberle thought it was risky to begin construction of the ship before the prototype had been tested, but he had to admit that "the Commission" (i.e., Rickover) was taking an even greater gamble in rushing the development of the reactor. All agreed that the Navy would be in an impossible position if the Commission had the reactor ready on time and the bureau had no hull available.[26]

The difference between Rickover's approach and that of the ship design division was more than a question of schedules. The fundamental issue was concurrent development. The division was seeking a conservative, evolutionary approach in which each step was based on the successful completion of the preceding one. It was the same philosophy that guided the ship characteristics board when it proposed that the first nuclear submarine be a test vehicle. For his part, Rickover was assuming that the design, development, and construction of a propulsion reactor were primarily matters of shrewd and sophisticated engineering. Success would depend upon his gamble that there were no unknowns—nothing in the laws of nature—that would make it impossible to build a small, high-powered reactor which a Navy crew could operate. This assumption involved something of a risk, as General Electric had discovered in its attempt to develop the power-breeder reactor. No one could overcome obstacles imposed by fundamental laws of the physical universe. If later development of the Mark I encountered any of these, all the work on Mark II and the ship would be wasted. And if that happened, one man would be clearly responsible—Rickover.

The Reactor for the First Submarine

When Rickover first proposed to include a nuclear submarine in the Navy's shipbuilding program in January 1950, he was not yet ready to commit himself to the type of reactor which would go into the one hull to be authorized for fiscal year 1952. By the time President Truman had signed the authorization bill eight months later, in August 1950, there was no longer any question that the water-cooled Mark I would be ready before the sodium-cooled Mark A. During those months Westinghouse had made excellent progress on the Mark I design while General Electric had only begun to organize the Mark A project at Knolls.

Table 2.   Navy Nuclear Propulsion Program in 1953

	Water-Cooled Reactor	Sodium-Cooled Reactor
AEC Field Office	Pittsburgh	Schenectady
AEC Contractor	Westinghouse (Bettis Laboratory)	General Electric (Knolls Atomic Power Laboratory)
Land Prototype	Submarine Thermal Reactor (STR) Mark I, National Reactor Testing Station, Idaho	Submarine Intermediate Reactor (SIR) Mark A, West Milton, New York
Nuclear Submarine	*Nautilus* SSN 571 STR Mark II	*Seawolf* SSN 575 SIR Mark B
Shipyard	Electric Boat Division, Groton, Connecticut	Electric Boat Division, Groton, Connecticut

Thus the plan was to start construction at once on the Mark I prototype of the submarine thermal reactor at the Idaho test site. Construction of the Mark A prototype of the submarine intermediate reactor would begin about six months later. The first submarine hull, which would contain the Mark II plant, would be laid down in the summer of 1951. The second hull, which would be included in the Navy's shipbuilding program for fiscal year 1953 and contain the Mark B plant, would be started in the fall of 1952. In consolidated form the scheduled completion dates were as shown in table 3.

Table 3.   Schedule of Completion Dates for Reactor Prototypes
                         and Submarines.

Prototypes	Mark I	Mark A
Preliminary design	September 1950	February 1951
Detailed design	June 1951	December 1951
Construction	December 1951	June 1953
Submarines	Mark II	Mark B
Preliminary design	November 1950	September 1952
Contract plans and specifications	June 1951	April 1953
Construction	August 1954	November 1954

The dates in the August 1950 schedule[27] would change several times; that was to be expected. The controlling date in Rickover's mind was January 1, 1955. There was no other way to meet this goal except by concurrent development.

The Role of Electric Boat

When Robinson signed the Westinghouse contract on February 23, 1950, neither he nor anyone else at Electric Boat could have had any real conception of the task the company faced. The very fact that the company would have a part in building two prototypes—the first one 2,000 miles away in Idaho—indicated that the new venture would be far different from any shipbuilding effort the company had undertaken in the past. Another novel and contentious feature would be the experience of working with Rickover and his organization.

Within a few months Rickover's incessant demands and the rapid proliferation of assignments began to upset established patterns of operation at Groton. Electric Boat had started with the task of preparing preliminary layouts of machinery and equipment for the Mark I. Before the end of June Code 390 had assigned the company the additional job of making detailed arrangement plans for the steam generating compartment, the main propulsion machinery compartment, and the piping system. Because Electric Boat as a submarine builder had no experience with steam systems, a few engineers from the Bethlehem Shipbuilding Company were brought in to assist. Another requirement was the construction of a full-scale mock-up in wood and cardboard of every piece of equipment in both compartments.

In the spring of 1950 Electric Boat had almost no engineering staff to handle this flood of requirements. During the lean years after World War II many experienced engineers had left the company, and of those remaining only a few had yet obtained security clearances for access to classified information on nuclear reactors. At that time the Commission was beginning a vast expansion of its production facilities, and thousands of construction workers and engineers were awaiting clearances for projects in all parts of the nation. As a stop-gap Robinson set aside a few rooms where the men who were cleared could start work. Thomas W. Dunn, an engineer who had been with the company before World War II, agreed to collect information on the design of the reactor compartment, while Frank T. Horan had a similar responsibility for the engine room. At Rickover's urging, Robinson required his key officials to attend a lecture series on nuclear technology given by members of the naval reactors branch and sent a few young engineers to the Oak Ridge school of reactor technology.

By September 1950 Electric Boat was almost inundated with design work. Now that General Electric was concentrating on the Mark A, the small design group at Groton had to begin to think about the second prototype. The greatest difficulty, however, came from the increasing number of design changes on the Mark I. Most of these were coming from Bettis or from other Westinghouse divisions which were providing steam components, but the Bureau of Ships was also at fault. An obvious remedy was to improve liaison between Electric Boat and Westinghouse on one hand and between Electric Boat and the bureau on the other.

Rickover set up meetings to clarify responsibilities. He discovered that the bureau and Electric Boat were preparing preliminary designs which differed markedly in certain dimensions. Although this kind of discrepancy was to be expected in the early stages of design, better coordination was necessary, particularly since construction had already started on facilities for the Mark I prototype in Idaho. Relations between the shipbuilder and Bettis were cordial, but Rickover thought they were far too informal. He complained that Bettis engineers sometimes sent plans to the bureau before they had been approved by responsible officials at Bettis or Groton. In many respects the situation resembled that which the naval reactors branch had faced in working with Bettis and Argonne in 1949.[28]

Early Construction in Idaho

In addition to the design work at Groton and Bettis, Electric Boat was soon involved in construction activities at the site near Arco, Idaho. As described in chapter 4, Rickover had already obtained a site at the National Reactor Testing Station and had resolved the delicate problem of delineating the responsibilities of his own contractors as opposed to those of the local Commission field office. While Westinghouse, Electric Boat, and Rust Engineering were still working out their initial construction plans, Rickover sent Commander Jack J. McGaraghan, an experienced officer in the Navy's civil engineer corps, to Arco to supervise the first activities on the site.

The size of the Mark I facility, which would have seemed impressive in a conventional industrial area, would be lost in the vast reaches of the southeastern Idaho desert. A mile west of the lonely macadam road which connected the test station's central facilities with the northern portion of the Commission reservation, the Navy site would be dominated by a large sheet-metal building which would house the prototype. Nearby would be spray ponds for dissipating the energy produced in the reactor and a limited number of shops, offices, and utility buildings.

From the beginning Rickover demanded speed and economy at Arco. On March 11, 1950, in reviewing Westinghouse's initial instructions for Rust Engineering, he insisted that the plans make possible enclosing the reactor building before the onset of winter. He directed that the site contain no more than the Mark I plant and those supporting facilities urgently needed. Buildings for the construction project should be designed for easy conversion to other uses later. Rickover told McGaraghan that if his group needed more office space, he could subdivide existing structures. Knowing that his project would be compared to others at the testing station, Rickover made every effort to hold down the cost of facilities and the number of men necessary to operate the plant when it was completed.[29]

Rickover preached economy, but he did not mean to stint on Mark I. In fact he thought a Spartan operation would strengthen the project. Although he was already planning other more advanced reactors, he would not let them interfere with the Idaho project. As he wrote in February 1952, "The success of the Mark I will determine the extent of the support we receive from the Atomic Energy Commission and the Navy."

That same month Rickover went to Idaho to explore ways of speeding up the work. He made clear that prime responsibility for constructing the Mark I

8. The naval reactor installation at the
National Reactor Testing Station in Idaho
in December 1953. The submarine
thermal reactor, Mark I, is in the
large building.

rested with Electric Boat as the subcontractor to Westinghouse; all the other organizations were to serve Electric Boat at Arco. The organization was complicated enough, however, to require special means of communication. A summary of decisions at the weekly production meeting at Bettis was to be teletyped to the Idaho site. At the end of each week the Westinghouse contingent at the Mark I site would send Bettis, Code 390, and Electric Boat at Groton a teletype covering procurement problems and their effect on the construction schedule. In addition, McGaraghan was to institute a weekly "gripe" letter. These were to go only to Rickover and were to be kept in his personal file. McGaraghan could use them to bring to Rickover's attention any situation he thought necessary, but he was expected to have something to report every week.[30]

Many aspects of Mark I construction were similar to those encountered in any project under a tight schedule. Sooner or later the delivery of equipment would be delayed, forcing a readjustment of schedules. Such delays occurred on the Mark I, but Rickover's reaction was far from typical. Instead of just patching up the schedule, Rickover was interested in finding the root cause of management failure. To minimize delays on the Mark I project, Rickover set up a production control section under John F. O'Grady, a former naval reserve officer whose competence as an expediter had impressed Rickover during the war. O'Grady helped the project in many ways, from badgering suppliers on delivery dates to cajoling labor leaders into releasing components in strike-bound plants.[31]

In other respects the Mark I posed difficulties uncommon to most construction projects. One of these was the exceptional cleanliness required of all components to be installed in the plant. Fabricators had to follow special procedures during manufacturing and inspection to insure that no foreign matter was introduced. Code 390 imposed special wrapping and tagging regulations unfamiliar to most industries. When circumstances warranted, Rickover sent teams of Westinghouse and Navy personnel to manufacturers' plants to inspect cleanliness procedures. At the Idaho site careless handling during installation could render useless all the precautions taken during manufacture and shipping. In March 1952 Rickover made a special trip to Idaho to discuss the subject. He insisted that McGaraghan take personal responsibility for the cleanliness of every component installed in the prototype. McGaraghan was to compile check-off lists, file reports, and recommend improvements in existing procedures. Back in Washington Rickover used such data to follow every step of McGaraghan's work.[32]

By the autumn of 1952 McGaraghan found his job coming to an end. He could handle the familiar construction activities in the early phases of the project, but as reactor components began to arrive in Idaho, he found it more difficult to make decisions on his own. More than ever he had to rely on instructions from Washington, principally from Kintner, who was soon to succeed Dick as the Mark I project officer after the latter's death following a brief illness in January 1953. As Mark I became a reactor, new talents would be needed to bring it into operation.

Mock-Ups for Mark I

Mock-ups in one form or another were not new to naval shipbuilding, particularly in submarine construction. As he did in so many instances, Rickover took an existing technique and exploited it in new ways. He had Electric Boat mock up in wood and cardboard every pipe, valve, electrical panel, and large motor in the reactor and machinery compartments. The full-scale mock-up had a special fascination for Rickover. During visits to Groton he would climb through the simulated compartments in the drab shed-like structure near the river bank, studying the configuration from several angles to make certain that there was enough space for men to maintain and replace equipment at sea and to make sure that a valve handle would not project dangerously into a walkway. Rickover was convinced that full-scale mock-ups provided information that even the most experienced shipbuilder could not gain from drawings or quarter-scale models.[33]

Westinghouse had a different kind of testing structure for the Mark I in one of the high-bay buildings at Bettis. In this instance, the purpose was not to simulate the physical layout of the shipboard plant in wood and cardboard, but to duplicate the operating conditions of the primary coolant system in pumps and pipes. The reactor vessel at Bettis contained a dummy core with the pumps, valves, and piping identical to those in Idaho. In every respect except the important one of nuclear operation, the mock-up could produce the conditions which would occur in Mark I. So closely did the Bettis mock-up resemble the Mark I that some of the equipment originally fabricated for the mock-up was actually installed in Mark I. As a result, the reactor was in operation before the mock-up, but the Bettis facility still proved valuable for testing and trouble-shooting.

As the various laboratories, shipbuilders, and industrial contractors could rightfully claim, the Mark I prototype as it took shape in the shops at Bettis,

9. A full-scale wood and cardboard mock-up of the crew's quarters aboard the *Nautilus*. Similar mock-ups of the machinery areas were used to make certain that components were accessible for maintenance and repair. They also helped to train workmen in installation procedures.

10. The West Milton, New York, site in the summer of 1953. The spherical containment vessel, 225 feet high, dominates the site. The hull section for the submarine intermediate reactor, Mark A, is being assembled outside the sphere in the right foreground.

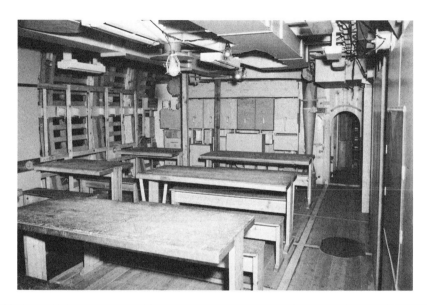

9

10

11. President Truman about to place his initials on the keel plate of the *Nautilus* on June 14, 1952. Behind the president to the left in the light suit is John Jay Hopkins. To the right behind the president is O. Pomeroy Robinson. Behind Hopkins are Mrs. Rickover and the Rickovers' son, Robert. Captain Rickover is partially obscured behind the two naval officers in the center of the photograph.

12. Westinghouse and Electric Boat officials at the *Nautilus* keel-laying ceremony. Left to right: Latham E. Osborne, executive vice-president of Westinghouse; John Jay Hopkins, president of General Dynamics; Gwilym A. Price, president of Westinghouse; Charles H. Weaver, manager of the Westinghouse Atomic Power Division; O. Pomeroy Robinson, general manager of Electric Boat.

11

in the mock-up at Groton, or in the plant in Arco was the product of many disciplines, industrial traditions, and crafts. It reflected the special talents of electrical and mechanical engineers, welders, steamfitters, metallurgists, physicists, and master carpenters. It also bore the indelible stamp of Rickover and the naval reactors branch.

Construction at West Milton

Although General Electric and Electric Boat did not begin any significant design work on the Mark A prototype until the summer of 1950, they were able to draw upon Knolls's extensive research on the power-breeder. For more than two years the laboratory had been studying the basic design of an intermediate reactor using a sodium coolant. The Genie heat transfer system, which General Electric had been developing since 1946, had first produced steam at the company's Alplaus, New York, plant in April 1950. Before the end of the year Knolls had brought a preliminary mock-up of the Mark A reactor core to criticality in a zero-power test facility.[34]

General Electric wanted to build the Mark A prototype at West Milton, New York, a small village twenty miles north of Schenectady. Originally recommended in July 1948 as the location of the power-breeder, the site was approved for purchase by the Commission in September. Site planning at West Milton had started just one year later, and only the Commission's increasing reservations about the power-breeder had delayed a full-scale construction effort in the closing months of 1949.[35]

The Commission not only had purchased the site but also had obtained from its reactor safeguards committee approval to build the power-breeder at West Milton, provided the entire reactor plant was enclosed in a huge sphere capable of containing any radioactivity that might be produced in a reactor accident. The decision to build the Mark A at West Milton required a new evaluation of safety hazards. This could not be accomplished until General Electric and Electric Boat had completed a detailed design of the plant in 1951. In January 1952 the reactor safeguards committee approved the plan to place the Mark A in a sphere 225 feet in diameter constructed of 1-inch steel plates.[36] Not until August 1952 was concrete poured for the foundations of the sphere which was to become the distinctive feature not only of the West Milton site but also of many other power reactors.

By early 1953, as the Mark I was nearing completion at Arco, construction at West Milton was just getting into full swing. Rust Engineering had just been selected as the general contractor for the project and would take over construction of all general facilities at West Milton. The first six courses of steel plates had been set around the lower portion of the sphere, and just outside Electric Boat had assembled the submarine hull section in which the Mark A system would be placed. Mark A was running two years behind Mark I, but the project was beginning to pick up momentum.[37]

Keel Laying

Rickover's strategy of concurrent development and his January deadline meant that construction of the ship had to begin long before the prototype was completed. In June 1952, while the Electric Boat team was installing the Mark I pressure vessel, steam generators, and primary coolant piping in the hull section at Arco, shipyard personnel at Groton were fabricating hull sections and preparing for keel-laying of the *Nautilus.*

A better name for the world's first nuclear submarine would have been hard to find. In 1801 Robert Fulton had called his experimental undersea boat the *Nautilus,* and Sir Hubert Wilkins had given the same name to the craft he had used in 1931 in his daring attempt to penetrate beneath the Arctic ice. The United States Navy had assigned the name twice to submarines, first to the H-2 boat in 1913 and then to the SS-168. Launched in 1930, the Navy's second *Nautilus* completed fourteen war patrols before being decommissioned in 1945.

The most famous of all ships bearing the name was the fictional submarine created by Jules Verne. Finding an original edition of *Vingt mille lieues sous les mers* in the library, Roddis had been fascinated with comparing the specifications of Captain Nemo's famous craft and those of the new submarine. The nuclear ship would be somewhat longer, a little greater in beam, and of far larger displacement. Verne's creation, however, could travel at 43 knots with a cruising radius of 43,000 miles, somewhat in excess of the performance then planned for the Navy's new ship. Instead of nuclear power, Verne's craft relied on a sodium "Bunsen" apparatus. Whatever hazards this system possessed, presumably radiation was not one of them. Intriguing, too, was the pipe organ in the crew's lounge. The nuclear *Nautilus* could never match this.[38]

Choice of the name had been somewhat fortuitous, at least as far as Rickover knew. The Navy practice was to place the names of newly decommissioned submarines at the bottom of a list and then reassign those from the top as new submarines were built. Somehow or other the match was made in the Navy bureaucracy, and on October 25, 1951, Secretary Dan A. Kimball established the designation SSN for nuclear submarines and officially named the first ship the *Nautilus*.[39]

Rickover set out to make the keel-laying worthy of a famous name and an historic ship. It was not only a sense of history that stirred his imagination; he also saw a chance to win support for nuclear propulsion. Usually a keel-laying was not an occasion for ceremony in submarine construction because there was really nothing sacred about the date. Any one of the midship sections being assembled in the yard could be moved to the building ways when convenient.

Nothing could assure more attention to an event than attendance by the president. Rickover had enjoyed a session with President Truman at the White House in February 1952.[40] Pleased with Truman's interest in the project, Rickover thought the president would accept an invitation to speak. Rickover followed political protocol by approaching Senator Brien McMahon. Not only did McMahon come from the state in which the *Nautilus* was being built, but he was also the Congressional leader most closely connected with atomic energy. As chairman of the Joint Committee on Atomic Energy, McMahon was deeply concerned about national defense. He had met Rickover at committee meetings and admired his energy and ability. Then at the height of his political power but dying of cancer, McMahon had been mentioned as a possibility for the Democratic nomination as vice-president in the national campaign that year. The keel-laying could bring national attention to the accomplishments of the Truman administration in national defense. Unable to leave his sickbed, McMahon telephoned Truman. The president gladly accepted McMahon's invitation.

On June 14, 1952, Truman stood on temporary stands over the building way before a shipyard filled with spectators. Around him were important leaders from industry: John Jay Hopkins, president of General Dynamics, the recently formed corporation of which Electric Boat was a main constituent; Gwilym A. Price of Westinghouse; and Ralph J. Cordiner, the new president of General Electric. The Navy was impressively represented. Secretary Kimball was surrounded by admirals: William M. Fechteler, Chief of Naval Operations; George C. Crawford, commander of the Atlantic submarine force; Calvin M. Bolster, chief of naval research; Homer N. Wallin, chief of the Bureau of Ships; and Evander W. Sylvester, assistant chief of the bureau for ships. In the background and in civilian clothes was Rickover.

After the inevitable congratulatory remarks and introductory speeches, Truman began. He recalled the role atomic energy had played in his administration: the first nuclear detonation at Alamogordo and the bombs used against Japan which were, for most of mankind, the first knowledge of the power of the atom. The *Nautilus* was a military project, but the president saw the ship as a step toward peaceful uses of atomic energy. For her use new metals had been made and new machinery designed. Someday these could be used to produce electric power.

As he concluded, Truman raised his hand. A crane picked up a huge bright yellow keel plate and laid it before the stands. Truman walked down a few steps and chalked his initials on the surface. A welder stepped forward and burned the letters into the plate.[41] The public could have had no better demonstration that the *Nautilus* was under construction.

Forcing Improvements in Management

With two prototypes and one submarine under construction by the summer of 1952, the fate of the naval reactors project hung increasingly on the performance of Electric Boat. Within another year the keel of a second nuclear submarine would be laid at Groton, and the company would have even more responsibility. Rickover suspected that the load was getting too heavy and the pace he was setting too fast. He decided the remedy was not to ease the burden on Electric Boat but to demand better performance.

As he had done at Westinghouse and General Electric, Rickover started at the top. He was much impressed by Hopkins, the company's president. Hopkins was as fine an example of the new school of American industrial leadership as Robinson was of the old. A lawyer by profession, Hopkins had joined Electric Boat in 1937 as a director and had risen to vice-president in 1942 and president in 1947. Determined to diversify the company's activities, he had formed the General Dynamics Corporation in 1952 and was eagerly seeking major contracts in defense industries and atomic energy.

During the autumn of 1950 Rickover had seen an opportunity to bring new management talent to Groton. Carleton Shugg, the Commission's deputy general manager who had supported Rickover's effort since coming to Washington in 1948, had just been passed over for the position of general manager. A forceful administrator and experienced shipbuilder, Shugg had been looking for a new job. Rickover had mentioned this opportunity to Hopkins, who soon convinced Shugg to take charge of all construction for Electric Boat in 1951.[42]

By the summer of 1952 Rickover was convinced that more muscle was needed in management at Electric Boat. On August 23 he called Hopkins and Robinson to his Washington office for one of his now legendary Saturday meetings. With brutal frankness Rickover told his visitors that Electric Boat would have to reorganize. The yard was still operating as if it were turning out submarines of one basic type. Robinson was trying to meet the demands of the nuclear project by promoting men from within the company. He was adding design engineers and draftsmen, but in Rickover's opinion, far too slowly to produce the thousands of detailed drawings needed for prototypes and ships. Rickover compared Electric Boat's personnel policy with that of another shipbuilder, which used slack periods to weed out all but the best employees. Electric Boat, Rickover charged, retained most of its employees between jobs and therefore had little flexibility in hiring better men. Hopkins could see how far apart Robinson and Rickover were when Robinson offered to hire sixteen more engineers. Rickover declared the number completely inadequate. As the argument grew bitter, Hopkins intervened and accepted Rickover's demands.[43]

Robinson was clearly the victim of the feast-or-famine cycle which was the lot of all American shipbuilders. He was reluctant to add men to his payroll until he knew what they could do. Because constructing a nuclear submarine was something new, Rickover could not predict exactly how many men would be needed, but he was convinced Robinson would never meet the schedule if he waited until precise requirements were apparent. Then there would be no time for training or security clearances.

Robinson had scarcely begun to change his hiring policy when Rickover learned of further trouble at Groton. Reports from the bureau's submarine design group indicated a rash of changes in the Mark II. Although the facts at Groton were anything but clear, it seemed possible that rearranging the control rod drive mechanism and increasing the size of such major components as the turbogenerator sets would require an increase in the length and diameter of the *Nautilus*.[44]

Reacting angrily, Rickover declared that the incidents in the Mark I and the *Nautilus* were merely symptoms of more fundamental problems. With the help of field representatives in Pittsburgh and Schenectady, he compiled a list of management weaknesses at Electric Boat: vague assignments of responsibility, undefined lines of authority, inefficient follow-up procedures, illogical planning, and incomplete drawings and specifications.[45]

Before taking any action, Rickover sent Panoff to Groton to investigate. Panoff's common-sense investigation revealed several new facets of the problem. Changes were being made in the Mark II, but these resulted from improvements in design; the proposed modification of major components would not require an increase in hull dimensions. Some uncertainties could not be resolved until several drawings were completed. All in all, however, the status of affairs at Groton was far less serious than it had seemed.[46] It was unusual that Rickover in this instance had learned of difficulties at Groton from someone in the bureau outside his own organization.

Probably this incident alone did not cause Hopkins to reorganize the company, but perhaps it provided one more piece of evidence that reorganization was necessary. In a number of sweeping changes on November 1, 1952, Hopkins abolished the position of general manager, which Robinson held. Electric Boat became a division of General Dynamics with Shugg as the division manager in charge of the entire shipbuilding operation. Robinson remained a senior vice-president of the parent organization.[47]

A New Grip on the Bureau

Even before the reorganization at Electric Boat, Rickover was pressing the company to move up the schedule for completion of the *Nautilus* by as much as five months. Robinson had been willing to accept the proposed schedule, if only to avoid another fight with Rickover. With Robinson gone, Shugg faced the same question. He was at first reluctant to agree, but after studying the situation for several days, he decided that by using overtime and extra shifts, he might pick up four months, but not five.[48]

Rickover's pressure not only affected Electric Boat but also the Bureau of Ships. In the first place, Shugg would have to convince the bureau to approve the new schedule. Secondly, many of the components of the *Nautilus* were supplied not only by Bettis but also by private vendors under bureau contracts. The bureau also was responsible for the timely delivery of government-supplied items. In accepting the new schedule the bureau would be committing itself to provide on time the components for which it was responsible. It was perhaps indicative of Rickover's limited authority in the bureau that he chose to bring pressure on Shugg rather than his military superiors.

On November 25, 1952, Shugg presented the new timetable to Rear Admiral Sylvester, assistant chief of the bureau for ships. Shugg expressed his confidence that Electric Boat's experience in constructing the Mark I would make it possible for the company to move up the scheduled completion of the

Nautilus by four months. As for the propulsion plant, Shugg believed that the Commission and the bureau had only to meet their present delivery dates. Shugg also related the steps he had already taken to speed up work at Groton. A new building was being constructed for the design force. Already Shugg had increased that group by about ninety men and he intended to add 225 more.

Sylvester was on the spot, for he must have known that Rickover was behind Shugg's proposal. Sylvester's own reports from Groton suggested that Shugg was overly optimistic. Already Electric Boat had missed some target dates because the company lacked certain equipment. Even so, Sylvester recognized that the real trouble was the late delivery of items for which the bureau was responsible. Given the necessary sequence of trials required before the *Nautilus* went to sea, Sylvester was uncertain that the bureau could meet the existing schedule, let alone a new one. A quick check confirmed his doubts: with no change at all, the bureau could miss the present goal by as much as six months. After studying delivery dates, manpower curves, and schedules, one of Sylvester's men thought that there was a real possibility that the hull would not be ready to receive some of the main propulsion plant equipment.[49]

Sylvester at once took steps to meet this challenge. He reassigned some work on other submarines to other yards; he delayed the installation of some gear not required for the sea trials; he requested help from other organizations in the Navy and authorized additional overtime and subcontracting at Electric Boat. Within the bureau, Sylvester warned his branches that there would be no changes in the *Nautilus* design unless they were essential to the safe operation of the ship, and he ordered prompt action on all plans and requests from Groton.[50]

This was not the first time that Rickover had maneuvered the bureau into accepting his demands, but always before he had acted in his role as a Commission official. Now Shugg could help him exercise some control over Electric Boat, and he could use that power to bring the bureau to terms.

"Trans-Atlantic Voyage"

By early 1953 construction of the Mark I prototype was nearing completion at Arco. Most of the buildings had been erected, and attention now focused on the bizarre structure resembling a section of an exhumed automobile tunnel that filled the central floor area of the large reactor building. The curved

Chart 5. The Nuclear Power Division was placed under the Assistant Chief for Ships and received a new code (490). The Argonne Laboratory had dropped out of the project.

THE NAVY NUCLEAR PROPULSION PROJECT IN NOVEMBER 1952

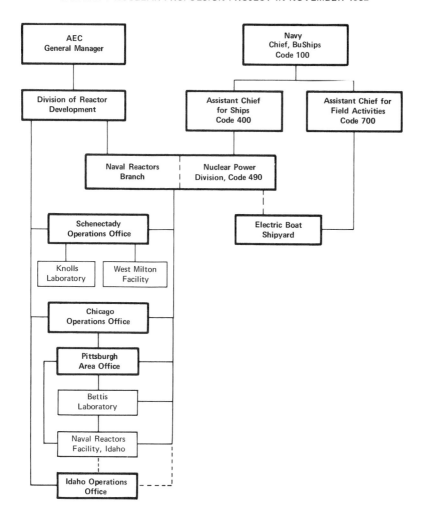

vertical plates which comprised the water tank outside the hull indicated the location of the reactor compartment. Once inside, many people would find it hard to believe that they were not aboard an operating ship. Only in the after end of the hull where some of the propulsion equipment was mocked up in wood did the plant depart obviously from reality.

As Mark I began the transition from a construction project to an experimental reactor, Rickover moved new talent to the scene. His own representative was Kintner, while the senior Westinghouse engineer was John Simpson, second only to Weaver in the Bettis hierarchy. Both competent and efficient, Kintner and Simpson worked well together. By becoming almost inseparable they provided the kind of coordination the project needed in its final stages. To make certain that he understood clearly what was going on at the site, Rickover insisted that the two men be on the telephone when they talked to him.

By the middle of March 1953 the Idaho team had completed the thousands of checks and adjustments necessary before the reactor could be operated. Rickover was on hand when the first tentative withdrawal of the control rods began during the closing days of the month. As the rods were inched out of the reactor, engineers checked the hundreds of instruments which indicated conditions in every part of the plant. More than once as the reactor neared the point of criticality something would trip the safety rod and the reactor would automatically shut down, or "scram." Then came a painstaking analysis to see what had caused the shutdown—whether there was a serious fault or whether the instruments were adjusted too finely. To the relief of everyone, the instruments were the cause most of the time. Finally, at 11:17 p.m. on March 30, 1953, the Mark I went critical. According to plan the power level attained was under .01 horsepower. It was sufficient to obtain physics data and shielding information.[51]

The next two months were filled with data gathering. The number of safety circuits—those which caused the reactor to scram—was reduced fourfold. As preparations were completed for bringing the reactor to power, Rickover flew in from Washington, bringing with him Thomas E. Murray, the first engineer to serve as a member of the Atomic Energy Commission. Murray was particularly concerned with the military applications of atomic energy and had become one of Rickover's staunchest supporters in Washington. To Murray fell the honor of opening the throttle which for the first time fed steam generated by nuclear power into the turbine. On May 31, 1953, the Mark I generated several thousand kilowatts of power.

Test operation continued day and night as the operators gradually increased reactor power in 5-percent increments. After each increase came a thorough analysis of plant conditions. The Mark I behaved well, the heat-transfer curves followed predictions, and there were no indications of over-heating. Particularly reassuring were the data collected on radiation levels: conservative designs had yielded levels significantly below those which had been calculated.

The Mark I had not yet reached full power. First plans called for a 48-hour run as essential for gaining crucial physics data. Later calculations showed that the information the physicists needed could be obtained in twenty-four hours. Rickover learned of the twenty-four-hour decision on his way to the site on June 25 and countermanded it. He determined upon a 100-hour run, one which would not only give the nuclear data, but also thor-oughly test the plant components. Not everyone agreed, but Rickover over-ruled all objections. That night on a cot in the Quonset hut, he thought of posting a chart of the North Atlantic in the control room, so that at the end of every four-hour watch the crew could mark the position of the "ship." Rickover had to return to Washington while the run was still some hours from ending. At the Mark I site all went well until about the sixtieth hour, when troubles began to accumulate: motor generators started to act up; nu-clear instrumentation became erratic; and a large reactor coolant pump de-veloped strange noises. Accepting full responsibility, Rickover in Washington refused all requests to shut down the Mark I. In the control room the officers watched the progress on the chart. When their reckonings showed that the "ship" should have reached Ireland, they shut down Mark I in accordance with established procedures.[52]

The track on the chart was graphic evidence to naval strategists that nu-clear power could soon revolutionize naval warfare. Even more impressive to engineers, the demonstration had come in the first stages of operating the prototype. Certainly one of the most remarkable features was the source of the difficulties occurring toward the end of the 100-hour run. They occurred in the steam plant and mechanical equipment. Nothing had gone wrong with the control drive mechanism, or with any of the nuclear components which had received so much agonizing attention. Rickover had disregarded the advice from Idaho to shut down the reactor because he saw that it was oper-ating well. He realized that those who wanted to terminate the test run earlier were not concerned about safety but about the risk of damaging steam plant components. As long as there was no danger to the reactor, Rickover saw no reason for not pushing the plant to the limits.

The test run had been an extraordinary achievement, with a significance extending far beyond the Navy and the United States. The Mark I was the world's first fully-engineered nuclear reactor capable of producing practical amounts of energy on a sustained and reliable basis. That fact represented the combined skills of Rickover and his organization, the Bureau of Ships, Westinghouse, Electric Boat, and hundreds of contractors. But as the leader of the project Rickover deserved and received the largest share of individual credit. The Mark I alone was enough to establish Rickover as an authority in nuclear technology.

Rickover and the Navy

Under most circumstances the striking accomplishments which the Mark I represented might have been expected to open new professional opportunities for the leader of the project, but Rickover had no such illusions in the spring of 1953. It was ironical that at this moment of achievement he faced the almost immediate termination of his career as a naval officer and thus as head of the naval propulsion project.

To provide promotion opportunities for officers in all grades, the Navy had developed over three decades a personnel system which required officers above certain ages in certain ranks to retire from active service if they were not promoted within a specified number of years. Formal selection boards established under Navy regulations determined which officers were to be promoted and which, in effect, were to be retired.[53] Rickover, who had been a captain since 1942, was now fifty-three years old. Having already been passed over twice for selection to rear admiral—the second time in July 1952—Rickover was faced with retirement on June 30, 1953.

Why Rickover, or any other officer for that matter, failed to be promoted by the selection board was a difficult question to answer. From the Navy's perspective, there were always more officers well qualified for promotion than the Navy could support. The selection process was, therefore, to some extent arbitrary, but most officers believed the selection board system established in 1916 was a vast improvement over the old method of promotion on the basis of seniority in grade.[54] Instead of leaving promotion to chance or perhaps political influence, the selection board process placed the difficult task of evaluation in the hands of professional naval officers who seemed best able to determine whether those of lower rank possessed the experience and ability needed for assignments at higher rank. Promotion, therefore, was not an award for accomplishments but recognition of an officer's capacity for greater responsibility.

13. The submarine thermal reactor, Mark I, at the Idaho test site as it appeared in 1954. The reactor is located within the portion of the hull surrounded by the water tank.

13

Precise regulations governed the methods of selecting members of boards, the composition of each board, and its functions. The Chief of Naval Operations and other senior admirals selected members of boards considering promotions to rear admiral. When the board was evaluating engineering officers, the membership included three admirals with an engineering speciality. To facilitate the frank exchange of opinion in evaluating candidates, the proceedings of the selection boards were secret. No official records were kept of their deliberations, and members were honor-bound not to reveal what they had discussed. Selection by the board was tantamount to promotion, but the civilian authorities—the Secretary of the Navy, the president, and the Senate —had to approve. These procedures were designed to provide some measure of objectivity and the exercise of professional judgment in promotions.

The selection board process, however, was subject to criticism. Its most vulnerable feature was the secrecy surrounding deliberations. Secrecy always left the boards open to the accusation, particularly by those officers who had been passed over for promotion, that capriciousness or favoritism entered into the selection. Such charges were impossible to prove, but many people in the Navy believed that an officer's social standing, his ability to get along with his superiors, and even the social graces of his wife could be just as important as his technical or administrative ability in attaining flag rank.

Given the system as it operated, one could only speculate why two boards had failed to select Rickover for rear admiral. There was no question that in terms of achievement he was one of the outstanding officers in the Navy. Even before Mark I began operation he was a prominent figure. In their September 3, 1951, issues *Life* and *Time* magazines featured the Navy's nuclear project and its leader. The press had noted the Navy's failure to promote Rickover in 1952. The magazine section of the *New York Times* on October 26, 1952, ran a favorable article on the project and spoke of the recognition Rickover had won from the Commission and Congress. Secretary Kimball in July, 1952, almost a year before Mark I reached full power, expressed the opinion that "Rickover had accomplished the most important piece of development in the history of the Navy."[55] Rickover was also well known to Congressional leaders and important figures in industry.

In many respects, however, Rickover's official position in the Navy and the Commission did not reflect either his achievements or his reputation. He had been a captain for more than a decade. The nuclear power division which he directed now reported to the assistant chief of bureau for ships (Code 400) rather than to the assistant chief for research and development, a change that suggested nuclear power had been recognized by the bureau as more than a research possibility. But Code 490, as the division was called, was still buried deep in the bureau organization. Likewise, in the Commission, Rickover occupied only a modest rung as a branch chief in the division of reactor development.

Among the arguments against Rickover's promotion the most obvious were those involving the personality clashes which had studded his career. An "outsider" since his days at the Naval Academy, Rickover had deliberately ignored and even ridiculed traditions that were part of the naval officer's world. He had offended some of his superiors by bluntly speaking his mind, but as a young officer he was protected by men many years his senior precisely because he got results. As he moved up in the Navy, his refusal to compromise on technical matters involved him in disputes with his near contemporaries. By 1953 those senior officers such as Mills, who had supported Rickover despite the antagonisms he created, had left the scene. Not every officer opposed Rickover. Some admired him for his accomplishments; some respected his ability but opposed his promotion to rear admiral; others welcomed the prospects of his retirement.

Unquestionably some of the opposition in the Navy to Rickover's promotion was based on personal animosities, spite, and even religious prejudice,[56] but underlying these emotional forces was a fundamental issue: the role and responsibility of officers in the modern Navy. The conventional view was that the naval officer had to be a well-rounded man who acquired special skills and administrative talent by filling a variety of assignments of increasing responsibility during his career. Rotation was expected to keep officers from going stale in routine jobs. It could bring new insights and fresh experience in practical situations to important staff assignments. It provided a number of officers with experience in each technical position so that the Navy, as one Chief of Naval Operations put it, would never have to rely on one person in a speciality.[57] Diversity of experience was also considered an important qualification for higher command. The advantages of the rotation system had been obvious in earlier days when it was possible for an individual officer to master all aspects of his responsibility. As naval technology rapidly became more complex in the years after World War II, it was less certain that rotation provided the kind of officer the Navy needed.

Whatever its merits, the rotation system had important implications for the operation of the Navy. Because no officer occupied any position in the Navy hierarchy for more than a few years, the system could not operate unless each officer recognized that authority rested in the title and not in the personal qualifications of the incumbent. Using the rotation system to provide broad experience and the selection boards to weed out the less efficient, the Navy tried to produce officers who could exercise authority with responsibility. Particularly in the higher ranks where authority was broad, these qualities seemed more important to many officers than highly specialized technical skills. Rear Admiral Homer N. Wallin, chief of the Bureau of Ships, was to state the traditional view before a Congressional committee in March 1953 when he declared: "The nuclear power billet in the Bureau of Ships is presently a Captain's billet, and we now have on hand a number of Engineering Duty Captains who are well qualified to assume this post."[58]

Statements such as this were senseless to Rickover and his organization because they saw the role of the naval officer in an entirely different context. Rickover insisted that the true basis for authority was demonstrated competence relevant to the position and not a record of broad experience or military rank. Neither military rank nor civilian status had any place in Code 490. Rickover tried to assign authority on the basis of competence and effectiveness. From this perspective it was ridiculous for Wallin to suggest that any other captain in the Navy could approach Rickover in his qualifications to head the nuclear power project. Rickover believed that he and his assistants were the best qualified men in the nation for their assignments. Bureaucratic devices such as selection boards and rotation systems in their opinion would deny the program the technical excellence it required.

There were officers in the Navy who might have been willing to accept Rickover's approach as it applied to a technical project of limited scope, but they could not accept the extension of that idea to the Navy at large. Rickover's highly disciplined (some would say obsessive) concentration upon a single objective largely explained the early success of the Mark I. It had also convinced many of Rickover's superiors both in the Navy and in the Commission that he was not "broad" enough for higher responsibility, say as chief of the bureau or as director of reactor development. This sort of criticism referred not to Rickover's intellectual attainments but to his tactics. His insistence upon the highest priorities for his own projects indicated to his superiors a narrowness of view, an inability to appreciate the value of other activities supported by the Navy or the Commission. Even if Rickover and his associates had been able to admit this weakness, they would have put it in moralistic terms—that they would not compromise their integrity by cooperating with what they believed to be inefficient or useless projects in hopes that the favor would be returned. Rickover's tactics were to approach his assignment with ruthless determination and as a project manager to fight to the last for everything he needed to attain his goal. These tactics, admirable in a project manager, were precisely what appeared to many to disqualify Rickover for "broader" responsibilities.

The Fight for Survival

With the regular channels to promotion blocked, Rickover would have to consider other strategies if he were to remain as head of the nuclear project. One possibility, which Admiral Wallin suggested, was that Rickover retire as a passed-over captain and then accept a recall to active duty in the same billet. This solution would keep him on the job, but Rickover feared that it would undermine his authority in the Navy. As the number of nuclear submarines increased, he would not be able to exert sufficient influence over other parts of the naval establishment. For similar reasons Rickover also rejected the idea of a special presidential nomination to rear admiral. Other officers also objected to this idea because it would tend to threaten the integrity of the Navy's promotion system.

The last alternative was for the Navy to keep Rickover on active duty after June 30 so that his name could once again be presented to the selection board in July. In the normal course of events there was little reason to believe that the board would select Rickover for promotion, but some of his staff believed that they could bring pressure on the board to make a favorable decision. Obvious sources of support were Congress and the press. Many Congressmen, especially members of the Joint Committee on Atomic Energy, were impressed by Rickover's accomplishments and saw his continued service as vital to national defense. The press saw the opportunity for a good story in Rickover's challenge to the Navy bureaucracy.[59] Rickover's staff stirred up some public interest, but many of the expressions of support from both government and industry were spontaneous.

Rickover's staff helped prepare the material that Congressman Sidney R. Yates read into the *Congressional Record* in early February 1953. Yates was a member of the House Armed Services Committee and represented the Illinois district from which Rickover had been appointed to the Naval Academy. The material consisted of magazine articles and the hyperbole delivered on ceremonial occasions by President Truman, Senator McMahon, and Chairman Gordon Dean of the Atomic Energy Commission. The central theme in

Yates's presentation was that the Navy was retiring its best "nuclear scientist" and risking the future of nuclear propulsion simply because the admirals on the selection board did not like Rickover personally. This complaint, which the Navy had often heard from disappointed officers, was coupled with a far more serious charge—that any selection system capable of such a gross error must be faulty.[60] When Yates introduced a bill proposing to restructure the selection process by adding civilians to the board, the Navy began to take the Rickover issue more seriously.

Yates's speeches, however, proved only the opening salvo of the attack. The main blow was to be delivered by Henry M. Jackson, the promising young senator from Washington. While in the House of Representatives, Jackson had been a member of the Joint Committee on Atomic Energy. He had met Rickover briefly at committee hearings, but he had not come to know him well until the two men found themselves on the same airplane headed for the nuclear weapon tests in the Pacific in the fall of 1952. While waiting through the interminable hours for the test to occur, Rickover and Jackson had struck up some lively conversations. Jackson was intrigued by Rickover's candor and intensity and listened with rapt attention to Rickover's accounts of his incessant battles with the Navy bureaucracy over nuclear power.[61]

Jackson, who had also been alerted by Rickover's staff, now jumped into the fray. After discussing the situation with three of Rickover's men, Jackson told the press that he intended to write Senator Leverett Saltonstall, chairman of the Senate Armed Services Committee, to raise a question about the Navy's failure to promote Rickover. At the time, the committee was considering the nominations of thirty-nine captains who had been recommended for promotion to rear admiral by the same board which had passed over Rickover's name in July 1952. As a member of the Joint Committee, Jackson had been behind the security barriers of the atomic energy program and was reported to have said that he "knew the full story of the Rickover case."[62] Jackson was also able to provide material to challenge the point that Wallin would make: that the Navy had several engineering captains who had been in the effort from the beginning and who were competent to fill Rickover's shoes. The committee announced on February 26 that it was withholding action on the thirty-nine nominations pending an investigation of the selection system.[63]

Now the Navy had no choice but to find a graceful way of promoting Rickover. The capitulation came in the form of a letter from Secretary Robert B. Anderson to Saltonstall announcing that the Navy was convening "a selection board to recommend engineering duty captains for retention on active duty for a period of one year with a requirement in the precept that one of those recommended for retention be experienced in the field of atomic propulsion machinery for ships." Should Rickover be selected, he would not be forced to retire in June and would be eligible for consideration by the selection board in July 1953, which would work under the same requirement in choosing captains for promotion to rear admiral.[64]

The promotion issue was now settled for all practical purposes, but some officers, particularly in the Bureau of Ships, refused to surrender. The March selection board retained several other captains apparently for the sole purpose of maintaining the appearance of routine procedure. The engineering officers on the July board refused to select Rickover even though such an action by the entire board would certainly have aroused strong criticism in Congress and threatened the whole selection process. To avoid this danger, the line officers on the board broke tradition by casting the majority vote and Rickover became a rear admiral.[65]

The successful operation of Mark I and his promotion to rear admiral gave Rickover a new stature in the Navy. He had gone to Oak Ridge in 1946 as an engineering officer little known outside the Navy. Seven years later he emerged as an authority on nuclear engineering and as one of the most influential officers in the United States Navy. During those years Rickover and his associates had developed a new style of engineering administration. They had created the industrial laboratories and organized the industrial team which had proved highly effective in directing and controlling the complex operations of modern technology. They had built powerful alliances outside the Navy with the Congress, the Atomic Energy Commission, and some of the best industrial companies in the nation. The future would hold many bitter struggles, but Rickover could now move with a new confidence and a new sense of independence toward building a nuclear Navy.

In some respects the first six months of 1953 had been highly encouraging for Rickover's organization. The sparkling performance of the Mark I plus Rickover's victory in his fight to retain his position in the Navy suggested that the 1946 dream of a fleet of nuclear-powered ships might not be far from reality. But Rickover could see danger in these very successes. The virtually trouble-free operation of the Mark I might lead some Navy officers to believe that building nuclear ships would be simply a production process. Furthermore, Rickover would pay a high price for the promotion which made it possible for him to remain as head of the project. The appeal to Congress for support and the implied threat to the selection board system had won him hostility and isolation as well as some degree of independence in the Navy.

Introducing nuclear propulsion into the fleet, even after the accomplishments of Mark I, would not be easy. Although the Navy was eager to have nuclear ships, Rickover suspected that few officers understood the impact which the new technology would have on conventional Navy activities ashore and at sea. As he reminded his superiors in April 1953, nuclear power for naval vessels was still in the development stage. The first two nuclear submarines would be custom-built and not production models with proven propulsion systems. Although he had already asked the laboratories to study other types of reactor plants, neither the laboratories nor Electric Boat had begun to make the transition from engineering development to the type of production operations necessary to build a nuclear fleet.

Looking to the Future

Rickover had concentrated the energies of his organization and contractors on the reactors for the first two submarines, but he was always interested in new designs which might offer advantages either in submarines or surface ships. His insistence upon considering engineering development as a learning process enabled him to see countless opportunities for improving both the physical design of reactor plants and the development process itself.

New designs were a frequent theme in Rickover's discussions with his staff even as early as 1949, but the first general consideration of that subject appears to have occurred in February 1950—months before ground had been broken at the Idaho site for the Mark I and before General Electric was fully committed to the Mark A. At this early stage of development the possibilities for applying nuclear power in the Navy seemed almost infinite, but in submarine design Rickover and his staff could see two probable courses. One was to produce a number of plants similar to the first two prototypes but in-corporating improvements gained through operating experience. The second was to develop an entirely new submarine reactor which would deliver three or four times as much horsepower and drive the submarine at much higher speed. For a more powerful plant, Rickover was particularly interested in using two reactors rather than one. His own experience with conventional submarines had convinced him that the extra margin of safety was worth the cost.[1]

The early development of the *Nautilus* had convinced Rickover that pro-ducing new designs of reactor plants would require many years of work. He thought it wise to begin studies of a faster submarine even before the *Nautilus* was completed.[2] By taking the initiative, Rickover might be able to discour-age the Navy from requesting a variety of reactor designs, each intended to meet a special need. There was also the very practical question of workload. As the first two propulsion plants neared completion, the laboratories would require new assignments, or their experienced scientists and engineers would be transferred to other projects.

In 1951 Rickover began the task of generating a Navy requirement for a submarine substantially faster than the *Nautilus* would be. On October 22, 1951, the Chief of Naval Operations signed a memorandum drafted by Rick-over's group requesting the bureau to "establish design criteria for an im-proved SSN, embodying very high submerged speed as its principal feature." Before the end of the year Rickover presented the idea at Bettis and Knolls. He proposed that the laboratories make a six-month survey in which they would not limit themselves to existing approaches.[3]

The call for new studies had come at an opportune time for General Elec-tric. The company had already informed the Commission of its desire to re-sume work on the power-breeder as a secondary effort to the Mark A and B. The Navy's interest in the new submarine opened still another opportu-nity. As in 1950, General Electric was reluctant to drop any possibility, but both Hafstad and Rickover wanted to prevent the company from spreading itself too thin. Rickover especially was determined to keep Knolls concen-trated on Navy work. First he made certain that the laboratory was involved in the new reactor studies. Then he applied pressure through the Commission to force General Electric to withdraw its proposal for the power-breeder.[4]

Through 1952 General Electric supported studies of the design criteria at Knolls with existing Commission funds. Two dozen scientists and engineers were all that were needed, and most of the initial work did not require costly experimental facilities. At first the Knolls group investigated the comparative advantages of water-cooled and sodium-cooled systems and tried to understand the engineering implications of a two-reactor plant. Because all previous reactor studies at Knolls had been focused on sodium systems, the laboratory found it necessary (and quite easy) to obtain expert advice on water-cooled systems from Argonne. Rickover ordered Knolls and Electric Boat to explore ways in which a two-reactor plant could be incorporated into a submarine hull.[5] By the end of the year Rickover was ready to move beyond studies into large-scale development. The future of the submarine advanced reactor would rest with the Eisenhower administration, which was about to take control of the executive branch.

Interest in a Nuclear-Powered Carrier

In 1950 Rickover's group could predict several reasons why the Navy would be interested in building a number of fast nuclear submarines in the future, but it was much harder to see how nuclear-powered surface ships could offer advantages worth the cost. Furthermore, a reactor for a surface ship would require a very large amount of uranium fuel, which was still in short supply. Rickover and his assistants concluded that "we should not, at this time, do anything toward pushing nuclear propulsion for surface vessels."[6]

About the time Rickover's group arrived at this position, Admiral Forrest P. Sherman was reaching a different conclusion. Sherman had become Chief of Naval Operations in November 1949, at a critical time for the Navy. The abrupt cancellation of the supercarrier *United States* soon after its keel-laying earlier that year had precipitated the "Admirals' revolt" and shattered the morale of the Navy. As Sherman's firm hand restored confidence in the service, he gathered strength as an influential force in Navy planning. A naval aviator, he appreciated the value of the carrier in a naval task force. In August 1950, after the invasion of Korea, Sherman asked the Bureau of Ships "to explore the feasibility of constructing a large carrier with an atomic power plant, and to determine time factors, cost factors, and characteristics."[7]

Despite their earlier skepticism about nuclear power for surface ships, Rickover and his group responded quickly to Sherman's request. Acting on the assumption that a nuclear-powered carrier could be built, Rickover asked Argonne, Knolls, and Oak Ridge to prepare feasibility studies of a reactor plant. The naval reactors branch had the reports within two weeks, and ten days later Rickover had a memorandum to the Joint Chiefs of Staff ready for Sherman's signature. Although Rickover's draft proposed completing a land-based prototype in 1953 and a shipboard plant in 1955, the final version mentioned only the land prototype.[8]

In limiting his recommendation to the prototype, Sherman might have been recognizing the strong reservations which both the Joint Chiefs and the Commission had about the project. Under a presidential directive the Commission was expanding its capacity for producing special nuclear materials for fission and thermonuclear weapons.[9] It did not seem feasible to give the carrier reactor a high priority without disrupting the expansion program. Instead the Commission urged in August 1951 that Westinghouse be requested to make an engineering study of a carrier reactor. Even worse, from Rickover's perspective, was the Commission's decision to assign responsibility for the Westinghouse study to another group in the division of reactor development rather than to the naval reactors branch. Hafstad feared that Rickover, in the interests of speed, would simply scale up the submarine design without considering other alternatives. It was imperative in Hafstad's opinion to consider the carrier reactor in terms of the Commission's long-term goals for building a reactor capable of generating electricity and for minimizing the diversion of fissionable material from nuclear weapons.[10]

Growing more impatient, Rickover attempted to force a faster pace on the carrier project, first by trying to push through a formal military requirement. Without the support of Admiral Sherman, who had died in July 1951, Rickover made slow progress. Marshalling new forces to his cause, Rickover established a close relationship with Commissioner Thomas E. Murray, who shared with Rickover the practical approach of the engineer and a consuming interest in national defense. Another source of support was Senator Brien McMahon, chairman of the Joint Committee on Atomic Energy. This alliance with key Commissioners and with the Joint Committee would be essential to Rickover's success during the next two decades. It may have encouraged the Joint Chiefs of Staff to establish a formal requirement for the carrier reactor in November 1951.[11]

With some prodding from Rickover and the Commission, Westinghouse completed its study of alternative designs in January 1952. The 133-page report analyzed the advantages and disadvantages of six reactor designs, of which five were found feasible for a carrier. Because the Westinghouse report did not include any recommendation, that task fell to Rickover's group. Within a few days the naval reactors branch concluded that the most promising design for immediate development would use ordinary water as the moderator and coolant. Both the Navy and the Commission endorsed Rickover's choice, and Hafstad assigned development of a land prototype of the carrier reactor to Rickover and to Westinghouse. Rickover estimated that it would take four years and as much as $150 million to build the land prototype.[12]

Despite the Commission action, the land prototype was in a weak position as the Eisenhower administration came into office in 1953. Part of the vulnerability lay in the narrowness of its initial support. Not many officers felt that a nuclear carrier held sufficient advantages to justify the expense, particularly if other parts of the Navy budget would suffer. Moreover, the timing of the project from its earliest stages had been unfortunate. Circumstances surrounding the atomic energy program in 1950 and 1951 had led to hesitation and delay, so that the project was just at the point of hardware authorization at the time of the 1952 presidential election. No matter which candidate was elected, it was likely that the new and expensive project would receive close scrutiny. Finally, the Joint Chiefs' requirement to combine the functions of land prototype, plutonium production, and electric power—no doubt intended to make the project appeal to several interests—could also be interpreted as a failure to agree on goals.

The Administration Acts

There was no doubt that the new administration would scan the Commission's program in a search for savings. President Eisenhower made that point clear in his State of the Union message in February 1953, when he declared that a reduction of federal expenditures and a balanced budget were indispensable to the economic health and military strength of the nation. The next day Joseph M. Dodge, the new director of the Bureau of the Budget, asked the Commission along with all other federal agencies and departments to cut their budgets for the coming fiscal year to the minimum necessary to maintain essential services.[13]

The Commission reacted by cutting back its civilian power program to only one reactor project, a sodium-cooled, graphite-moderated reactor, which was then considered among all the Commission's experiments the one nearest to actual operation. The Commission soon learned, however, that the administration's demand for retrenchment was not confined to civilian projects. At a meeting of the National Security Council on March 31, 1953, Lewis L. Strauss, now serving as Eisenhower's special assistant on atomic energy, suggested that eliminating the Air Force's nuclear aircraft project and the Navy's nuclear carrier would save more than $200 million annually. As a retired admiral in the naval reserve, a former Commissioner, and a staunch supporter of military programs, his opinions carried great weight.

There were several other reasons for cancelling the carrier project. For one thing, relatively little money had yet been spent on it. For another, the Joint Chiefs' requirements did not present a strong case. Although the Navy had requested the land prototype, it admitted that a requirement for the carrier itself would be premature. If elimination of the reactor would not be an immediate blow to defense, the principal value of the reactor would be its ability to produce plutonium and electric power. While the Eisenhower administration agreed that early development of nuclear power was important, Strauss and others argued that the goal "should be attained primarily by private and not government financing." Finally, the abridged project was inconsistent with the administration's intent to draft legislation that would permit private industry to own and operate nuclear power facilities, buy or lease fissionable material, and enjoy more liberal patent rights than were available under the existing Act.[14]

Rickover's efforts to save the carrier project had little impact on the administration. At a meeting of the National Security Council on April 22 Deputy Secretary of Defense Roger M. Kyes recommended indefinite postponement. Both the president and Strauss urged quick action on all projects to be terminated so as to reduce cancellation costs. Chairman Gordon Dean argued that the action would virtually kill the Commission's efforts to develop civilian nuclear power, especially since the administration had now cancelled the sodium-graphite reactor. Eisenhower replied that he would consider any recommendation the Commission might wish to make for converting the carrier reactor to a civilian power effort. All that the security council had determined was that the carrier reactor was not required for national security.[15]

In Rickover's view the carrier reactor was the victim of half-hearted support in the Navy. The administration's decision not to cancel the Air Force project to build a nuclear-powered bomber tended to support Rickover's opinion, although a comparison was difficult to make because the two projects were in different stages of development. Within the Bureau of Ships there had been no firm backing for the carrier. The ship design coordinating committee in 1952 had questioned the feasibility of obtaining sufficient fissionable material for a fleet of nuclear submarines and surface ships, and warned against seeking authorization of a large surface ship before operating experience had been gained on even a small ship. Rear Admiral Wallin, chief of the bureau, had pointed out to the Secretary of the Navy in August 1952 that there was no reason to believe that the first nuclear-powered aircraft carrier would be capable of speeds surpassing the most modern conventional carrier. Nuclear power would have some clear advantages for surface ships, but Wallin did not expect it to have revolutionary effects deserving exceptional priorities. In December 1952 the bureau concluded that no decision should be taken on laying down a nuclear-powered carrier until the first nuclear submarine was in operation and the land prototype of the carrier had been tested.[16]

Since the death of Admiral Sherman no senior officer on the staff of the Chief of Naval Operations had backed the nuclear carrier. The reason was that some officers, at least, saw in the project a threat to plans for building conventional carriers. The shipbuilding program for fiscal year 1952 had included the *Forrestal,* a large aircraft carrier which was the first major ship to be constructed for the Navy since the end of World War II. The Navy hoped that the *Forrestal* would be the first of a series, one of which would be authorized each year until defense requirements were satisfied. Too much talk about the promise of nuclear energy for carriers in the future could jeopardize the present effort. Admiral Donald B. Duncan, Vice Chief of Naval Operations, frankly had warned the bureau that "too rosy a picture" might undermine efforts to win Congressional support for more ships of the *Forrestal* class. In his view, it had made more sense to support the carrier reactor only as a land prototype which might also be useful in producing electric power.[17]

How much loss of the carrier project could be laid to the Navy was hard to say. Rickover had fought hard for the project and was disappointed both by the result and by the failure of the Navy to unite behind it. The fundamental error, however, was the commingling of several requirements in a single reactor at the very time those requirements were subject to review by a new administration. As it later turned out, all the requirements were met but with separate reactors. As we will see in chapter 8, the administration in the summer of 1953 approved the Commission's proposal to build a civilian power reactor in place of the carrier reactor. Just a year later the administration reestablished the requirement for a land prototype of a nuclear-powered aircraft carrier. In the meantime, Rickover had won Commission support for the submarine advanced reactor to be developed by General Electric as the next generation of nuclear power plants for submarines.[18] Thus in the long run the cancellation of the carrier reactor was only a temporary disappointment. Ultimately it would enlarge rather than restrict the role of Rickover's group in developing nuclear power systems.

New Navy Leadership

During his campaign for the presidency, Eisenhower had promised a new look at the defense needs of the nation. To meet this pledge he submitted to Congress in April 1953 a reorganization plan which was aimed at strengthening the authority of the Secretary of Defense. During the same year the terms of the current members of the Joint Chiefs of Staff expired. By the end of June, when Congress accepted the proposed changes in the Department of Defense, Eisenhower had also nominated the new members of the Joint Chiefs of Staff. He selected Admiral Arthur W. Radford as chairman; General Nathan F. Twining to represent the Air Force; General Matthew B. Ridgeway, the Army; and Admiral Robert B. Carney, the Navy.[19]

Carney came from the command of the United States Sixth Fleet and the NATO forces in Southern Europe to take his oath as Chief of Naval Operations on August 17, 1953. His record included extensive service at sea during both World Wars, and he had held important positions in Washington. As a member of Admiral William F. Halsey's staff, he had spent much of World War II aboard battleships and carriers in the Pacific. His love, however, was destroyers. He could still speak warmly of the *Fanning* on which he was officer of the deck in November, 1917, when she sank the German submarine *U-58*.

The conflict in Korea, which had ended a few weeks before Carney was sworn in, had made a profound impact upon the Navy. When the fighting had broken out in 1950, the Navy had been reducing its fleet in order to adjust along with the other services to extremely tight budget restrictions. Five years after Hiroshima, the Navy was still uncertain as to the role of seapower in the age of the atom. At the time of the North Korean attack, the active fleet consisted of 671 vessels, including seven large aircraft carriers, seventy-two submarines, and one battleship. The North Koreans, with only a few combat ships, had offered the United States Navy little challenge except by mine warfare, but the Navy had played an effective part in the conflict by providing carrier-based strikes, bombarding coastal facilities from surface ships, and landing and evacuating troops. Although the fighting was localized in the Far East, the war itself created tensions around the world, a fact which required the Navy and other services to build up strength in other areas. By the time the cease-fire had been signed in Panmunjom on July 27, 1953, the Navy's active fleet stood at 1,129 ships. In every category an increase had been achieved, mostly by taking ships out of "mothballs." The number of large aircraft carriers had doubled, submarines had risen to 110, and even battle-ship strength had temporarily climbed to four.[20]

Although the Korean War had shown that control of the sea was impor-tant in limited war, Carney did not face an easy future. The place of the Navy in the military establishment was still uncertain. Eisenhower's proposal for reorganizing the Department of Defense was evidence of strains and tensions among the three military services, and of the growing sophistication of the unification struggle. The first Republican budget for defense set a new peace-time record. More significant to the Navy, however, was the fact that the Air Force appropriation was larger than that of any single year during the Korean conflict, while the Navy and the Army took substantial cuts.[21]

Turning to the Navy itself, Carney also faced difficulties. The Navy had fought off the Korean coast with ships of World War II vintage which had performed satisfactorily except against the surprisingly effective mine warfare waged by the North Koreans. In consequence the Navy had embarked upon a hurried development of new minesweepers far in advance of what it had possessed at the end of the war with Japan. In a way, the minesweeper inci-dent was indicative of the problem. Large ships, such as carriers, could be modernized fairly successfully and years added to their active life. For small ships, such as destroyers and submarines, the gains through modernization were more limited. These vessels too, were essential to the Navy, but most had been built during World War II and were showing their age.

Carney was convinced that the greatest threat to the United States came from the Soviet Union. The mines off Wonson were of Russian origin; the Soviet submarine fleet was active, large, and growing. To counter this threat, the Navy had to make use of new technologies. The fleet of World War II—serviceable off Korea—was rapidly becoming obsolete. Some of the lessons of that war were still applicable; the capital ships of the Navy were no longer the battleships, but the submarine and the aircraft carrier.

In submarine warfare the Navy had been forced in two directions. One was to develop better submarines for attack, the other was to devise improved means of destroying enemy submarines. One of Admiral Sherman's first acts as Chief of Naval Operations had been to seek authorization of a small high-speed submarine for use as a target to train surface forces. This function was also one of the missions listed by the ship characteristics board for the *Nau-*

tilus. In addition, submarines themselves might be used against underwater craft. Most promising were hunter-killer submarines, especially constructed for quiet operation and carrying sensitive detection equipment. Even as the first of these special ships was being built, the Navy decided to convert some of the fleet-type submarines to hunter-killers. The Navy also tried to use submarines for other purposes. A few were converted into oilers, cargo carriers, and transports.[22]

The need for better carriers stemmed from the increasing size, weight, and speed of aircraft and the introduction of jets with their voracious appetite for fuel. The carrier *United States,* cancelled in 1949, had represented one approach to providing better flight platforms. The funnels of the ship were to be flush with the flight deck and her bridge was to be retractable. Another solution was the British-inspired angled flight deck. In 1952 the *Antietam* was fitted with this innovation, which made possible recovering some aircraft while launching others. So successful were the tests on the *Antietam* that the Navy began converting its other large carriers. The *Forrestal,* laid down in 1952, incorporated all the major post-war modifications: the angled deck, the steam catapult (another British development which made possible launching heavier planes), and the closed-in bow.[23]

Improving the carrier, however, was not enough. The carriers needed a task force of smaller ships to defend against enemy submarines and aircraft. But as the carrier became larger and faster, she tended to outrun her escorts. Even destroyers built late in World War II could not keep up with a fast carrier, particularly in heavy seas. In an effort further to strengthen the carrier task force, the Navy decided in October 1950 to convert some of its fleet submarines to radar picket vessels. These would operate far in advance of the task force and, with only the radar antenna above surface, would warn of hostile aircraft, control friendly planes, and as the use of missiles developed, guide them to their target.[24]

Carney was convinced that the Navy had to be modernized. He saw the potential of nuclear energy but he had no real grasp of the new technology. Carney had known Rickover as the engineering officer on a destroyer in the 1920s when he himself was an engineering officer for the destroyer squadron. Years later Rickover recalled Carney's initiative and industry in writing a damage control manual. After their joint destroyer service, the two men seldom saw each other. In 1946 Carney had noted with wry amusement the effort of some officers in the Bureau of Ships to prevent Rickover from being assigned to Oak Ridge. Shortly after Carney became Chief of Naval Operations, Rickover called upon him with information on the general status of nuclear propulsion.[25]

Command changes in the Bureau of Ships also occurred at this time. Wallin was relieved as bureau chief and became commander of the Puget Sound Shipyard. His successor was Rear Admiral Wilson D. Leggett, Jr. With a background which included extensive submarine duty, Leggett at one time had been in charge of internal combustion engine development for all Navy purposes, and he had been instrumental in developing diesel electric propulsion. Rear Admiral Sylvester continued to serve as assistant chief of the bureau for ships.

By the fall of 1953, the outlook for nuclear propulsion had greatly improved. With the end of the war in Korea, the Navy under new leadership could plan to build a new fleet. Rickover's technical achievement with the Mark I made certain that nuclear energy would be given serious consideration, even if no reactor-driven ship had yet gone to sea.

Defining Submarine Requirements

Rickover's justification for the submarine advanced reactor had convinced the Commission that it should continue to support further development of Navy propulsion plants. In the process of winning Commission approval Rickover had also succeeded in tying the Knolls laboratory more firmly to his activities. These achievements, however, would mean nothing unless Rickover won Navy support for the new submarine project. By the time the Commission agreed to finance the submarine advanced reactor, there were signs that the basis for Navy interest was shifting.

The size of the *Nautilus* and the *Seawolf* illustrated the truism that no advances in naval architecture were made without penalties. The two nuclear submarines would have far more horsepower than the *Tang* class, the Navy's latest attack submarines, but they would be far larger, displacing while submerged about 4,000 tons compared with slightly over 2,000 tons for the conventionally driven ships. High-speed submarines using two advanced reactors would have a submerged displacement of close to 6,000 tons. Despite the potential advantages of nuclear propulsion, many officers were disturbed by the huge size of the vessel being considered for the new reactor plant. Rear Admiral Charles D. Wheelock of the bureau had this reaction after visiting the Mark I facility in Idaho. The soaring expectation and obvious engineering mastery exhibited by the group at Idaho made a lasting impression on Whee-lock. He reported to Admiral Wallin his conviction that nuclear power for submarines was closer to reality than some of "the doubting Thomases" in the Navy believed. At the same time Wheelock confessed that he "simply [could] not stomach" the idea of a very large high-speed, high-powered sub-marine. Such a vessel did not fit into current or even advanced tactical con-cepts. "It looks like grandstanding and has as its only purpose a technique for gaining the dollar support of AEC."[26]

Wheelock's appraisal had an element of truth. Rickover had committed the high-speed submarine reactor to Knolls. In February 1953 the Chief of Naval Operations had signed an operational requirement for the ship and in April the Commission had accepted formally the submarine advanced reac-tor as part of its reactor development program. If the Navy failed to support the project, Rickover would lose Knolls; the Commission would turn the lab-oratory's resources in another direction; and General Electric would inevita-bly seek an assignment in civilian power. But even more was at stake than the Navy's relations with the laboratory. The advanced reactor was the Navy's only remaining close tie to the Commission's laboratories and technical capa-bilities. If the high-speed submarine were rejected because of its size, Rick-over might find it hard to convince the Commission of the need for continuing support, and the Navy would soon find itself in the technological backwater from which it had struggled to escape in 1946. Finally, Rickover believed it was foolish to reject the advantages of high submerged speed and long en-durance in order to hold down the size of the vessel. This kind of thinking, he declared, was short-sighted and unimaginative.[27]

The trouble in the summer of 1953 was that Rickover had tied the future of nuclear propulsion to a project which had declining appeal to the Navy. He could argue that submarine officers did not really know what a very large submarine could do with the high speed and long endurance which nuclear propulsion would make possible. But, as several admirals insisted in a Penta-gon meeting on August 5, 1953, the Navy simply had no use for a submarine with a displacement of more than 4,000 or possibly 5,000 tons. The argument was that larger submarines would be hard to maneuver and would make too big a target for the enemy. Preliminary studies at Knolls and Argonne during the previous eighteen months had indicated quite convincingly that the most promising designs for the submarine advanced reactor would be too large for a submarine of such limited displacement if it were to attain the submerged speed set forth by the Chief of Naval Operations in October 1951.[28]

To make matters more complicated, the Navy was now demanding nuclear submarines with the size and maneuverability of conventional attack submarines even at a sacrifice of speed. For Navy purposes, a scaled-down version of the Mark II plant would probably be satisfactory, but such a reactor would not qualify with the Commission as a development project. In short, the Commission was willing to support the advanced reactor and the Navy would finance a small submarine, but neither agency could give very much weight to the special interests of the other.

Rickover, caught between conflicting interests, had to move cautiously. As he explained to his superiors on August 10, 1953, the Navy had to support some kind of advanced reactor to keep the Commission involved. An equivocal attitude on the Navy's part would lead to the kind of disaster that had overtaken the carrier project. Rickover suggested a strong, high-level expression of Navy interest in an advanced submarine reactor, with some subtle modification of the original requirement. In place of the heavy emphasis on a high-speed capability, the Navy could stress the need for a new plant with a weight/power ratio much lower than would be possible in the *Nautilus* or *Seawolf* and capable of some improvement in speed. Such a project would insure Commission participation.[29]

To meet the Navy's immediate needs, Rickover proposed to begin the design of a modified and improved Mark II for a submarine of approximately the same size as the *Tang* class. The development of the nuclear power plant for this fleet-type submarine would be largely a Navy undertaking, although Rickover undoubtedly expected to get some Commission support for work at Bettis.

Under the circumstances that prevailed in the summer of 1953, the Bureau of Ships and the Chief of Naval Operations were willing to accept a compromise. Rickover would develop the submarine fleet reactor, and the Navy would take a firm but rather general position on the need for the submarine advanced reactor in order to meet the Commission's requirements. Rickover proceeded to draft a letter for the signature of the Secretary of the Navy declaring the Navy's strong support for the advanced reactor. His staff worked closely with the Commission's reactor development and budget groups in drafting papers which the Commission approved on September 9, 1953.[30]

The Commission's action resolved a touchy situation. Rickover had perhaps overreached himself in stimulating a requirement for a high-speed submarine which would have been far larger and more expensive than line officers in the Navy would accept. Perhaps he believed the vision of a truly new type of submarine would overcome conservative opposition in the Navy. But in the end he had succeeded in fashioning a compromise which preserved the Navy's ties with the Commission's laboratories and promised the Navy a substantially improved submarine.

The Fleet Submarine

The second part of the compromise involved the Navy's decision in the autumn of 1953 to include at least one small nuclear-powered fleet-type submarine in the 1955 shipbuilding program, which was then being formulated in the Bureau of Ships and in the office of the Chief of Naval Operations. Rickover had already begun work on a propulsion system during the summer of 1953. He had asked Bettis to analyze the technical assumptions, development problems, and costs of what was to become known as the submarine fleet reactor or "SFR." With this study in hand early in September 1953, he had officially notified Hafstad that the Navy was awarding contracts to Westinghouse and Electric Boat for the design of the new submarine. Although the Navy would defray the cost of design, development, and construction, Rickover needed Commission authorization to use Bettis personnel and facilities, even though the Navy had not yet determined the size of its shipbuilding program.[31]

Within the Bureau of Ships the prospects for a third nuclear submarine raised the question of where it would be built. Since no private company other than Electric Boat was constructing submarines in 1953, Admiral Sylvester proposed to assign the new ship to the Portsmouth Naval Shipyard, which had been building submarines for years and had worked closely with Electric Boat. Located in New Hampshire on the Maine border, Portsmouth was not far from Groton. Another advantage of its location was that Portsmouth could count upon the congressional delegations of two states. Sylvester thought Portsmouth should take on the third submarine while Electric Boat provided design and consulting services.

Rickover accepted the need for greater shipbuilding capacity, but he thought Sylvester's plan was premature. Rickover had vivid memories of the troubles he had in building competence at Electric Boat. The assignment was even more challenging because the reactor for the small submarine would be built without the benefit of a land prototype. Portsmouth had no knowledge of nuclear technology, and in Rickover's opinion the yard's experience in constructing conventional submarines would be of little value. Furthermore, Electric Boat did not have the resources to serve as a design consultant. Against this background Rickover offered a counterproposal. The 1955 program should include two fleet-type nuclear submarines. One could be built at Portsmouth, the other at Groton. If the two projects were phased so that at each stage Electric Boat was some months ahead of Portsmouth, the new yard could be trained with the least burden on the old.[32]

The possibility of more submarines also brought up the question of the role the other bureau codes would play. They were already heavily involved in the design and construction of the nonpropulsion portions of the *Nautilus* and *Seawolf,* but the boundary between the responsibilities of Code 490—as the nuclear power division was now called—and the other codes was not clear. As Bettis was drawing up its first reports on the submarine fleet reactor, the bureau's ship design division was investigating the feasibility of a nuclear submarine with either a single propeller or twin screws, and with or without a conning tower. The machinery design division was exploring several possible turbine propulsion systems: direct drive, electric drive, and a reduction gear. Before the end of September Leggett had reports from these divisions. On October 13, 1953, he decided that if a nuclear submarine was included in the 1955 program, Portsmouth would build it. Code 400, headed by Sylvester, would direct all design work and administer the contracts. Leggett's orders did not mention Rickover or Code 490, but because Code 490 was part of Code 400, presumably Rickover would have a part in technical matters.[33]

The absence of any mention of Code 490, however, did suggest that Leggett was attempting to broaden competence in nuclear technology in the bureau as well as the shipyards. From his vantage point it was logical to start with the third submarine. Even if it was powered by the new fleet reactor the ship would require less development than either the *Nautilus* or the *Seawolf.* Consequently the other codes could take up their traditional work in the more routine aspects of ship design and construction, while Rickover and his organization could concentrate on truly innovative designs such as the submarine advanced reactor. Whatever Leggett's motives, he was acting on the assumption that the bureau and Portsmouth could do the job, and he was pointedly ignoring Rickover's warning that the assignment would prove too much for both.[34]

Leggett and Rickover were grappling with a major feature in any successful development effort: the transition from the first-of-a-kind to production types. Some demarcation between Code 490 and the rest of the bureau was necessary and inevitable; the question was the terms of the arrangement. After several conferences, the bureau leaders on November 19, 1953, reached an agreement which covered the division of responsibilities for design studies for the submarine fleet reactor and the submarine advanced reactor. In the case of the submarine fleet reactor, all hull and ship design and all steam machinery would come under Sylvester, but Rickover would retain responsibility for the reactor. For the submarine advanced reactor, which was a developmental effort, Rickover would control the entire propulsion system from the reactor to the propeller shaft.[35]

The solution was logical. Rickover's responsibility for an entirely new propulsion plant was the outgrowth of his experience and that of the bureau with the *Nautilus* and the *Seawolf*. Although the November 19 agreement covered only the design studies, it contained the seeds of the permanent arrangement between Rickover and the bureau. He would always have charge of the reactor, regardless of whether or not it was for the first ship of a class. For the following ships of that class the bureau would be responsible for the steam plant. Changes which might affect reactor operation or specification would require Rickover's approval.

The Nuclear Power Division

The impact of these new responsibilities on Rickover's nuclear power division (or the naval reactors branch as it was still called in the Commission) was difficult to measure primarily because Rickover refused to accept the rigidities of a formal organization based on functional assignments. Rather than create a structure which divided responsibility in a formal way along functional lines, Rickover preferred to assign tasks according to the abilities of his staff, regardless of whether these assignments made sense on organization charts. For this reason, there were few major reorganizations within Code 490. One systematic effort to reorganize had occurred in the autumn of 1953 when the number of projects was about to increase from two to five. Rickover suggested that his senior staff find some remote location where they could examine the organizational impact of the new projects, away from the daily preoccupations of the office. Panoff offered the use of a canoe club on Sycamore Island in the Potomac, and the group consisting of Roddis, Kyger, Laney, Panoff, Mandil, and Rockwell assembled there on September 10, 1953.[36]

The Sycamore group concluded that it would not be feasible to have five project officers working independently with Bettis and Knolls. All activities in each laboratory would be coordinated in Washington by one engineer who would be known as a laboratory officer. These two men—one for each laboratory—would monitor the activities of the full-time project officers at Bettis and Knolls. To relieve the laboratory officers from budget and administrative responsibilities held by the present Washington project officers, the Sycamore group recommended the creation of a budget and reports group.

Rickover did not think the new organization would work, but he was willing to try it. Laney became the laboratory officer for Knolls, and Panoff took that position for Bettis. Lieutenant Commander Vincent A. Lascara, a Navy supply officer, set up the new budget and reports group. Whether Rickover's judgment was correct or whether he saw to it that his prediction came true, the laboratory officer system was not successful. The project officers in the field were never formally designated, and within a few months Rickover was making other assignments which undercut the laboratory officers. By the summer of 1954 the old project officer system, plus Lascara's budget and reports office, was reestablished. Panoff was the project officer for Mark I and II and would later take over the submarine fleet reactor at Bettis as well. Laney had gone to Pittsburgh as Geiger's operations officer, leaving three men from the 1953 postgraduate class at the Massachusetts Institute of Technology as project officers: Lieutenant David T. Leighton for the submarine advanced reactor, Lieutenant Commander Robert A. Hawkins for the Mark A, and Commander James C. Cochran for the large ship reactor. Commander Joseph H. Barker, Jr., from the Navy civil engineering corps, had been selected to serve as project officer for the new civilian power reactor.

This roster (or any other during the middle 1950s) offers but a momentary glimpse at an ever changing pattern of names and positions. Of the four officers from the 1953 MIT class, only Leighton served more than a few years on the Washington staff. For a time he was stationed at Mare Island, but both before and after this assignment Leighton was one of the inner circle upon whom Rickover relied heavily. The original Oak Ridge group, which had provided Rickover's senior staff in the early years, was all but gone in August 1954. Dick was dead; Libbey had left the Navy; and Dunford was temporarily on another assignment. Only Roddis was still in Code 490 with responsibility for officer assignments and administrative liaison with the Navy. Of the civilians in the Oak Ridge group, only Emerson was still working in Code 490, on reactor containers and pressure vessels.[37]

Most of the original staff of the technical groups was still in Code 490 in the summer of 1954. Kyger continued to serve as Rickover's principal advisor on physics, although Radkowsky now had an important role. Rockwell and Mandil now had heavy responsibilities in nuclear technology and reactor engineering, respectively. Kintner was at the moment in charge of advanced design, one of the more important assignments among the technical groups, while Marks was responsible for ship applications. Condon, Wilson, and Kerze continued to serve as senior engineering specialists. Among the many newcomers in the technical groups, there were several who would have a lasting impact on Code 490. Milton Shaw, a Navy civilian engineer, took the Oak Ridge training course before coming to Washington to direct work on systems for the submarine advanced reactor. Later he would become a project officer and a member of Rickover's senior staff. Jack C. Grigg, already Rickover's specialist on control and electrical systems, would occupy that position for two decades. Robert W. Dickinson and Robert F. Sweek, two officer graduates of the MIT course, and Theodore J. Iltis, a civilian engineer, held responsible technical positions in Code 490 for several years.

In its relations with the Bureau of Ships, Code 490 occupied essentially the same position in 1953 that it had in 1949. Still a division within the office of the assistant chief of the bureau for ships (Code 400), the nuclear power division reported through Sylvester to Leggett. In July 1954 the division's designation was changed from Code 490 to 590, but this modification reflected only a larger reorganization in the bureau and not a shift in Rickover's responsibilities.[38]

The Concept of a Nuclear Fleet

Although the Navy had no firm commitments to more than three nuclear-powered submarines in the autumn of 1953, Secretary Robert B. Anderson, Admiral Carney, and Rickover were all considering the possibilities of building a fleet of both nuclear submarines and surface vessels. To some extent the outcome would depend upon Carney's assessment of the needs of the Navy, Anderson's concern over budget implications, and congressional willingness to appropriate funds.

For Carney, the question of nuclear propulsion had to be weighed carefully. On the one hand, Carney felt it was urgent to modernize the fleet, and he was convinced that nuclear power would be an important element in rebuilding the nation's naval forces. In December 1953 he asked Leggett for data on a nuclear-powered destroyer and a large submarine. Carney and Anderson were both thinking of the large submarine as a carrier for the *Regulus,* a surface-to-surface air-breathing missile which had been test-fired from a conventional submarine in July 1953.

On the other hand, Carney saw that modernization of the fleet required more than nuclear-powered ships could provide. When he testified before a House appropriations subcommittee on February 9 and 10, 1954, on the 1955 shipbuilding program, Carney asked for one *Forrestal*-class carrier, five destroyers, five destroyer escorts, and three submarines—one of which would have nuclear power. He characterized the proposal as only the first step in a systematic effort to build a modern fleet. A deliberate plan would avoid the extra expense of a large construction effort as well as the ultimate danger of letting a large part of the fleet become obsolete at the same time. Although he did not explicitly make the connection, Carney's plan for a long-range approach was clearly consistent with the Eisenhower administration's intent to build for the "long haul." Carney declared he was certain that nuclear power would revolutionize the fleet. "But until we know exactly where we are going I would not recommend a large and precipitate building program."[39]

Carney had not yet decided whether the Navy should ask for more than one nuclear submarine. The day after his second session with the subcommittee he and a group of admirals went to Rickover's offices in the temporary buildings on Constitution Avenue for a briefing on nuclear propulsion. The staff had carefully prepared a series of charts covering all aspects of nuclear propulsion, including its application to various types of submarines and surface ships, costs, and the number of shipyards required. Well aware that Carney was troubled by the cost of nuclear submarines, Rickover pointed to reductions already achieved and the promise these held. He described the existing projects and his relations with the Commission, the Navy, and the contractors.[40]

From these details Rickover turned to the future. He thought the 1956 plan should include three more nuclear-powered ships: two fleet submarines and one guided-missile submarine. Operational experience could make it possible for the Navy to determine the number it needed in the 1957 program to replace the conventional ships. An effort of this magnitude Rickover believed would require at least two more shipyards to take over the general design while Bettis and Knolls concentrated on developing new and improved reactor cores. He also recommended that another yard be selected to build nuclear ships. As for a nuclear-powered carrier, Rickover believed that no construction funds should be requested in the 1955 program, because the studies recently authorized at Westinghouse and Newport News were enough to keep the project alive.

Carney was looking for fresh ideas. Even before his congressional appearance he had established a committee on shipbuilding and conversion to canvass the Bureaus of Aeronautics, Ordnance, and Ships for new proposals. The same day as the Rickover briefing, Carney announced his intention of undertaking a long-range study of the Navy's needs. Stating that the Navy could no longer exist on World War II ship conversions, he declared that the service was entering a new era which would see the application of nuclear energy to weapons and propulsion. These would not alter the need for controlling the sea, but Carney believed they would change the character of fleet operation and the types of ships and aircraft.

How far the bureau had come in its thinking could be seen in the reply which Leggett sent on February 17, 1954, to Carney's committee on shipbuilding and conversion. Leaning heavily on a draft Rickover's group had prepared, Leggett proposed laying down one conventional carrier each year until the Navy had built up its strength. Turning to nuclear matters, Leggett saw no benefit in placing a nuclear propulsion plant in a battleship or a cruiser. As for submarines, he thought conventional units should be added until the nation had a capability of building nuclear submarines in large numbers. As a minimum, Leggett believed two nuclear submarines should be laid down each year beginning with the 1955 program. For the 1956 effort he would add the submarine with the advanced reactor.[41]

Leggett's reply showed Rickover's influence, but there was an additional significance. Pointedly absent was any reference to the closed-cycle submarine, an omission which must have caused some heartache in the bureau. On March 12, 1954, the ship design committee warned that nuclear power for submarines and surface ships was being given far too much emphasis prior to any shipboard operating experience. The committee asked that work on the closed-cycle continue and that one such vessel be included in the 1955 program. Leggett rejected the advice on the grounds that the Chief of Naval Operations had declared that there was no requirement for the ship and other officers in the bureau believed the approach was obsolete. Leggett's reply showed that the future of submarine propulsion lay with nuclear energy.[42]

Formal action implementing the decisions of February and March came quickly. On April 20, 1954, Carney sent Leggett a requirement for a radar-picket submarine which would use the submarine advanced reactor plant.

Tentatively, the ship would be in the 1956 program. Six days later in Leggett's office, Rickover learned that Anderson and Carney, reacting to strong pressure from the Joint Committee on Atomic Energy, had decided to add a second fleet reactor submarine to the 1955 program. Electric Boat would build one and Portsmouth the other.[43]

Although these decisions followed his recommendations, Rickover viewed the results with some concern. He was worried that the Navy might be concentrating too much on the production of reactor plants at the expense of developing them. He believed he could best counter this trend by outlining the future of nuclear power in the Navy and by explaining his plans. A memorandum to Carney in May 1954 set forth his view of the future course of nuclear power in the Navy. The proposal was to develop a family of five reactors covering a wide range of shaft horsepower so that they could be used in every class of ship for which nuclear propulsion seemed feasible. Rickover carefully pointed out that he was not suggesting the immediate application of nuclear power to a large number of ships, but rather a deliberate process of thorough testing and operational evaluation. Of the five reactors, two—the submarine fleet reactor and the submarine advanced reactor—were already under development. Not yet active were projects for a reactor for a small submarine, a twin-reactor plant for a large destroyer or cruiser serving as a carrier escort, and a multiple reactor plant for a cruiser or a large attack carrier.[44] The five-reactor plan assumed that nuclear energy could meet the main propulsion needs of the future Navy: for submarines designed for a variety of missions and for various ships in a carrier task force.

By this time Rickover had formally revived the carrier project. He had completed negotiations of a study contract with the Newport News Shipbuilding and Dry Dock Company for the "large ship reactor" or "LSR." Bettis already had several engineers working exclusively on the design of the new reactor, and the effort would grow rapidly during the remainder of 1954, especially after receiving formal approval from the administration. On July 23 President Eisenhower approved the action of the National Security Council rescinding the decision of the previous year. A few weeks later the Commission approved research and development for a land-based prototype of the large ship reactor at an estimated cost of $26 million over a five-year period.[45]

The 1955 shipbuilding program, the five-reactor proposal, and the action on the large ship reactor were evidence that nuclear propulsion was passing from a purely developmental effort to one which would play a key part in

the Navy's operations. The transition was that which followed every technological innovation as it moved from the laboratories and engineering shops
to the production floor. Carney at once saw the possibility of using nuclear
energy in the new fleet of ships which he expected would modernize the Navy.
If Rickover had not from the beginning insisted that the Mark I resemble a
shipboard plant, Carney would not have been able at that time to include
nuclear propulsion in his plans for the new fleet. It was unfortunate, however, that the submarines to be driven by the fleet reactor would represent
something less than the optimum in design and performance. In its anxiety
to have nuclear submarines at the lowest possible cost, the Navy had been
willing to settle for inferior performance. Years later Rickover would look
back on the decision as a lack of imagination on the part of officers who were
operating the submarines. These men, Rickover would contend, could not
see what lay before them.[46]

Although the 1955 program with the submarine fleet reactor foreshadowed
conflicts which Rickover would have with other parts of the Navy, the effort
had immediate implications for his own activities. He did not accept the assumption, common in the Navy, that it would be comparatively easy to modify the Mark II for the fleet type reactor, or to "educate" (to use Leggett's
term before the House appropriations subcommittee) other yards to build
nuclear submarines. Neither task would be easy, and they would fall largely
upon the organization which Rickover had created to build the *Nautilus* and
the *Seawolf*. He would have to transform that structure into one capable of
producing nuclear ships in numbers as well as of developing new types. The
transformation would require the reorganization of the laboratories, the contractors, and his own staff. Nuclear power in the Navy was entering a new
phase.

The *Nautilus*

The Navy in 1954 had embarked upon an ambitious plan to build nuclear
submarines. This ready acceptance of a new technology was somewhat surprising because no vessel had yet been propelled by nuclear energy. Admittedly the Mark I was compiling a successful operating record, but the plant
was still only a prototype in the desert. Not until the *Nautilus* was at sea,
driven by the Mark II, could Rickover demonstrate beyond any doubt that
naval warfare stood on the brink of a revolution.

The responsibility for transforming the *Nautilus* on the ways at Groton into a sea-going Navy ship rested with the officers and crew led by Commander Eugene P. Wilkinson. Born in 1918, Wilkinson had grown up on the West Coast and had graduated from the University of Southern California in 1938. For a few years before joining the Navy in World War II he had taught chemistry and mathematics in a California high school. After receiving a commission he had served one year aboard a heavy cruiser before transferring to submarine duty. Before World War II ended, he had completed eight war patrols. Deciding to stay in the Navy, Wilkinson was serving in the office of the Chief of Naval Operations when Rear Admiral William S. Parsons sent him to Oak Ridge to learn nuclear technology with a group of other experienced submarine officers. From Oak Ridge Wilkinson went to Argonne and Bettis, where he worked on the early development of Mark I. Impressed by Wilkinson's ability, Rickover had urged him to transfer to engineering duty, but Wilkinson had preferred sea duty on submarines. When Rickover was ready to select a commanding officer for the *Nautilus,* Wilkinson was his choice for the assignment. There were other candidates for this obvious prize, but Rickover with some difficulty prevailed.

The paramount importance which Rickover attached to training applied just as firmly to the officers and crew of the *Nautilus* as to the naval reactors staff. Rickover personally interviewed each of the officers recommended by the Bureau of Naval Personnel and accepted only those he thought could master the intricacies of nuclear technology. Both the officers and the crew, who were selected from the best men in the submarine force, were required to complete a gruelling one-year course in mathematics, physics, and reactor engineering at Bettis. They arrived at Arco in time to be an integral part of the team that brought Mark I to criticality and to full power. With their hard-won qualifications stemming from more than a year of training, the ship's company as it began to arrive in Groton late in 1953 was able to assume a key role in testing plant systems and components. The officers and men worked aboard the ship or in the yard alongside experienced engineers and technical specialists from Electric Boat, Westinghouse, and the supply contractors.[47] Never before had officers and crew of a new type of submarine come aboard with such a detailed knowledge of the propulsion plant and its components.

Only the traditional launching ceremony in January broke the steady pace of around-the-clock activity aboard the *Nautilus* during 1954. In September the discovery of faulty steam piping threatened the effort to complete the ship by January 1, 1955. Electric Boat ripped out the bad pipe and struggled to make up the lost time. With a last-minute burst of effort, the team of ship's engineering officers and crew and Electric Boat and Westinghouse engineers brought the Mark II to criticality on December 30. On the second day of the new year, the reactor briefly supplied steam for the ship's electrical system. Later that day steam was fed into the turbines, and the propellers turned over slowly as the ship lay at the pier. On January 3, 1955, the Mark II reached full power. As far as the propulsion plant was concerned, all that remained before the sea trials were more tests and some insulation around the piping.[48]

The cramped compartments of the *Nautilus* were unusually crowded on the morning of January 17, 1955. Joining Wilkinson and the crew were Rickover and a few of his staff, Carleton Shugg and other officials from Electric Boat, several contractor representatives, and officers from various Navy commands. At 11:00 a.m. the crew dropped the mooring lines and Wilkinson, on the bridge with Rickover, gave the command to back. When the ship was scarcely clear of the pier, the engineering officer in the maneuvering room reported to Wilkinson on the bridge that there was a loud noise in the starboard reduction gear and that he had switched to electrical propulsion. Under normal circumstances Wilkinson would have returned at once to the dock, but in full view of the press boats and other small craft attracted to the scene, Rickover was determined not to terminate the trial unless it was necessary. While the ship proceeded down the river on the port propeller alone, Panoff and the engineering officer inspected the noisy gear. It took but a few minutes to replace a loose locking pin on a retaining nut, and Wilkinson shifted back to steam propulsion. As the *Nautilus* slipped down the Thames past the breakwater into Long Island Sound, a signalman on the submarine blinked to the escort tug *Skylark:* "Underway on nuclear power."

This and subsequent trials exceeded expectations. During the first trial, while the *Nautilus* was confined to surface runs, the ship ran into seas heavy enough to make her roll violently. Many of the crew and technicians aboard became seasick as they struggled to measure the performance of the ship and its propulsion plant, but both operated perfectly. Submerged tests a few days later were more comfortable. Again the nuclear propulsion plant functioned faultlessly. To some officers the performance of the *Nautilus* was almost unbelievable. No longer did submarines need two propulsion systems—electric for submerged runs and diesel for surface operation. No longer was split-second timing necessary in crash dives in order to close air-intake valves for the diesels. No longer, indeed, would there be the familiar throb of the diesels. In some parts of the ship it was hard to tell whether the *Nautilus* was on the

14. The *Nautilus* (SSN-571) was launched at the Electric Boat yard, Groton, Connecticut, on January 21, 1954. The ship was less than a year from sea trials.

15. The world's first nuclear-powered ship ready for initial sea trials. The *Nautilus* was about to leave her dock at Groton on January 17, 1955.

14

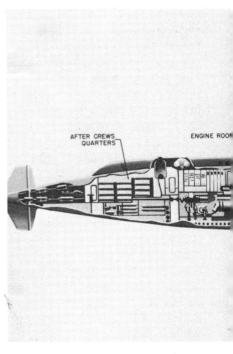

15

16. A cut-away drawing showing the main compartments of the *Nautilus*. The arrangement of the reactor compartment is only schematic.
U.S. Navy

17. Commander Eugene P. Wilkinson, the first captain of the *Nautilus*, at the periscope.

16

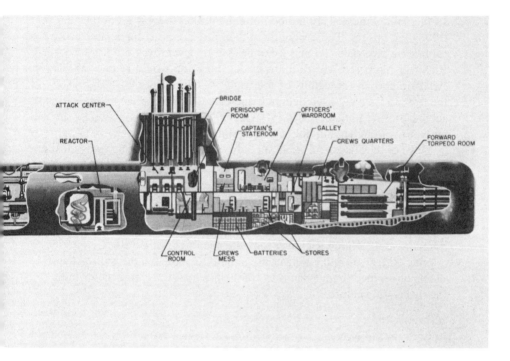

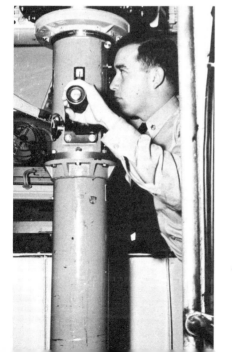

17

surface or submerged. In a mood of exhilaration the test group put the ship through her paces. There were more than fifty dives, setting a new record for submarine sea trials. To make certain that the event was properly recorded, Rickover discussed with Wilkinson the report he would send to the Navy. After describing the details of the trials, Wilkinson wrote: "The results of the tests so far conducted definitely indicate that a complete re-evaluation of submarine and anti-submarine strategy will be required. Its ultimate impact on Navy warfare should not be underestimated."[49]

News of the successful trials spread rapidly. Press coverage was most extensive in the East and in Pittsburgh, Groton, and Schenectady where there was local interest, but the event caught the attention of newspapers in other parts of the country as well. The wire services picked up the accolade "outstanding" bestowed by Secretary Charles S. Thomas. Carney told the House Appropriations Committee that the initial tests had gone off better than he had dared hope. In March Rickover took the members of the Joint Committee on Atomic Energy on a day-long cruise aboard the now-famous ship. Advertisements blossomed in magazines as companies boasted of their contribution to the successful venture.[50]

Impact on the Navy

It was easy for the public to see the *Nautilus* as the harbinger of the nuclear age, and engineers and scientists who had some experience in nuclear technology could appreciate the accomplishment which the new submarine represented. As promising as the ship's early performance was, however, the Navy had to reserve final judgment until an extensive series of trials was completed. Carney did what he could to speed the trials and was even willing to anticipate favorable results. He had already approved construction of three attack submarines using the fleet reactor being developed at Bettis: the *Skate* (SSN-578), to be built at Electric Boat; the *Swordfish* (SSN-579) to be constructed at Portsmouth; and the *Sargo* (SSN-583), scheduled for the Mare Island Naval Shipyard.[51]

Carney hoped to include additional nuclear submarines in the 1956 shipbuilding program; but, as he reminded the House Appropriations Committee in March 1955, the Navy's commitments were widespread and demanded modern vessels of several types. Thus his 1956 request included five conventional submarines as well as three nuclear-powered ships: two more submarines of the *Skate* class and a large radar-picket submarine using the ad-

18. Members of the Joint Committee on Atomic Energy aboard the *Nautilus,* March 20, 1955. Seated around the wardroom table clockwise from the foreground are: Representative Carl Hinshaw, Senator Bourke B. Hickenlooper, Senator Albert Gore, Representative Carl T. Durham, Senator Clinton P. Anderson, Senator William F. Knowland, Representative James E. Van Zandt, Representative James T. Patterson, Senator John O. Pastore, Representative W. Sterling Cole. Standing, left to right: Joint Committee staff members Corbin A. Allardice, Walter A. Hamilton, J. Kenneth Mansfield, Edward L. Heller, Representative Herbert C. Bonner, Admiral Rickover, Rear Admiral Frank T. Watkins, Commander Eugene P. Wilkinson.—*U.S. Navy*

vanced, two-reactor system. Carney saw his proposal as a prudent and orderly transition to nuclear power, particularly since industry had not yet demonstrated that it could build reactor plants in significant numbers.[52]

Almost as Carney was speaking, the *Nautilus* was proving the surpassing superiority of nuclear propulsion in demonstrations of performance completely beyond the capability of conventional submarines. Since the first trials, the *Nautilus* had been an unparalleled success, despite some deficiencies. Because of faulty design, the hull vibrated excessively under certain conditions. When the Navy took possession of the ship on April 22, 1955, temporary restrictions made it impossible to reach design speed. After a few days' delay because of a steam leak, the *Nautilus* began a shakedown cruise to San Juan, Puerto Rico. In eighty-four hours the *Nautilus* steamed 1,300 miles submerged, a distance greater by a factor of ten than that previously traveled continuously while submerged by any submarine. It was the first time that a combat submarine had maintained such a high submerged speed—about 16 knots—for longer than one hour. It was the first passage between New London and San Juan by any submarine, surfaced or submerged. Later the *Nautilus* did even better, going from Key West, Florida, to New London, a distance of 1,396 miles at an average speed of over 20 knots.[53]

Most impressive of all was the performance of the *Nautilus* in operations with the Atlantic fleet. In July and August 1955 the *Nautilus* and some conventional submarines of the *Guppy*-type simulated attacks on an antisubmarine force consisting of a carrier with its aircraft and several destroyers. Even against the *Guppies* the task force was hard pressed, but the *Nautilus* was almost invulnerable. At great ranges the nuclear submarine could locate the hunter-killer group, but the surface ships could not detect the *Nautilus*. Because the ship did not have to surface, it was almost immune to air attack. With its high submerged speed, the submarine could overtake a surface force making 16 to 18 knots and, in certain conditions, even evade the standard torpedo attack. To those officers taking part in the exercises and evaluating the first data, one fact was clear: in combat one nuclear submarine was worth more than several conventional ones.[54]

Since the Korean War the Navy had built some new types of conventional submarines and modified others for special missions. One idea was to build a small hunter-killer submarine, a slow but very quiet ship which could lie in wait for enemy submarines at strategic points. Rear Admiral Frank T. Watkins, commander of the Atlantic submarine force, saw in nuclear propulsion an exceptional opportunity to improve the effectiveness of the hunter-killer. The unlimited endurance of nuclear propulsion would enable the hunter-killer to proceed to its station submerged and thus undetected over the entire route. Watkins was also greatly impressed by the advantages of high speed which nuclear power offered in attack submarines. Watkins had concluded from fleet exercises with the *Nautilus* that as attack submarines, the ships in the *Skate* class would be too slow to avoid underwater detection.[55]

Nuclear power also seemed to have growing promise for a guided-missile submarine. Early in 1955 a committee under James R. Killian prepared a report on missiles for the National Security Council. The committee urged developing intercontinental and intermediate-range ballistic types, and especially recommended shipboard launching. The Navy already had modified two conventional submarines and several surface ships to carry the *Regulus I*. Under development since 1948, the *Regulus* looked like a small airplane, but it was a surface-to-surface air-breathing missile with a range of 575 miles and a speed of 600 miles per hour. There was some Navy interest in building a nuclear submarine to carry a *Regulus* missile. There was even some talk of converting the *Nautilus* and *Seawolf* to this purpose and using the two-reactor plant Rickover was developing to power a large, fast submarine carrying several *Regulus* missiles. Navy proponents saw the nuclear-powered guided-missile submarine as a reasonable step in technology and an excellent challenge to the missile activities of the Army and Air Force.[56]

In June 1955 two powerful members of the Joint Committee on Atomic Energy took up the idea. Congressman Melvin Price, chairman of the subcommittee on research and development, found it ironical that the *Nautilus* —the world's most advanced naval ship—should be armed with conventional torpedoes, a weapon which had not changed much since World War I. Price called for a nuclear submarine which would carry a missile armed with a nuclear warhead. Senator Clinton P. Anderson, chairman of the Joint Committee, had similar views. In lauding Rickover's achievements, Anderson remarked that these had been attained in spite of the Navy rather than because of it. Anderson suggested a reorganization of the nuclear propulsion program as a step toward defining and establishing responsibility and authority for the development of a complete nuclear propulsion and weapons system. Anderson's words were perhaps intentionally vague, but he seemed to be suggesting that the Navy give Rickover responsibility for developing nuclear weapon systems as well as propulsion plants for nuclear submarines.[57]

Probably in response to such pressures, Secretary Thomas proposed including a nuclear-powered guided-missile submarine in the next shipbuilding program. Far from being intimidated by his superior, Carney stuck to the position he had taken before congressional committees earlier in the year, with only a minor modification to recognize the new interest in hunter-killers. In his opinion the attack submarine still deserved the highest priority for nuclear power, followed by the hunter-killer and the radar-picket submarine. Carney maintained that he did not oppose the application to guided-missile submarines in principle, but he doubted that nuclear power would add much to their effectiveness. He also thought the rapid evolution of missile technology made it prudent to delay until missiles better than the *Regulus* had been developed.[58]

Admitting that it was not his job to establish shipbuilding priorities, Rickover was concerned about the impact of Carney's ranking on the nuclear program. The high priority Carney gave to the hunter-killer was particularly significant to Rickover, who understood the difficulty of fitting a reactor plant into the small hull of this submarine. Anticipating this new interest, Rickover had already begun the design of a reactor for the hunter-killer. After preliminary studies by his own staff, Rickover had convinced the Commission to fund the construction of a prototype under a 1953 requirement. In July 1955 the Commission had selected Combustion Engineering, Incorporated, to develop and build the prototype at Windsor, Connecticut. As for a missile-carrying nuclear submarine, Rickover agreed that it would take longer to bring an improved missile into operation than most people thought. For this reason it was important to continue the development of the submarine advanced reactor at Knolls. He believed it could be used for either radar-picket or guided-missile submarines.[59]

If the tentative priorities established by the Navy's leaders during the summer and fall of 1955 did not seem entirely orderly or consistent, it was worth remembering that the *Nautilus* had been operational for only a few months. It would take time to understand the full impact of nuclear propulsion and to sort out and evaluate potential applications. The Navy itself was a rapidly changing organization. One group of officers on Carney's staff or in the Bureau of Ships might find one application impelling; a few months later their successors might favor another. The *Nautilus,* however, did make one fact clear to everyone. As Admiral Jerauld Wright, Commander in Chief of the Atlantic Fleet, put it after reading the final report of the antisubmarine exercises with the *Nautilus:* "It is urgent that countermeasures be developed for the true submarine and that no future combatant submarine be built that is not nuclear powered."[60] At the very least, the *Nautilus* had brought the submarine into the nuclear age.

Nuclear Power Beyond the Navy

The decision of the Eisenhower administration to abandon the carrier reactor was a severe blow to Rickover's hopes for a nuclear-powered surface ship. Yet it provided the base for a new effort which would carry Rickover and his organization far beyond their familiar world of Navy bureaucracy, engineering laboratories, and shipyards.

The following pages describe how Rickover and his associates fashioned the remnants of the carrier project into a vast and organizationally complex undertaking to build the world's first full-scale electrical generating plant using nuclear energy. The civilian power project, which produced the reactor plant at Shippingport, Pennsylvania, brought Rickover's methods and philosophy into a new area of technology which private industry would develop in the future. The patterns created, not only in administering technical work but also in designing power reactors, would influence nuclear technology in the United States for decades to come. The role of Rickover's group in this accomplishment is the subject of this chapter.

Nuclear Power and the Carrier Reactor

The dream of using nuclear energy to generate electric power had influenced the Navy project from its very beginnings. Plans to develop the Daniels reactor in the spring of 1946 had provided the impetus for sending the Navy team to Oak Ridge and gave the Navy its first effective entrée into the world of atomic energy. The very size of the carrier reactor, its potential capacity as a power generator, in turn, readily suggested its application to electrical power production.

Robert LeBaron, chairman of the Military Liaison Committee, had raised this possibility in February 1951 when Rickover briefed the committee on his ideas for the carrier reactor. A chemical engineer, LeBaron had become deeply involved in the efforts of the military services to increase the production of nuclear weapons. One idea which LeBaron advanced was to design the carrier reactor so that it would serve both as a land-based prototype and as a producer of electric power, which was in short supply during the Korean War. Any large power reactor would also produce substantial quantities of plutonium, which under certain circumstances could be used for weapons if it could be recovered from the reactor. It was tempting to try to meet these three needs with one project, but Rickover warned that these would be conflicting requirements. For example, a carrier reactor should be designed to

operate as long as possible without refueling, but that would only delay plutonium recovery and would create a form of plutonium of limited use in weapons. Furthermore, a civilian power reactor would not need the operational
flexibility or rugged construction of a sea-going propulsion system. Rickover
looked at the carrier reactor with the single-mindedness of a project director.
To him, it made sense to concentrate on the carrier if that was what the Navy
wanted. But the urgent need for both plutonium and power and the growing
interest of American industry in nuclear power systems made a combination
of goals almost irresistible.[1]

As described in chapter 7, Rickover had been able to sidetrack the idea of
a multi-purpose reactor during 1951. By the time the Westinghouse study
appeared a year later, the land prototype was to be designed specifically for
shipboard application. But interest in a multi-purpose reactor had not died.
Since 1951 the Commission had been considering the possibility of building
one reactor which would produce plutonium for weapons, power for defense
industries, and engineering data for designing naval propulsion plants. The
prospects of such a reactor had attracted more interest among the Commissioners than among the reactor experts. Zinn already had his hands full with
the plutonium production reactors which Argonne was developing for the
Commission's new Savannah River plant. General Electric was equally preoccupied with its assignment at Hanford and with work on the Mark A and B
submarine plants. Hafstad would have been the last to underestimate the challenge which military requirements had placed upon the Commission for reactor products. During the spring of 1952, despite the Commission's interest
in multi-purpose reactors, Hafstad had insisted upon concentrating the efforts
of the reactor development division and its contractors on plutonium production. For studies of power reactors Hafstad was content to rely for the moment on private industry.[2]

With Hafstad's encouragement, private industry had already demonstrated
a strong interest in building power reactors. In 1951 the Commission had
permitted four study groups comprised of private power companies and industrial manufacturers to have access to classified information which would
enable them to evaluate the economic prospects of nuclear power. Encouraged by the preliminary results of these paper studies, the Commission early
in 1952 had invited these groups and other companies to submit proposals
for the actual design and construction of power-and-plutonium reactors. By
the summer of 1952 private industry was considering at least one proposal
for a joint venture with the Commission. A private utility would provide the

site and build the entire plant. The Commission would pay for and run the reactor and recover plutonium for nuclear weapons. The company would own the generating equipment and would market the electricity produced.[3]

To some extent the interest of private industry in power reactors could be expected to take some of the pressure off the Navy to make the carrier reactor a significant power producer, but Rickover could not escape that requirement entirely. Despite the efforts of Hafstad and others to discourage the idea of building multi-purpose reactors, the Commission had not given up the idea. In September 1952, in reviewing criteria which Rickover and his staff had developed for selecting a site for the carrier prototype, the Commission had moved power production from last to first priority. After safety, land acquisition, technical features, and administrative arrangements, Rickover had mentioned the power aspects in terms of selecting a site which would have access to a public utility and which would take into consideration the sale value of the power. In contrast the Commission assigned the highest priority to a site "in an area of high cost power."[4]

In some respects the growing interest of private industry in nuclear power biased the carrier project in the direction of power production at the expense of its function as a land prototype for a nuclear carrier. There were strong elements within the Congress and the executive branch, particularly in such agencies as the Department of the Interior and the Tennessee Valley Authority, which proposed to see nuclear power developed strictly as a government enterprise in order to prevent a private monopoly of a new energy source. For such individuals the enthusiasm of private industry enhanced the importance of the carrier project as a bastion against private monopoly. Whether he liked it or not, Rickover found by the end of 1952 that the carrier project had become a pawn in the old battle between public and private power interests in the United States, a fight that went back at least to the origins of the Tennessee Valley Authority and would continue long after the carrier prototype had been forgotten.[5]

Rickover made no attempt to exploit these forces in trying to reverse the decision of the Eisenhower administration against the carrier reactor early in 1953. He preferred to attack the project primarily as a misdirected effort to achieve economies in federal spending—misdirected because those criticizing the project did not understand the technical issues involved. This conviction was clearly evident in Rickover's efforts to revive the project even after the National Security Council's firm decision on April 22, 1953. Refusing to give up, Rickover had enlisted the support of John F. Floberg, Assistant Secretary of the Navy for Air, who arranged a meeting with LeBaron and Roger M. Kyes, the Deputy Secretary of Defense, on April 30. LeBaron expressed his conviction that private industry was prepared to undertake and finance the development of large power reactors, a contention which Rickover could not substantiate from his experience with major contractors like Westinghouse and General Electric. Neither did he accept LeBaron's claim that the project was on too long a time scale to be worthy of government support. The chief argument Kyes and LeBaron advanced against the project was that the technical approach was wrong. They contended that it should be possible to adapt the submarine reactor for this purpose. Rickover presented many arguments intended to refute this idea, but he was unable to convince Kyes. The deputy secretary was determined that there would be no reprieve for the carrier reactor.[6]

Industry and Nuclear Power

In his last-ditch effort to save the Navy project, Rickover chose to ignore new developments which already anticipated the demise of the carrier reactor. Commissioner Murray, who had long been Rickover's principal source of support within the Commission, had realized the project was dead after Strauss had advocated eliminating the carrier at the National Security Council meeting on March 31, 1953. Murray had suggested to Chairman Dean that the Commission consider combining the carrier project with its civilian power work in one effort which might meet both aims at less cost. Having discussed this possibility informally, the Commission did not overlook the opportunity which the president had offered Dean on April 22 to redirect the carrier project to a nonmilitary objective. Murray took the lead in drafting the Commission's proposal and in Dean's absence became the spokesman for the idea in a meeting with Eisenhower on May 4.[7]

The Commission's proposal cited the administration's high priority for "the early development of nuclear power by the United States." The Commissioners were convinced that the pressurized–light-water reactor offered a promising approach to civilian power. The redirected project was especially important after the administration's decision to cancel the Commission's only other experimental power project, the sodium-cooled, graphite-moderated reactor. The Commission also agreed to make maximum use of private industry. This would be possible if Congress adopted the Commission's proposed revisions in the Atomic Energy Act of 1946. But even then the Commission believed private industry would not assume the financial burden of long-term development of a power reactor.[8]

During the same week, the Joint Committee on Atomic Energy was also beginning to move in the direction of civilian power. In a hearing on May 1, 1953, with Kyes, LeBaron, and Floberg, Chairman W. Sterling Cole and his colleagues heard Kyes explain the decision to cancel the carrier reactor, an action which Floberg considered "a tragic error." At first concerned over the fate of the Navy project, the committee soon began to give more attention to the idea of a civilian power reactor, especially after Admiral William M. Fechteler, the Chief of Naval Operations, showed little enthusiasm for the nuclear carrier in a hearing on May 6. By that time the committee also had a copy of the Commission's letter to the president urging the civilian reactor project and a statement from Walker L. Cisler, president of the Detroit Edison Company, that a group of companies under his leadership was prepared to undertake development and construction of a power reactor. Murray countered this claim with a statement that none of the twenty-seven companies which had expressed an interest in participating in the atomic energy program would be able to develop a large power reactor at its own expense. The committee concluded that Kyes had not understood this point.[9]

Cole was determined to see that the administration did not postpone development of a power reactor in the hope that private industry would do the job. He was pleased to learn a few days later that the National Security Council on May 6 had approved the inclusion of $7.9 million in the Commission's budget for fiscal year 1954 for research and development of a power reactor based on the carrier reactor design, but he deplored the administration's failure to approve any funds for construction. In a strong letter to the House Appropriations Committee, Cole complained: "I believe that it is not only ridiculous but dangerous beyond our ability to foretell that we should now appropriate over a billion dollars for continued atomic weapon supremacy and yet allow nothing to start building at least one atomic power plant for peacetime use. This is not an act of economy but of folly." Cole urged the committee to add construction funds to the Commission's budget if the administration would not do so.[10]

The Joint Committee itself supported the power reactor in no uncertain terms when it reviewed the Commission's budget on May 18. Cole and other Republican members of the committee did not hesitate to disagree with the administration on this issue and gave Dean and Murray full opportunity to make the point that private industry was not prepared to make a large investment in nuclear power. Democratic members, led by Congressman Chet Holifield, warned against monopolistic control of a new power source by a few large companies and recommended a government project as a way of preventing it.[11]

By the middle of May 1953 it seemed likely that the Commission could save the project by converting it from a naval to a civilian activity. The staff was already developing plans for the new effort. Hafstad had two approaches open to him. One, which Rickover had prepared at Murray's request, was to continue the existing Navy project with the shipboard features deleted. This approach would undoubtedly produce an operating reactor in the shortest possible time, but it had distinct disadvantages from Hafstad's point of view. The methods Rickover had developed on the submarine plants allowed no room for exploratory studies of reactor systems by the Commission's laboratories or for independent development projects by private companies or groups, such as Cisler had proposed. Hafstad thought Rickover would make the minimum number of changes necessary to modify the design for civilian application only and then would proceed to build the reactor under the tight controls he exerted in the Navy program. This approach in Hafstad's opinion would give the Commission a working reactor, but he feared it might not be the best example of the efficiency or economy of nuclear power systems. As for industrial participation, Rickover would be willing to go no farther than to permit industrial representatives to work on the project much as they had on the Daniels reactor in 1946; there was no thought of sharing responsibility beyond Westinghouse and the private utility that would operate the plant.[12]

The second approach, recommended by Stuart McLain, chief of the production reactor branch, seemed more sound to Hafstad. McLain had been involved for some months in exploring the design of large plutonium production reactors, which opened up a variety of possibilities for power systems. Under McLain's guidance, Argonne had been spending close to $2 million per year investigating reactors using both light and heavy water as the moderator-coolant. Other laboratories and Commission contractors had studied slightly enriched reactors, which amounted to variations on the design Rickover had proposed for the carrier project. McLain thought it would make sense to spend a few months evaluating the various possibilities before reaching a decision. It was quite possible, for example, that these studies might lead to a reactor capable of generating much more power than the amount Rickover then proposed in modifying the carrier reactor. Only the larger reactor would have any chance of generating electricity at a cost close to that of conventional power plants.

McLain also favored a more open and flexible organization than would be possible under Rickover's control. In the submarine project, Rickover had gradually transferred development from Argonne to Westinghouse so that by early 1952 Argonne had few responsibilities left for naval reactors. Rickover had built a lean and hard team at Bettis, a group almost independent of the rest of the Commission and its contractors. To give Rickover the power reactor would leave the Commission's laboratories and private industry virtually no opportunity to participate. McLain was convinced that development of a good power reactor would require a broad spectrum of talents from the national laboratories and industry.

Hafstad had no difficulty in selecting McLain's approach. He conceded that Rickover had done an excellent job in building the *Nautilus* prototype, but the complexities of designing a power reactor were something else again. The Navy group, with its stress on producing hardware, was not likely to have much patience with new ideas that would require months and years of testing. And without such an approach, how could the Commission hope to develop a power reactor that was really based on the latest technology? Hafstad had support for his position in his own staff and especially from Robert P. Peterson, an engineer whom he had hired as chief of the industrial power branch.

Marion W. Boyer, the Commission's general manager, was inclined to agree with Hafstad and Peterson. Boyer, on leave from his position as vice-president of Standard Oil of New Jersey, believed in team enterprises and the industrial approach. At the same time he perhaps understood better than his subordinates the pressure that was building up on the Commission to produce the power reactor. Cole's letter to the House Appropriations Committee suggested that the Joint Committee had a special interest in the project and would be looking to the Commission for positive results. Murray was so intensely concerned about the project that he had asked the Commission and staff to make no commitment while he was absent from Washington during the last week of May. Without making any formal recommendations himself, Boyer sent Hafstad's paper to the Commission.[13]

The situation was touchy enough after Murray returned that the Commissioners decided on June 16, 1953, to consider the matter in executive session without any of the staff, even Boyer, present. Thus there was no record of the discussion, only the conclusion that the new project, now called the pressurized-water reactor, would follow the carrier reactor design and that it would be assigned to the naval reactors branch under Hafstad's general supervision. The Commissioners also approved a directive which Rickover's group had prepared requiring that the reactor use slightly enriched uranium and pressurized light water. The directive also specified the net power capacity and steam pressure. Within a week Rickover had also drafted for Hafstad's signature instructions to the field reflecting the Commissioners' decision.[14]

Technically the decision at the executive session on June 16 merely provided policy guidance for the staff; it did not constitute formal approval of the project. By the time Hafstad had completed a revised paper, the Commission had a new chairman, a fundamental change that gave those opposing the Navy approach a second chance. On July 1, 1953, Lewis L. Strauss, who for six months had been Eisenhower's special assistant on atomic energy, succeeded Dean as chairman. Strauss was closely associated with the administration's decision favoring the civilian power project and with his experience as a financier and his close ties with industry could be expected to support industrial interests. As the staff soon learned, the fact that Strauss was a retired reserve admiral did not mean that he would automatically favor a Navy project. In fact, Strauss was sensitive about his former Navy connections and intended to lean over backwards to avoid any appearance of being a Navy "stooge."

With a new chairman in office, a full-dress review of the June 16 decision was inevitable. Hafstad, Peterson, and others who hoped to reverse the decision would have to rely on the new chairman and Commissioner Henry D. Smyth, a physicist and author of the Smyth report on the Manhattan project. Smyth followed the Commission's research and reactor development programs closely and could be expected to appreciate the kind of broad investigative approach which Hafstad and McLain were advocating. Hafstad's first revision of the original paper presented only the bare specifications which he had sent to the field on June 23. Both Strauss and Smyth found it hard to believe that these data did not stem exclusively from the naval application. Only after Rickover had expanded the statement and discussed it with Smyth were the two Commissioners convinced that all special features of the carrier reactor had been deleted.[15]

Peterson again made a plea to the Commission on July 9 for a reactor which would be large enough to have a chance of being economical. Although the Commissioners understood his point, they believed that construction of a higher powered reactor represented too great a step in technology. Rickover asserted that existing technology would not permit construction of a pressure vessel for a plant capable of such a large power rating.

The issue seemed pretty well decided as the Commission discussion proceeded, but Rickover noticed that Hafstad and Peterson were nervously watching the door of the conference room as if they were expecting someone. Suddenly a courier arrived with a special message from the Joint Committee. The letter informed the Commission that Cole's appeal to the House Appropriations Committee had apparently been successful and that the Commission's budget would probably include $7 million for the start of construction of the pressurized-water reactor. As Cole pointed out, the provision established "a program *initiated* by the Congress." The Joint Committee would have "a more than usual interest" in how the Commission proceeded. The committee, according to Cole, would be concerned about "too heavy emphasis on the Navy aspects," which would result from Navy direction. Cole asked the Commission to inform the committee of "the specific administrative and organization plan" the Commission intended to follow before it was put into effect.[16] The letter was a last-minute attempt to reverse the decision assigning the new project to the naval reactors branch.

From earlier conversations with leaders of the American power industry, Murray suspected that Cole's letter represented a ploy to keep the project out of Rickover's control and thus assure industry a free hand in building the nation's first civilian power reactor. Strauss believed it would be wise to postpone a decision until the Commission could study the situation, but Murray was adamant. The Commission had spent a month investigating the project and was ready to act. In his mind, the question was whether the Commission or the Joint Committee was going to run the nation's atomic energy program. Once Murray put the issue in those terms, the Commission voted to reaffirm its June 16 decision. A few weeks later the Commission sent Cole a letter which Murray had drafted informing the committee that Rickover had been assigned full responsibility for the pressurized-water reactor.[17]

Murray soon discovered that the issue had not been fully laid to rest. During the next several weeks Rickover and Murray's own staff brought him reports of industry attempts to overturn the decision. There were several oral attacks on Rickover and new expressions of private industry's interest in building a power reactor. Murray found it necessary to warn some of the Commission staff against attempting to undermine the decision, and he insisted on substituting his own draft for the staff's proposed reply to the Joint Committee. His draft made the Commission's decision sound final and enthusiastic rather than tentative and reluctant.[18]

New threats to Rickover and the power project loomed up in August 1953. When the General Advisory Committee met in Washington, some members complained that the Commission had acted without consulting the committee.

Hafstad had also assembled representatives from the national laboratories to explain the work they were doing on power reactor systems. Although they were careful not to criticize the Commission's decision directly, the presentations left the clear implication that the laboratories considered the decision unwise. Approaching the issue with the scientists' experimental outlook, the laboratory leaders opposed investing virtually all the funds available for power reactor development in a reactor that did not seem to offer any impressive advances in design. Another factor was certainly the animosity which Rickover's aggressiveness had generated in the laboratories. Murray was furious at this attempt to line up the General Advisory Committee against the Commission and demanded that Hafstad be reprimanded. The Commission was not willing to go that far, but it did not reopen the question.[19]

The last salvo against the decision came from the Navy itself. Secretary Anderson told Strauss on August 20 that the Navy wanted no part of the project now that it had no relation to military requirements. The Navy, Anderson said, was ready to cooperate with the Commission in any way, including the transfer of Rickover to another assignment if his continued presence in the naval reactors branch caused difficulties. Murray had no trouble convincing his colleagues to reject the idea. By the end of August all sides seemed convinced that the Commission would not change its mind.[20]

The Commission's action represented a substantial victory for Rickover. It was first of all a clear vote of confidence in the system he had established for building naval propulsion systems. The successful full-power operation of the Mark I prototype that same month was living proof of his claim that he could build a submarine reactor on a time scale others thought was impossible. His promotion to rear admiral (described in chapter 6) showed that the Navy could no longer overlook his accomplishments. Even more important, the decision would have lasting effects on reactor development in the United States. It committed the nation at least for the moment to virtually exclusive development of light-water reactor systems. It also meant that the Commission would base its primary effort in reactor development on Rickover's hardheaded engineering approach, not on the research-oriented techniques of the national laboratories. Finally, the decision would require American industry to accept, at least for this initial effort, the technical standards and administrative controls which Rickover and his organization imposed. The decision would color American reactor technology for decades to come.

Organizing the New Project

As the Commission expected, Rickover and his group lost no time in making plans for the pressurized-water reactor. Before the end of July, they had sent Hafstad a revised cost estimate for the plant. In the rush to pull together a proposal for the National Security Council, Hafstad had guessed that the whole project would cost close to $100 million, roughly $75 million for construction and $25 million for development. Rickover now thought construction would run no more than $55 million, including architect-engineering, purchase of the site, all the buildings, the reactor, generating equipment, and the initial fuel loading. The schedule called for completing the project in the fall of 1957, and the naval reactors branch was proceeding on the basis of the new cost figure and this completion date.[21]

The only guidelines the branch had were the broad specifications which the Commission had set forth in the executive session on June 16. The Commission's directive made clear that time and money were limiting factors. To build the reactor quickly and within the cost limitation, the branch would have to depend upon the technology developed for the carrier reactor. There was no doubt in Rickover's mind that his branch would have to maintain the same strong centralized technical and administrative controls it had exercised in the submarine projects. The division of responsibility would be essentially the same, with Westinghouse having complete authority for the reactor and the primary coolant circuit and another contractor (either a construction company or a private utility) for the steam machinery and generating plant. This latter function would be comparable to Electric Boat's work on Mark I.

The design of the reactor itself would be the responsibility of Westinghouse. A supplement to the Westinghouse contract, signed on October 9, 1953, provided that the company would design, fabricate, assemble, and test the reactor and the primary heat-transfer system at an estimated cost of $19.5 million during fiscal year 1954.[22] Although the new project for the time being permitted cancellation of all work on the carrier reactor, it promised to increase the workload at Bettis substantially. As early as the spring of 1952, when Rickover assigned the carrier project to Westinghouse, the relatively unstructured organization which had grown up around the submarine project was beginning to show signs of age. Rickover complained that too many activities were scattered about the laboratory without proper coordination and direction. At his suggestion Weaver established four departments—for reac-

tors, power plants, physics, and materials. To make certain that the departments met the needs of both the submarine and carrier projects, Weaver
established a small project office under the direction of John W. Simpson,
who was now serving as assistant manager of the atomic power division.
Simpson, who had studied reactor engineering at Oak Ridge in 1946 as a
young engineer, had emerged as one of the most capable technical managers
in the Bettis organization. Weaver gave Simpson responsibility for planning,
scheduling, and checking on all activities related to the projects with the assistance of two project engineers.[23]

Simpson and his small staff tried during the following year to apply the
project approach at Bettis, but by the summer of 1953, when the civilian
power reactor and the submarine fleet reactor were added to Bettis' responsibilities, the assignment again outgrew the organization. Early in 1954, under
Rickover's pressure, the reactor and power plant departments announced
new alignments of functions which, among other things, fixed responsibilities
for project activities within the departments. The last step in the transition
was completed on September 1, 1954, when Rickover insisted that Bettis reorganize entirely on a project basis. The central structure of the organization
followed the four reactor projects—for the large ship reactor, the civilian
power reactor, the submarine fleet reactor, and the Mark II reactor. Most of
the technical departments by this time had been dissolved and their personnel
assigned to each of the four projects. Thus each project was essentially self-
sufficient in all the scientific, engineering, and technical capabilities needed
to design and build the propulsion plant.[24] In adjusting to new requirements,
Bettis was evolving from a small development laboratory for a single submarine project into a complex organization capable of designing and building
several types of reactor plants.

Selecting an Industrial Partner

Westinghouse had a vital role in designing the civilian power plant, but an
equally important element of the project was finding an industrial partner,
presumably one or more power companies, to build the electrical generating
portion of the plant and operate it. The Commission had justified the project
to the Eisenhower administration as an opportunity for private industry to
participate in developing nuclear power, and industry had expressed a strong
interest in such a partnership. The Commission's proposal was that Westinghouse would design and build the reactor, while a utility would finance, build,
and operate the electrical generating portion of the plant.

This was a reasonable expectation on paper, but as Murray pointed out, it was not likely to be fulfilled unless private industry was aware of the opportunity. More than a month after the Commission's decision to give Rickover and Westinghouse the assignment, Murray complained that the administration still had not announced its decision to build the pressurized-water reactor. Strauss accepted the need for an announcement, but he thought the matter of sufficient importance to reserve for the president. At the time, Eisenhower and his advisers were attempting to settle on a draft speech in which the president would give the American people a candid appraisal of the Cold War. Murray thought this an excellent idea, especially after the revelation that the Soviet Union had successfully detonated a thermonuclear weapon device on August 12. The United States, in Murray's opinion, could achieve a stunning propaganda advantage by announcing the civilian power project as a response to the Russians' warlike gesture. As that opportunity slipped away during the following weeks of inaction, Murray grew more impatient. Finally he extracted from Strauss permission to make the announcement in a speech before a group of public utility executives in Chicago on October 22, 1953.[25]

In his Chicago speech Murray maintained that a civilian nuclear power plant was as important as the thermonuclear weapon in the nation's grim race with the Soviet Union. Unless the United States took steps to develop nuclear energy for the power-hungry countries of the world, the nation would not be able to count long on foreign sources of uranium ore, on which the growing stockpile of nuclear weapons depended. He acknowledged that private industry could not yet accept the full financial burden for the power reactor, but he was convinced that a strong government-industry partnership would be essential in reaching the goal of economic nuclear power.

Murray's speech was designed to arouse industrial interest, but it did not state specifically how industry might participate. Only those companies which had joined the industrial study groups in earlier years could surmise from Murray's statements that it would be appropriate to submit proposals for financing and building a portion of the plant. By the end of November 1953 only two such proposals had been received, one from a utility company in South Carolina and the other from the Nuclear Power Group, consisting of four large power companies and a major construction contractor.[26] Murray thought that a formal invitation from the Commission might elicit additional proposals, some of which might offer greater advantages to the government. At Murray's urging the Commission issued a formal invitation for proposals to be submitted by February 15, 1954.[27]

Of the nine offers received by the February deadline, the one from the Duquesne Light Company of Pittsburgh was clearly superior. The company offered to build a new plant on a site it owned at Shippingport, Pennsylvania, on the Ohio River twenty-five miles northwest of Pittsburgh. At no cost to the government, Duquesne offered to provide the site, build the turbine generator plant, and operate and maintain the entire facility. The company also agreed to assume $5 million of the cost of developing and building the reactor, which Westinghouse would design and the Commission would own. For the steam delivered by the reactor the company was willing to pay the equivalent of 8 mills per kilowatt-hour, a comparatively high price. The Commission staff estimated that over the course of a five-year contract Duquesne's contribution would be more than $30 million, compared to $24 million for the next most attractive proposal. Also under the Duquesne offer the Commission could cancel the contract at any time without incurring termination charges. It was abundantly clear from the proposal that Duquesne had much more than a casual interest in the Commission's invitation.[28]

The source of this interest lay in the chairman of Duquesne's board of directors, Philip A. Fleger, a lawyer and business executive who had brought a firm hand to Pittsburgh's power company. Although he made a point of keeping his stockholders happy, Fleger was not afraid of new ideas. The Commission's call for industrial studies of nuclear power in 1950 had attracted his attention, and he had ordered his best engineers and executives to join him in taking a course in the principles of atomic energy at a local university. In the spring of 1953 Fleger had proposed to the Commission a joint venture with another company to explore the feasibility of nuclear power.[29]

It was Murray's speech, however, that really sparked Fleger's interest. He was startled by Murray's statement that the nation's continuing access to foreign sources of uranium would depend upon the early achievement of economic nuclear power. He agreed with Murray that industry had a vital part in that effort, and he feared that the government would do the job alone if industry did not offer attractive terms for partnership.[30]

Equally important in Fleger's mind was the exceptional opportunity which the Commission's invitation offered to a relatively small power company. Duquesne, with its limited resources, could not accept an open-ended commitment to a nuclear power plant, much less hope to build one without government assistance. In this instance, Duquesne could limit its financial commitment to the terms of its proposal. Fleger was also willing to gamble that, after the original contract expired, the Commission would be willing to underwrite operating costs so that the company would pay no more for nuclear power than it would for energy from a conventional plant. If the project in fact imposed no long-term economic disadvantages, would not the publicity alone, which the presence of the world's first full-scale power reactor would bring to Duquesne and Pittsburgh, be worth the investment? Fleger was convinced it was.

The Commission had no trouble accepting the Duquesne proposal. It made possible the government-industry partnership which the Commission had been trying to establish for years. Furthermore, the terms of the agreement, with the $30-million contribution from Duquesne, would satisfy an economy-minded administration. Westinghouse was pleased to have the plant close to Pittsburgh and under contract with one of its oldest customers. From the perspective of the Commission and Westinghouse, the arrangement could hardly have been better.[31]

The Contractor Team

Even if Rickover accepted these premises, he was more impressed by the obstacles he faced in directing such a complex partnership. In the first place, he was working outside the familiar territory of the Bureau of Ships and its long established relationships with Navy suppliers and shipbuilders. Rickover and his staff had little direct experience in large construction projects and even less with the power industry. Secondly, the complexity of the new organization would make it difficult for the naval reactors branch to exercise the kind of controls it had imposed on the submarine projects. Although Westinghouse and Duquesne would be the only prime contractors, they could not build the plant themselves. Westinghouse had already taken steps to find an architect-engineer to design the plant as a whole, and both Westinghouse and Duquesne would need subcontractors to construct their portions of the plant. This arrangement would require five contractors, and beneath them would be hundreds of fabricators and suppliers.

The contractor selection process itself was becoming much more complicated. The days were gone when Rickover and his staff could investigate a number of potential contractors informally, pick the one they thought best qualified, and then obtain a rubber-stamp approval from the Commission. The selection of Westinghouse for the power reactor project in the summer of 1953 had provoked some mild criticism from the Joint Committee, presumably generated by companies hoping to take part in the enterprise. Beginning with the Duquesne contract, the Commission required Alfonso Tammaro, as manager of the Commission's Chicago operations office, to appoint contractor selection boards. The boards issued formal invitations for proposals, evaluated them in accordance with Commission regulations, and then submitted a written evaluation and recommendation to Washington.[32]

Rickover considered the board procedure cumbersome and slow, but it did not frustrate his ultimate purpose to obtain the best contractors available. His own staff in Washington and Geiger in Pittsburgh were close enough to the selection process to assure that competent contractors were chosen. A board appointed by Tammaro in the autumn of 1953 selected the Stone & Webster Engineering Corporation of Boston as architect-engineer in April 1954. More than a year later a board with James T. Ramey of the Chicago office as chairman selected the Dravo Corporation of Pittsburgh to install piping and other equipment in the reactor portion of the plant. For the Duquesne portion Fleger selected Burns & Roe, Incorporated, and two associates in February 1955 without consulting the Commission.[33]

On such critical items as the pressure vessel, steam generators, heat exchangers, and reactor materials the naval reactors branch exercised the same controls over contractor selection and performance it had maintained in the submarine project. Rickover and his assistants were never a force to be ignored. They would demand realistic scheduling, good performance, and high-quality work. They would warn the contractors about the exceptional standards required in a nuclear plant. But Rickover knew from experience that such warnings were not likely to be accepted at full value by those who had never suffered the agonies of building a reactor.

Design Philosophy

In drafting the brief specifications for the reactor plant which the Commission had accepted in June 1953, the naval reactors branch had drawn heavily on existing technology. The use of pressurized water as a moderator and coolant had been tested extensively in submarine propulsion systems, and the proposal to use slightly enriched uranium as fuel rested on a substantial amount of research at Bettis for the carrier reactor. The idea of using saturated steam at 600 pounds per square inch originated in studies for the carrier reactor which showed that figure to be close to the practical maximum for the reactor plant being proposed. A civilian power plant, however, would require some specifications novel to Rickover's group, such as the use of concrete as shielding and adaptation to the kinds of turbines, electrical generators, and auxiliary equipment found in conventional central-station power plants. The June specifications also called for the simplest possible control system, the longest possible life for fuel elements, and the shortest possible refueling time.[34]

Although these initial specifications established the general characteristics of the plant, they did not begin to provide the degree of guidance which Rickover intended to give Bettis. If his branch was to have any effective influence on Bettis, as Rickover insisted, it would be necessary to formulate a fully articulated "design philosophy." That process required the branch to assemble all the technical information and experience from the naval projects and to determine how these could be applied in the basic conception of the new plant. Next, in a series of extended discussions, first with Rickover and then with Bettis engineers, the senior staff would hammer out the specific elements of the design philosophy, a process which involved the usual questioning, probing, arguing, and rethinking which went into Rickover's technical decisions.

The design process included the definition of general plant requirements. Set down in order of importance these included such prescriptions as: "(1) Safety must be an overriding feature. (2) The reactor coolant must be retained in a sealed system. (3) The materials in contact with reactor coolant must be corrosion resistant. . . . (9) Reactor decay heat must be dissipated without the use of an external source of power." In this generalized form the requirements were deceptively obvious, but that fact did not render them insignificant. The requirements represented the considered judgment of Rickover's group. Once agreed with, they could not be changed by an individual in the group or at Bettis without Rickover's approval. Every detail of the design had to conform literally to these requirements.

Formulating the objectives of the reactor design was an even more difficult process which involved the naval reactors branch in weeks of consultations with scientists and engineers at Bettis. The objectives, although more detailed than the requirements, were still general principles:

(1) That the core be designed so that it could be instrumented to measure actual fuel temperatures, coolant temperatures, and coolant flow at various states of its life, for comparison with calculated results. . . .

(2) That the reactor incorporate a system to detect and locate failed natural uranium fuel elements. . . .

(4) That the failure of a fuel element must not cause adjacent fuel elements to fall. . . .

(9) That the reactor pressure vessel design make a significant contribution to the technology of designing and fabricating large high pressure reactor vessels. . . .

(12) That the reactor be capable of shutdown with at least one control rod stuck in its uppermost position.

Like the requirements, however, the objectives were to be absolutely binding on all later design work. They determined the contractor's initial approach to design and provided definitive criteria against which all detailed designs would later be measured. Thus the formulation and codification of the design philosophy was an essential element in Rickover's system of technical management. The process demanded the services of experienced and talented engineers in the government organization. It focused clear responsibility for success or failure of the project on the government organization. At the same time, however, it enabled the project director to retain firm technical control over contractor activities.

Engineering Development

For Shippingport, Rickover relied on the kind of organizational structure he had created for the submarine projects. The several technical groups within the naval reactors branch worked directly with their counterparts at Bettis in designing specific portions of the plant. Shaw, Mandil, Rockwell, and Panoff served as Rickover's principal assistants. Rickover gave Shaw responsibility for the entire primary coolant system including the main coolant pumps, primary piping, various types of valves, steam generators, and related equipment. Mandil coordinated all aspects of reactor design including the pressure vessel, fuel core, control rod drives, and the refueling system. Rockwell was responsible for shielding design, coolant technology, and safety features of the plant. Panoff, as laboratory officer for Bettis projects, was directly involved in design decisions. Marks, Grigg, Radkowsky, and other veterans in the headquarters group brought their special talents to the project.

As usual, Rickover exercised his right to approve the assignment of key personnel at Bettis. Initially, in 1953, Weaver designated William R. Ellis, an engineer with experience in the company's commercial power division and in the carrier project, to coordinate work on the new power reactor. In September 1954, when Rickover pushed Bettis to reorganize entirely on a project basis, Simpson became the project director, and Ellis was appointed manager of the power plant division. Thus Ellis and Alexander P. Zechella were

Shaw's counterparts at Bettis on the primary coolant system, auxiliary systems, and general plant layout. Mandil's counterparts in reactor design were Philip G. DeHuff and Nunzio J. Palladino. Sidney Krasik of Bettis worked with Radkowsky on physics and Benjamin Lustman with Richard C. Scott and William H. Wilson on developing fuel elements. Dozens of other engineers both in headquarters and Bettis were similarly involved in engineering development.

To meet the tight schedule imposed by the Commission, the naval reactors branch had to make major commitments on plant characteristics long before most of the design philosophy had been formulated. Even before Duquesne proposed the Shippingport site, Washington and Bettis had reached some tentative decisions about the general layout of the plant and the size and design of the principal components. The primary coolant plant surrounding the reactor would be built largely underground in three huge airtight tanks which would contain any radioactive vapor which might be released in a rupture of the pressure vessel or primary steam lines. This added safety feature was intended to permit construction of the plant in a relatively populated area, should such a site be selected.[35]

Many of the major components were essentially scaled-up versions of equipment developed in designing the Mark I submarine plant. The Commission's decision to take advantage of the Mark I technology contemplated just this sort of advantage in designing the power reactor. Scale-up itself involved much more than just putting new dimensions on old blueprints. The reactor pressure vessel, for example, would be of impressive size, towering almost 35 feet in height with a diameter of more than 10 feet and a weight of 264 tons. Fabricating a vessel of this size would push existing technology to its limits and generate new engineering problems. The same could be said for the huge canned motor pumps, hydraulic valves, and steam generators needed to control 225 megawatts of thermal energy. Procurement of these components would not be an easy task, but the use of proven concepts had a distinct advantage. It shifted the heaviest load of responsibility from the already overburdened design forces to component fabricators. Although Washington found it necessary to ask Westinghouse to build the main coolant pumps, Westinghouse was authorized to negotiate contracts with Combustion Engineering, Incorporated, for the pressure vessel and with the Foster Wheeler Corporation and Babcock & Wilcox Company for the steam generators.[36]

These arrangements permitted Bettis to focus its design and development resources on the reactor core. The unprecedented power capacity of the plant would require an immense core assembly, almost 7 feet in diameter and 6 feet high. Within the core would be almost 100,000 fuel elements each meticulously encased in zirconium and welded into assemblies. The enormous investment of money and time required to fabricate these assemblies demanded the utmost care in design. In order to reduce these costs to a minimum, the fuel assemblies had to be built to withstand long irradiation (at least 3,000 megawatt-days) without failure. Another novel feature of the core stemmed from the decision to use slightly enriched rather than highly enriched uranium fuel. This innovation posed a host of uncertainties for Rickover's staff and the Bettis design group.[37]

The idea of using slightly enriched uranium in the core had originated in the carrier reactor project as a way of avoiding the commitment of a large inventory of weapon-grade material. Although several schemes for such a core had been proposed even earlier, no one had ever built a reactor of this type, and Mandil had explored several ideas with Bettis. One of the most promising had been proposed in 1953 by Radkowsky. Instead of attempting to achieve a uniform distribution of fuel elements containing slightly enriched uranium throughout the core, Radkowsky had suggested the possibility of using a "seed" of highly enriched uranium surrounded by a much larger "blanket" of natural uranium.

The idea seemed to have several advantages. It offered the possibility of refueling the reactor merely by removing the small seed rather than the entire core. More than half the total power output could be obtained from the natural uranium, and the formation of plutonium in the blanket might greatly extend the period of its useful reactivity. Adoption of the seed-and-blanket design would also enable the Bettis engineers to proceed with design of the blanket even before the amount of enrichment had been determined. Working closely with Radkowsky, Krasik and his physics group at Bettis devised a number of improvements in the design, such as placing the seed material in an annular arrangement rather than as a central cylinder within the core in order to get a better distribution of power. By the summer of 1954 Westinghouse was giving priority attention to the seed-and-blanket design, and Radkowsky's idea was to become a permanent feature of the reactor.[38]

The Commission's requirement to have the Shippingport reactor operating by the end of 1957, now less than three years away, forced Rickover to adopt some exceptional and costly methods. Contrary to the more leisurely and deliberate design studies which Argonne had proposed, Rickover insisted that

19. A full-scale model of the nation's first civilian power reactor at Shippingport, Pennsylvania. The pressure vessel itself was almost 35 feet high, more than 10 feet in diameter, and weighed 264 tons.

20. President Eisenhower in Denver, Colorado, uses a neutron source in a "radioactive wand" to activate a bulldozer at Shippingport, Pennsylvania, as part of the groundbreaking ceremonies for the pressurized-water reactor plant on September 6, 1954. The administration considered the Shippingport project a key element in the Atoms-for-Peace program.

20

19

Bettis fix the design of all major components quickly. This consideration was particularly important for the core, which would require long lead-time for production. One of the major uncertainties Rickover's group faced in 1954 was the form of uranium to be used in the core. The highly enriched seed material would certainly be a uranium alloy, but metallic uranium did not appear attractive for use in the blanket, where high integrity over very long periods of irradiation was required.

One possible solution was to find a uranium alloy that would be resistant to corrosion in high-temperature water. In 1953 Rickover ordered Bettis to begin an exhaustive study of many uranium alloys. This research led to the conclusion that the most promising materials were those containing up to 12 percent by weight of molybdenum and niobium or up to 3.8 percent of silicon. Special loops were built so that these alloys could be studied under the effects of hot water and radiation in a research reactor. These tests revealed that corrosion failures in the alloys were likely to be severe, and that the molybdenum alloys would have a large appetite for neutrons at energies for which the reactor was being designed.[39]

A second solution was to use uranium dioxide as the blanket material. Bettis had started some research on this material, but the oxides received much less attention than the alloys until the disadvantages of the former began to appear in 1954. By the end of the year Bettis was moving toward a decision in favor of the oxide fuel, although the arguments for and against such a decision were anything but clear. The Bettis engineers knew more about the alloys, but Rickover feared that equivalent study of the oxides would in time reveal as many disadvantages as work on the alloys had produced. He was convinced that the alloys offered the best solution.

An important consideration was that fabrication of alloy elements would have to begin in July 1955, while manufacture of oxide elements could be delayed until the end of the year. To be certain their decision was made on time, Rickover's group and Bettis began to summarize the advantages of the two designs during the spring of 1955. The oxide elements looked better, but all the evidence was not yet in. Under these circumstances Rickover was prepared to make the choice and take responsibility for it. In a meeting at Bettis on April 26, 1955, he carefully listened to the arguments on both sides, made a trans-Atlantic telephone call to check one technical point with a British metallurgist, and then reversed his earlier position by announcing that the blanket elements would contain uranium dioxide. Thus all the research on the alloys would contribute nothing to the Shippingport project, but it did provide valuable data for the future.[40]

Construction: A National
Priority

From the very beginning the pressurized-water reactor had been much more than a power engineering project. The Commission had first seen the new reactor as an impressive demonstration of the feasibility of nuclear power. The Eisenhower administration looked upon the idea as a way of bringing private industry into the new field of atomic energy. It therefore was not surprising that Chairman Strauss was able to enlist Eisenhower's participation in the ground-breaking ceremonies at Shippingport in September 1954. The new reactor would serve as a glittering example of what the president had anticipated in his Atoms-for-Peace address before the United Nations General Assembly the previous December. By the spring of 1955 the pressurized-water reactor had taken on new significance. Strauss, having learned of British progress on the first of a series of dual-purpose power-and-plutonium reactors at Calder Hall, looked to Rickover's project as the nation's only hope for earning the distinction of placing the world's first full-scale nuclear power plant in operation.[41]

For six months after the president's radioactive wand had set the first bulldozer charging into the Ohio River bank nothing much happened at the Shippingport site. Under relentless prodding from Rickover's group, Westinghouse and Duquesne had more than they could do just in assembling the team of contractors and developing a construction schedule. To give the project greater strength in nuclear engineering, Rickover had urged Weaver to appoint Simpson as the Westinghouse project director. On the Duquesne side, Rickover had been just as insistent that Fleger place his work under a man with firm grounding in nuclear technology. Rickover recommended John E. Gray, who had served for more than a year as a materials administrator in the naval reactors branch before going to the Commission's new Savannah River site. In Simpson and Gray, Rickover had two aggressive engineers who understood his "system" perfectly. Simpson was one of the brightest young stars in the Westinghouse organization. Gray had built a reputation both in industry and in government as a young man who would run away with a project if his superiors would let him.

With the nuclear aspects in firm hands, Rickover also made certain that he had experienced managers in large-scale construction work. This ability was unusually important because Rickover's organization had no special talent in this area. For this ability Rickover had turned to the Navy's civil engineer corps. In December 1952, when plans were maturing for the carrier reactor, Rickover had arranged for the transfer of Commander Joseph H. Barker, Jr., to his staff. He had known Barker, an experienced engineer, since 1937, when both were serving aboard the battleship *New Mexico*. Later, during World War II, Rickover had run into Barker at Okinawa. Rickover had sent Barker to Pittsburgh in 1953, where he worked with Geiger on construction plans for both the carrier and the power reactor. In October 1954, with the start of actual construction impending at Shippingport, Rickover brought Barker to Washington to serve as his project officer for the pressurized-water reactor. To take over Barker's duties at Pittsburgh, Rickover had two other officers from the Navy's civil engineer corps, Lieutenants Donald G. Iselin and Edward T. DiBerto. They worked closely with the Washington staff and with Geiger and Laney, who was serving as Geiger's assistant and as the Commission's technical representative at the Pittsburgh office.

In the first weeks of 1955 Barker struggled with Simpson and Gray to devise a consolidated schedule covering all the work by the four principal contractors and the dozens of subcontractors on the project. The schedule, drafted in final form for Rickover's approval by March 15, called for the installation of the last component just twenty-four months later. Considering the fact that design of the plant was only 15 percent complete and that only preliminary grading had been done at the site, the schedule seemed incredible. So intricately dovetailed was the scheduling that the late arrival of a single component could be a cause for concern.[42]

Although Westinghouse as the prime contractor was theoretically responsible for coordinating all construction and procurement, Rickover knew that he could not simply apply pressure at the top, but would have to intervene directly at all levels if the plant was to be completed on time. Westinghouse was not equipped to manage such a diverse operation, particularly in the area of construction. On the Mark I plant, where Westinghouse had a similar responsibility, Rickover had found it necessary to rely on Electric Boat for major support on construction. He still believed it necessary for Westinghouse ultimately to have control over all the technical specifications and quality of workmanship in the plant, but this kind of arrangement obviously complicated construction management.

Rickover had to find some way of coordinating the activities of the various contractors so that he would be able to intervene instantly when trouble occurred. He decided to put Barker, Simpson, Gray, and Iselin on a coordinating committee which would uncover problems and propose solutions. Barker, as his project manager, would serve as chairman. Simpson, for a time, and Gray would represent their organizations along with men with comparable authority from Stone & Webster, Dravo, and Burns & Roe. Meeting at first every two weeks and less frequently in 1956, the committee convened thirty-six times between March 1955 and April 1957. Between meetings, the members were in daily contact by telephone; only those items of special difficulty ever reached the committee's agenda.[43]

Useful as the coordinating committee was, it in no sense rendered Rickover's authority superfluous. His constant presence behind Barker could not be ignored for a moment. When Barker was in Pittsburgh, he called Rickover almost daily. Simpson, Gray, and Iselin checked with Rickover several times each week. He received detailed written reports from Westinghouse, Duquesne, and the Pittsburgh office, and he visited the Shippingport site at least once a fortnight. His usual schedule was to take the afternoon plane to Pittsburgh, ask penetrating questions in the car on the way from the airport to Shippingport, climb through those portions of the plant where problems had been encountered, and continue to discuss the job in the car all the way back to Pittsburgh, where he boarded the midnight sleeper for Washington. By nine the next morning he might be on the telephone in his Washington office calling Iselin about some detail.

The constant threat of Rickover's intervention was always a stimulus at Shippingport. Although construction forces and vendors believed they were already straining for the utmost in performance, Rickover considered the work shoddy and the effort less than full commitment. Orders to tear out equipment which did not meet specifications always met resistance. The human tendency was to see whether the literal specification was really necessary or to postpone action in the hope that an easier solution would show up. But Rickover had no patience with such delays. The members of the coordinating committee soon learned that somehow, some way, they would have to find immediate answers that would not compromise the quality of the plant.

Rickover's unyielding demand for responsive action often proved helpful when the trouble rested with one of the vendors or suppliers. In January 1956 Barker found that he could make no headway at all in obtaining a large order of structural steel from a subsidiary of one of the nation's largest steel companies. When the delay became serious, Barker mentioned it to Rickover, who immediately picked up the phone and called the president of the parent company. Before he could get back to his office, Barker had a call from the top expediter of the company wanting to know what all the fuss was about. The problem was settled that day, and the company revised its delivery schedule.[44]

In dealing with local construction problems in Pittsburgh, however, Rickover was not so successful. In fact, most members of the coordinating committee considered his efforts less than helpful. Rickover, by his own admission, did not understand the construction industry or the labor unions. He was accustomed to demanding superior performance from contractors over whom he held effective control. He refused to accept the traditional ways of the American construction industry with its independent and often inefficient methods. He did not understand that a vituperative tirade directed to a construction supervisor or a union steward would not produce the results it would have on a prime contractor whose future depended on Navy contracts. Barker, Gray, and Iselin moved cautiously trying to head off jurisdictional disputes, trying to get more craftsmen on the job, trying to find supervisors who could stop loafing. They held their breath during Rickover's visits to the site, for fear that a blunt question to a foreman or a sharp reprimand to a worker would cause a walkout. The members of the coordinating committee measured their success in terms of how infrequently Rickover had to make special trips to Shippingport. From Rickover's perspective, that motivation was as good as any other if it got the job done.

Completing the Plant

As the year 1956 began, the Shippingport plant still seemed a long way from completion. The project had been a grueling test of men and organizations since the spring of 1954. Slowly but inexorably time had outstripped the best efforts of those at Bettis and Shippingport. Steel shortages had delayed for three months the completion of the cavernous underground chambers which would house the reactor and steam generating equipment. Labor troubles and fabrication difficulties had caused the delivery date for the huge reactor vessel to slip into late 1956. Strikes plagued progress on the turbogenerator which Westinghouse was fabricating in South Philadelphia. There seemed little chance that the turbine could be delivered before February 1957.[45]

Delays in completing the underground enclosures for the reactor and steam generators had the greatest impact on the Dravo Corporation, the Pittsburgh construction firm which would install equipment for Westinghouse in the nuclear portion of the plant. Dravo had expected to start work in April or May, but by summer not all the steel for the enclosures was yet in place. Under the circumstances Rickover and Barker thought it all the more important for Dravo to be prepared to move quickly when the underground chambers were ready. They urged the Dravo management to get an experienced project director on the site and to begin training welders for stainless steel work. Dravo's response did not indicate to Rickover that the company fully recognized the magnitude of its assignment. By July Rickover was concerned enough to insist that Dravo assign a senior executive at the site. He also arranged to have several planning specialists from Electric Boat help Dravo in organizing the project. Finally Rickover decided that he would have to ask the Commission to authorize a sixty-hour work week for Dravo.[46]

In requesting overtime for Dravo, Rickover revealed another indicator of trouble at Shippingport. By the summer of 1956 Westinghouse had completed enough of the design of the plant to make a new estimate of what construction costs might be. The company concluded (and Rickover agreed) that the Commission should increase the construction authorization for the nuclear portion of the plant from $37.7 million to $45.0 million. Of the $7 million increase, about $2 million would pay the overtime costs for Dravo.

For Rickover to go to the Commission for help, the situation had to be bordering on the critical. Striving always to meet his commitments on schedule and within authorized costs, he considered an appeal to the Commission something of an admission of failure. Yet in the face of troubles besetting the Shippingport project in July 1956 he had no other choice. Without more money he could never have the plant ready for initial tests by March 1, 1957, as the schedule required.

As if his own difficulties were not sufficient, Rickover learned that the British had completed their first plutonium production reactor at Calder Hall in May 1956. The station not only supplied power for Britain's atomic energy plants but also began delivering 60 megawatts of electric power from two reactors to the national distribution system in October 1956. It was probably just happenstance that the British plant was producing power at the capacity designed for Shippingport.[47]

Westinghouse, Duquesne, and the construction contractors stepped up their efforts during the autumn of 1956 to get Shippingport back on schedule. Heroic efforts produced some improvements, but the outlook was, if anything, worse by the beginning of 1957. Because of late deliveries of equipment during the fall Dravo's overtime forces had not been kept busy, with the result that the company was further behind schedule than ever. Rickover still contended that Dravo did not have enough experienced supervisors on the job and that the company had been slow in recruiting and training welders for work on stainless steel. He could point to the fact that fewer than 25 percent of the pipe welds had been completed in a plant scheduled to be ready for testing in eight weeks. Convinced that Dravo could not complete all the work required in a reasonable length of time, Rickover demanded that Westinghouse find another contractor to take over all pipe installation in the large section of the plant which would handle radioactive wastes.[48]

The schedule crisis in early 1957 inevitably had financial implications. Westinghouse reported in January that costs for the nuclear portion of the plant were now likely to be $55 million, including a contingency of about $2.5 million. This apparent increase of $10 million in six months sounded alarming, but it did not represent a sudden escalation in costs across the board. Estimates for plant components, even on such complex and unusual items as the reactor core, had not risen inordinately. The greatest increases involved construction and services, the most glaring example being the increase in the estimate for installation costs from $3.2 million in 1954 to $12.8 million in 1957. The 1954 figure did reflect a substantial underestimate on the part of Westinghouse and the naval reactors branch, but by far the larger portion of the increase was attributable to Dravo's inability to gear up to the fast pace of the project.[49]

From the broadest perspective the cost increases were not excessive, especially when they were compared with those experienced in other Commission reactor projects. In developing a new technology it was never possible to predict costs accurately; one of the purposes of building reactors like the one at Shippingport was to discover what new technical problems cropped up and how good the original design and cost estimates were. At Shippingport there was the added difficulty of a very tight or perhaps even unrealistic schedule. It would be a real accomplishment if the pressurized-water reactor was completed on time, even with the cost increases. In February 1957 it was evident that costs would be above the original estimates but not badly out of line for a project of this nature.

What seemed of greater concern was the failure of Westinghouse and the naval reactors branch to exert effective financial controls over the subcontractors. At Rickover's insistence technical controls had been tight; but in the haste to complete the project on schedule, not enough attention had been paid to accounting and the administration of finances. A review committee appointed by the Commission's director of reactor development concluded that Westinghouse had not really assumed full responsibility for financial controls. The committee recommended and Rickover agreed that in the future large construction projects should be assigned to a single prime contractor and not subcontracted through a research and development organization like Westinghouse.[50]

This experience was perhaps a useful lesson for the future, but it did not help Rickover's predicament in the late winter of 1957. According to the original schedule, the entire plant was to be completed by March 1 and the reactor core installed for testing by July 1. By March, construction of the primary system was only 70 percent complete, and it did not seem likely that testing could begin before September. Deeply concerned, Strauss asked Rickover to do everything possible to have the plant in operation before the end of the year.[51]

Although it hardly seemed possible, Rickover increased the tempo of the project during the spring and summer of 1957. He replaced the coordinating committee with a new operations committee which had the job of identifying problems and working out solutions on the spot. The new committee met every Wednesday and compiled long lists of items to be corrected. As components finally began to arrive at the site, Joseph C. Rengel, who had replaced Simpson as project director, moved his Westinghouse office to Shippingport, and brought whole divisions of engineers to the site from Bettis so that they would be instantly available when the equipment they had designed was installed. As autumn approached, Gray, Rengel, and Iselin were virtually living at Shippingport. Mandil, Shaw, and Barker were coming out from Washington about twice a week. To keep up with the accelerating pace of events, Rickover asked that reports be sent to him by teletype.[52]

Time was now fast running out, but the added pressure merely forced greater performance. While extraordinary efforts were made to complete the reactor core and instrumentation, Westinghouse tested every valve, every switch, and every inch of pipe and electric cable on the site. Pipes were flushed with demineralized water until every trace of dirt had been washed away. Hundreds of valves and instruments already installed were found to be defective, were ripped out, and were rushed back to the manufacturer for repair or modification. On October 6 Westinghouse installed the reactor core. Then the head was bolted and welded in place; the control rod drives and the final instrumentation were installed.

The reactor was now ready for operation, but Rickover faced one procedural issue that he was determined to resolve before he would permit Duquesne to start up the reactor. He insisted upon the right to assign to the Shippingport plant a personal representative who would have absolute authority to shut down the reactor whenever he believed it was necessary. Fleger just as fervently held that granting Rickover this authority would be an infringement of his contract with the Commission. The climax of a series of disputes between Rickover and Fleger, this issue persisted down to the very day of startup, when Fleger finally conceded to Rickover's demand. Although the authority was rarely used, Rickover considered it an essential procedure in assuring safe operation of the reactor.[53]

The reactor first went critical early on the morning of December 2, fifteen years to the day after Enrico Fermi had achieved the first nuclear chain reaction in Chicago. Sixteen days later, on December 18, 1957, the turbine was synchronized with the generator, and Duquesne personnel took over operation of the plant. At 11:10 a.m. on December 23, just eight days before the end of the year, the pressurized-water reactor reached its full net power rating of 60 megawatts of electricity. Rickover had fulfilled his commitment to the Commission and the nation.[54]

The Significance of Shippingport

Although Calder Hall had earned the distinction of being the world's first operating nuclear power plant, Shippingport had a much greater impact on nuclear technology. Because Shippingport, unlike Calder Hall, had no military applications, every aspect of its design and operation could be declassified. During 1954 and 1955 Bettis, Duquesne, and the naval reactors branch organized four technical seminars for hundreds of engineers from Commission installations and private industry. With the basic information provided in these seminars, engineers throughout the world could begin to follow the Shippingport experience as Westinghouse made available thousands of technical reports on every facet of the project. Perhaps no engineering undertaking in history had been so thoroughly documented as the pressurized-water reactor. After the plant went into operation, Duquesne organized the first of a series of training courses in reactor safety and operation. Over the next six years more than a hundred engineers and technicians from the United States and ten foreign countries learned the rudiments of reactor technology at Shippingport.[55]

Not only the dissemination of knowledge but also the unchallengeable success of the pressurized-water reactor contributed to its enormous impact on the subsequent development of nuclear power. Following the initial power run in late December 1957, Rickover's group and Bettis began a series of tests and extended operating trials which continued until the first seed in the reactor core was exhausted on October 7, 1959. By that time the original core loading had operated for 5,800 equivalent full-power hours, compared to its design specification for 3,000 hours. During the same period the plant had generated more than 388 million kilowatt-hours of electricity. Even more significant than these statistics was the excellent performance of the plant as a power generator. Both in terms of stability of operation and flexibility in response to sudden changes in demand, the Shippingport plant had proved itself superior to conventional power plants. Only in the area of maintenance and operator skills did Shippingport impose demands not required in conventional operations. By the end of the decade the Shippingport reactor had clearly established nuclear power as a practical source of energy.[56]

To be sure, Shippingport had not begun to approach the goal of producing electricity at costs competitive with conventional plants. Westinghouse had never considered that goal within range for the Shippingport reactor, and the rapid escalation of costs during the last year of construction made any comparison almost meaningless. Rickover estimated that operating costs for the plant were about 64 mills per kilowatt of generating capacity, compared to about 6 mills for the average steam plant of that day. In several respects, however, Rickover's figure was not an accurate indicator of the potential for light-water reactors. The plant was the first of a kind, and it was not large enough to take advantage of economies in scale. In some respects Rickover's high cost estimate may have discouraged some industrial interest in nuclear power, but Rickover considered false optimism more dangerous than any exaggeration of the difficulties private industry would face.

In any case the excellent performance of the pressurized-water reactor did more to impress American industry than the most optimistic cost figures might have done. The Shippingport plant, following the success of the Mark I submarine reactor, showed that light-water reactors offered the best short-run prospects for economic nuclear power. Ten years later ten of the twelve central-station power reactors operating in the United States would use water as the moderator-coolant.[57] Almost as high a percentage employed slightly enriched uranium fuel in the form of uranium dioxide. Only the seed-and-blanket core design, which performed beyond all expectations at Shippingport, was not widely adopted by the nation's rapidly growing nuclear power industry. Presumably, high costs of design and development rather than any technical limitations prevented the widespread use of this type of core. In the

21. An aerial view of the Shippingport Atomic Power Station on the Ohio River northwest of Pittsburgh. The station, first operated on December 23, 1957, was the first full-scale nuclear power plant in the United States.

22. Central figures in the Shippingport project standing in front of the reactor control panel, probably early in 1958.

Left to right: Lawton D. Geiger, manager of the Commission's Pittsburgh office; Charles H. Weaver, Westinghouse vice-president for atomic power development; Rickover; James T. Ramey, executive director of the Joint Committee on Atomic Energy; John W. Simpson, manager, Bettis Atomic Power Division.

21

22

final analysis, it was not much of an exaggeration to say that the Shippingport plant served as a model for nuclear power development in the United States for more than a decade.

The Shippingport project also had an enduring effect upon the naval reactors program. The effort to build the pressurized-water reactor brought Rickover and his associates into a close and sometimes trying relationship with American industry. They now understood more clearly than ever before the limitations of American engineers and American technology in general. The report of the 1957 review committee had also made clear some of the lessons to be learned from the Shippingport project. This experience would stand Rickover's group in good stead during the last years of the 1950s, when the naval reactor program was moving from a small development effort on a single reactor into a vast enterprise to provide propulsion plants for a nuclear fleet.

Perhaps more important than anything else, Shippingport had given Rickover and his organization a new and commanding stature in the nuclear industry, the Navy, the Commission, and even the government at large. Just as Mark I had established Rickover as something more than an engineering duty officer in the Navy, so Shippingport had made him something more than a builder of military reactors. By 1958 Rickover had a national and even an international reputation. Thousands of visitors from all parts of the United States and many foreign countries flocked to the Shippingport site. A full-size replica of the pressure vessel for the Shippingport reactor dominated the American exhibit at the second international conference on the peaceful uses of atomic energy in Geneva, Switzerland, in the summer of 1958. A technical volume on the plant won an international prize at Geneva. The pressurized-water reactor, completely open for all the world to see, overshadowed all earlier power reactor experiments as the symbol of the peaceful uses of nuclear power, and Rickover, Westinghouse, and Duquesne were inseparably associated with that accomplishment in the public mind.

Rickover and his associates had led the world to new applications of nuclear power beyond the submarine. With that achievement came world-wide recognition and a new degree of independence in both the Navy and the Commission. Both would serve the Rickover team in the future as they pursued the goal of a nuclear Navy.

9

Propulsion Plants for the Fleet: Vertical Extension of the Navy Project

As we saw in chapter 7, the performance of the *Nautilus* during the spring and summer of 1955 had a profound impact on Navy attitudes toward nuclear propulsion. Nuclear power would soon become the standard propulsion system for submarines. The *Seawolf,* still two years from completion, was no longer a serious competitor for the *Nautilus.* The speed of the *Nautilus* in naval exercises showed such striking advantages that the new class of smaller and slower fleet-type nuclear submarines would be limited to a few vessels and would not be repeated. Even among line officers in the surface fleet there was a new interest in nuclear propulsion.

Although this new attitude was a source of some satisfaction to Rickover and others who believed that the future of the Navy depended upon nuclear propulsion, there were also some dangers involved. One was that the eagerness to add nuclear ships to the fleet would tempt the Navy to reduce the standards of quality and performance which Rickover had established for the *Nautilus.* Building a fleet of nuclear ships might also require the naval reactors branch and the laboratories to direct their efforts toward the production of propulsion plants rather than toward improvements in design. Rickover was especially concerned that the Navy not select the Mark II plant, the very first nuclear plant ever installed in a ship, as a production model for a nuclear fleet.

To some extent these fears would be realized before the end of 1955, both in terms of requirements for new types of nuclear propulsion plants and in the addition of nuclear submarines to the shipbuilding program. As this chapter will show, these two types of requirements, and particularly the multiple production of propulsion systems, required a vertical extension of the project system devised to build the *Nautilus* and *Seawolf.*[1] Rickover would extend his control over the development and manufacture of nuclear propulsion plants beyond the original laboratories at Bettis and Knolls to the hundreds of vendors and fabricators who would supply components of the propulsion equipment. How he was able to keep these new organizations virtually independent from the Bureau of Ships and the Navy's procurement system will be explained in this chapter.

At the same time, Rickover had to expand his organization in a horizontal direction. A specific project to develop a propulsion plant for a single ship like the *Nautilus* obviously ended with its completion. But the building of a fleet of nuclear ships required Rickover to establish essentially permanent working relationships with many other codes in the bureau and other organizations in the Navy. By insisting upon his continuing responsibilities for all new types of nuclear propulsion plants and for the safe operation of those plants in growing numbers of ships at sea, Rickover extended his influence on a permanent basis horizontally through many organizations in the Navy. Thus the nuclear propulsion project began to take on a never-ending and ever-broadening existence. Chapter 10 will explain how Rickover's group established a distinctive identity in all aspects of shipbuilding from the bureau codes to the private Navy shipyards. The ultimate extension of authority in the horizontal direction appeared in fleet maintenance and operation, described in chapter 11.

New Faces

The Navy's new interest in nuclear propulsion rested in part on larger forces on the international scene and on the striking performance of the *Nautilus,* but it also depended to some extent on new leadership in the Navy. By the spring of 1955 Secretary Thomas was actively seeking a replacement for Carney as Chief of Naval Operations. A strong leader, Carney was determined to run the Navy his way and to keep Thomas out of military concerns. He had pointedly rejected Thomas's suggestion that the Navy give high priority to a nuclear-powered guided missile submarine.[2] Despite Carney's strong voice for a modernized fleet, many officers and some civilians in the Department of Defense believed he was moving too deliberately.

Thomas passed over many senior officers to choose Rear Admiral Arleigh A. Burke, who at the age of fifty-four was one of the youngest admirals ever to be selected for the Navy's highest military position. After graduating from the Naval Academy he had studied engineering at the University of Michigan and had acquired a life-long interest in applying research and development to the Navy's problems. Early in World War II he had proved himself an aggressive, hard-hitting destroyer commander in the Pacific. After the Japanese surrender he had returned to Washington to serve in the office of the Chief of Naval Operations, but had chafed impatiently at the complacency, smugness, and unwillingness to pursue new ideas. He had organized a study of future naval warfare (see chap. 3) which had helped spark the establishment of a nuclear propulsion project in 1948. Neither Burke nor his fellow officers expected his selection as Chief of Naval Operations. Burke considered himself too young for the job, and he did not want to make it easy for Thomas to replace Carney.[3] In the end, however, Burke accepted the assignment as both a challenge and an opportunity. When he took the oath of office on

August 17, 1955, at the Naval Academy, Burke was determined to see that the Navy moved faster to get the modern ships and weapons it needed.

The Bureau of Ships was also under new and aggressive leadership. Rear Admiral Albert G. Mumma had become chief in April 1955. He and Rickover had clashed in the first days of the nuclear propulsion effort, and strong differences still remained. As bureau chief, however, Mumma was a better engineer and a more effective administrator than some of his predecessors. In July 1955 he issued a directive establishing the nuclear power division as a major unit of the bureau. As head of Code 1500 (as the new division was designated), Rickover acquired the title of assistant chief of bureau for nuclear propulsion. No longer subordinate to the assistant chief for shipbuilding and maintenance, Rickover would report directly to Mumma on all matters relating to nuclear ships.[4]

Looking to his relationships with the Commission, Rickover faced some worrisome if not vital changes in 1955. Among the Commissioners themselves there had been a number of replacements, but none of these had any real impact on the naval reactor program. Strauss and Murray, both of whom were staunch supporters of the Navy program, still dominated the Commission. Kenneth E. Fields, a retired brigadier general and former director of the Commission's weapon program, had succeeded Major General Kenneth D. Nichols as the Commission's general manager. Fields was an outstanding officer and civil engineer who had been associated with atomic energy activities almost continuously since Manhattan District days. With a good understanding of the Navy project and its relationships with the Commission, Fields seemed likely to support Rickover.

The big personnel shift had come in the division of reactor development, of which the naval reactors branch was still a part. After five years of intense activity, Hafstad had decided in 1954 to resign as director. As his replacement, Strauss had selected a young chemical engineer with a strong academic and industrial background. W. Kenneth Davis had done graduate work at the Massachusetts Institute of Technology, taught at the University of California in Berkeley, and served as a senior engineer with such large corporations as Ford, Bacon, and Davis, and the California Research and Development Corporation. Davis had some experience in nuclear technology but little with reactors. He joined the Commission in 1954 as Hafstad's assistant and moved up as director in February 1955.

Davis' background and interests suggested that he would be useful in promoting industrial participation in the development of nuclear power and the

Chart 6.   Rickover became Assistant Chief of Bureau for Nuclear Propulsion, reporting directly to the Chief, Bureau of Ships. The Portsmouth and Mare Island naval shipyards were preparing to build nuclear submarines.

THE NAVY NUCLEAR PROPULSION PROJECT IN JULY 1955

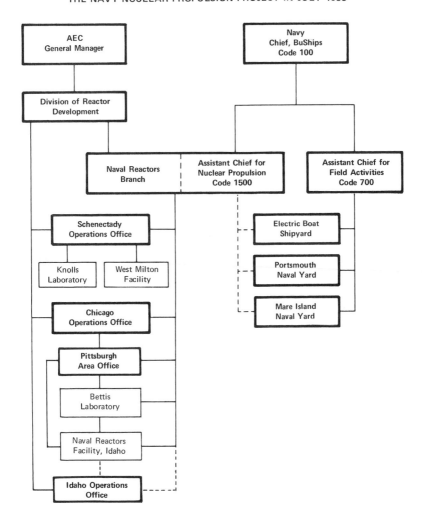

introduction of civilian power reactors into the American economy under the Eisenhower administration. Davis' ties would be much closer with industry than with other government agencies, a tendency that did not arouse any optimism in Rickover. It could not be said that Rickover had ever found himself on congenial terms with Hafstad, but his relationships with Davis were, if anything, even more strained. Lacking some of Hafstad's diplomacy, Davis was not reluctant to issue orders as well as requests to Rickover, and predictably the results were never pleasant and sometimes not constructive.

Rickover's own organization had continued to change both in personnel and assignments since 1953. Roddis, the last of the original Oak Ridge group, departed in 1955 to become Davis' deputy in the division of reactor development. Kyger had left the government for private industry, and Crawford had transferred to the Commission to succeed Dunford as Commissioner Murray's assistant. Later in the year Dunford would return to Code 1500 to take Roddis' job handling officer assignments and administrative liaison with the Navy.

More fundamental were the changes that had occurred among the project officers. With the *Nautilus* now at sea, Panoff could concentrate his attention on the design of the submarine fleet reactor at Bettis and on the construction of the *Skate*. He was also deeply involved with Rickover and others in formulating plans for new types of propulsion systems.[5] Soon after the initial sea trials of the *Nautilus* Panoff and others became convinced that the Navy would quickly appreciate the value of a high-speed submarine. Under Rickover's guidance they began to formulate design criteria for a new submarine propulsion plant. As an engineer with experience both in the technical groups and as a project officer, Panoff had become one of Rickover's most influential assistants.

Among the other projects, Lieutenant Commander Arthur E. Francis, a graduate of the MIT course in 1951, was now in charge of the Mark I. With the completion of the *Nautilus,* the prototype functions of Mark I were ended and the plant was now in the process of becoming a testing facility for new types of reactor cores and other components as well as a training facility. Sweek had moved over from the technical staff to become project officer for the Mark A and *Seawolf.* Leighton, as project officer for the submarine advanced reactor, would have an increasingly important role in Code 1500 when the reactor plant was tied to a military requirement for a new type of submarine. As in 1953, Cochran was still in charge of the large ship reactor project, and Barker was handling the construction of Shippingport. One new project officer, Edson G. Case, was responsible for the new propulsion plant being developed by Combustion Engineering for the hunter-killer submarine.[6]

With Kyger's departure, Rockwell, Mandil, and Shaw became the key members of the technical staff. In addition to his technical assignments, Rockwell had taken charge of Code 1500's rapidly growing responsibilities for the safe operation of submarine reactor plants and for training the increasing numbers of officers and men required for the nuclear fleet. Mandil, who had served under Kyger and Welch for several years, took over the reactor engineering group. Kintner had left to become Rickover's representative at the Mare Island Naval Shipyard.[7] In his new position, Mandil would have responsibility for initial development of new reactor types and for general design improvements in existing reactor systems. Shaw was heavily involved in power plant engineering both for the Shippingport project and for the submarine fleet reactor. When these projects were completed, he would transfer his attention to propulsion systems for surface ships.

As in the past both naval officers and civilians were leading the technical groups in Code 1500. Most of the officers were engineers who had taken the postgraduate course at MIT before coming to the nuclear project. After about three years' service in Washington they usually moved on to serve as Rickover's representatives in the shipyards or laboratories. A few, like Leighton, jeopardized their military careers by staying in Code 1500 well beyond the length of the usual tour of duty. When Leighton was eventually convinced that a naval career was not compatible with his work on Rickover's staff, he decided to resign from the Navy and continue as a Commission employee in the naval reactors branch.[8]

The civilian engineers did not face this particular problem, and some of them made a career of their work in Code 1500. Howard K. Marks, Jack C. Grigg, and Alvin Radkowsky were prominent members of this group. Many other engineers, however, remained only a few years before taking other assignments, and there was a constant demand for replacements. For a time Code 1500 had success in recruiting engineers from other bureau codes and from private industry, but these sources were limited and did not always provide acceptable personnel. In Rickover's opinion too many of these engineers had learned to work in the typical bureaucracy, which depended more on position and authority than on technical ability in making engineering decisions. Code 1500 found that engineers with this kind of background could be "re-educated" only with great difficulty. Beginning in 1956, therefore, Rickover began recruiting most of his technical staff from Naval Reserve Officer

23. Rickover and some of his senior staff in the summer of 1958. Standing at rear: Willis C. Barnes; from left to right around the table: Robert Panoff, Howard K. Marks, Milton Shaw, I. Harry Mandil, Jack C. Grigg, James M. Dunford, and David T. Leighton.

Paul Schutzer, *Life*

Training Corps units in the better engineering schools. By carefully screening each candidate individually, Rickover was able to recruit bright young engineers who were in Rickover's term "unspoiled" by exposure to the ways of either the Navy or industrial bureaucracy.[9] Not only did they lack prejudices which would have to be "unlearned," they were also likely because of their obligation to military service to accept their fate and adjust to it.

For all members of Code 1500, including Rickover, life was hard, and the pace relentless. The working day was long (normally ten hours for the professional staff), lunch hours short, and Saturday work a norm. Working space was always cramped in the shabby temporary buildings, and everyone was expected to be too busy or too concerned about his work to notice the inconvenience. Shirt sleeves were the uniform for men, as if to stress the informal, hard-working atmosphere. The constant pressure of responsibility left little time for friendly chats in the hallway, scarcely enough for the exchange of common courtesies. There was no time at all for fire drills, fund drives, office parties, or Navy employee programs. The few people in the quiet halls walked quickly about their business, those on their way to or from Rickover's office moving at a faster pace, sometimes on the dead run.

For some the intense concentration on technical detail, the incessant demands which overrode family life and personal interests, and the cold, mechanistic atmosphere, outweighed the advantages of training and education in the Navy project. But for many others, these same conditions were an irresistible attraction. The ambitious career-oriented engineer found in the naval reactors branch a priceless opportunity for learning and practicing his discipline. And for those with a touch of idealism there was plenty of incentive to try to accomplish great things while living by the highest standards of technical integrity.

New Priorities

In seeking ways to revitalize the Navy as a modern fighting force, Admiral Burke recognized the potential advantages of missiles and nuclear power. The achievements of the *Nautilus* suggested to some outside Code 1500 that the transition to nuclear power might be relatively easy. The 1956 shipbuilding program already included three nuclear-powered submarines—two of the fleet type and one radar picket. After conferences with Rickover and other officers of the Bureau of Ships in September 1955, Burke decided that two of the five conventional submarines in the 1956 program would be built with

nuclear power. One of these was to be the *Halibut,* the *Regulus* missile submarine which Carney had opposed. The other was the *Skipjack,* the first of a
new class of attack submarines. Burke also announced that all submarines in
the future would be nuclear powered. To speed the transition in the surface
fleet, he asked the bureau to advise him on the feasibility of installing nuclear
power on four types: a ship about the size of a frigate (another name for a
large destroyer) which would be able to escort carriers during high-speed
operations, a guided-missile light cruiser, a guided-missile heavy cruiser, and
a class of attack carriers.[10]

With the wealth of background material from previous efforts it did not
take Rickover long to provide an answer. The large ship reactor had already
been approved by the Chief of Naval Operations, and Congress had authorized the Commission to construct a land-based prototype which should be in
operation in 1958. The Navy had already submitted several possible schedules for a carrier using eight of the reactor plants. The design called for two
reactors to drive each shaft. A similar arrangement could be used with four
large ship reactors for a guided-missile cruiser. Nuclear propulsion for a frigate, however, was a different matter. Although various reactor types were
being investigated, there was at that time no nuclear propulsion plant which
could be installed on a ship as small as a frigate. To fill this gap Rickover
quickly established a new development project at Bettis.[11]

Burke's request for studies of two nuclear-powered cruisers showed his
deep interest in missile development. As Burke was considering a nuclear-
propelled guided-missile cruiser, the Navy was completing the conversion of
two cruisers to carry surface-to-air missiles and planning the installation of
Regulus on cruisers and carriers. Such efforts, however, did not begin to tap
all the possibilities which the Killian committee had seen early in 1955. In-
heriting from Carney a special study by the Naval Research Laboratory
favoring an immediate large-scale increase in Navy support for missile development, Burke arranged a joint project with the Army to augment work on
the *Jupiter,* a liquid-fueled ballistic missile with a range of about 1,500 miles.
To begin research on the Navy's launching system, Burke personally selected
Rear Admiral William F. Raborn to head a new organization called the Special Projects Office in the Bureau of Ordnance. Burke gave Raborn his choice
of any forty officers in the Navy and promised him the highest priorities. For
the moment at least, it seemed that the *Jupiter* with its huge tanks of liquid
propellant would be too large and too difficult to handle for submarine use,
but Burke was convinced that the new fleet ballistic missile would be almost

as effective if launched from a large surface ship. Hence his interest in developing nuclear propulsion plants for ships of this type.[12]

How far Burke had come was evident in his recommendations to the House Armed Services Committee in January 1956. He spoke of nuclear power as the most revolutionary innovation in the Navy since the introduction of steam. The *Nautilus* was in Burke's words "a major engineering achievement" but only the first step in the wide application of nuclear propulsion to ships. Nuclear propulsion for both submarines and surface ships was "not only warranted but mandatory." Burke asked the committee to authorize six nuclear-powered submarines, one nuclear-powered guided-missile cruiser, and preliminary design and advanced procurement on one nuclear-powered aircraft carrier. Of the greatest urgency were the submarines, all of which were to be of the new *Skipjack* class.[13]

The *Skipjack* Class

Rickover and his staff had foreseen that the first trials of the *Nautilus* would bring the Navy to appreciate the value of speed. They knew that Knolls would never be able to develop the submarine advanced reactor in time to meet the Navy's new requirements. The only alternative was to develop an improved version of the *Nautilus* plant at Bettis. Because the improved reactor would generate only about the same amount of power as the Mark II plant, the reactor alone was not enough to achieve the higher speed required in the new attack submarine.[14] To a large extent the Bureau of Ships was depending on a new hull design recently tested in the experimental submarine *Albacore*.

During the first half of the twentieth century, when submarines had only a limited submerged endurance, hull forms accommodated surface operation. Conning towers, guns, and other topside gear did not seriously hamper undersea performance at low speeds for short periods. Fleet-type submarines built by the United States during World War II had been long and narrow with a length-to-beam ratio of 11.5 to 1. Attempts after the war to achieve greater submerged speeds showed that the long, narrow hull was unstable. Experiments by the Bureau of Ships at the David Taylor Model Basin resulted in a new design. The new hull with a length-to-beam ratio of about 7.6 to 1 was short and wide, with nearly circular cross-sections, and streamlined like the body of a whale.[15]

The *Albacore* was an unarmed, experimental vessel driven by unusually powerful batteries. It had been commissioned late in 1953 to test the new

24. The *Skipjack* (SSN-585), the first of
a new class of submarines using the
S5W reactor plant and the streamlined
hull, is launched at the Electric Boat
yard at Groton, Connecticut, on
May 26, 1958.

24

25. From the *Nautilus* to the *Skipjack*. Scale models showing the evolution of the nuclear-powered attack submarine: The *Nautilus* (SSN-571), the first nuclear submarine; the *Skate* (SSN-578), in which the Navy incorporated nuclear propulsion in a hull about the size and configuration of a conventional submarine; and the *Skipjack* (SSN-585), in which nuclear propulsion and the high-speed hull form were combined to give maximum underwater performance.

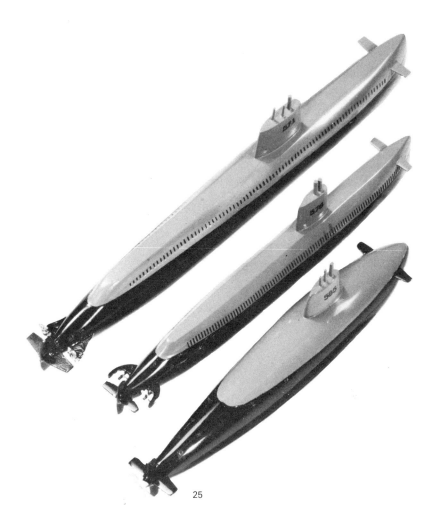

25

hull form. The Bureau of Ships used data from the *Albacore* trials in designing the last three conventional submarines for the fleet.[16] Although developed independently of nuclear propulsion, the *Albacore* hull would be an essential feature of the new *Skipjack* class of nuclear attack submarines.

Nomenclature

By the fall of 1955 the Navy project had so many reactors in operation or under development that the nomenclature was becoming confusing. In October Rickover's office announced a system for designating reactors which was quickly adopted and became permanent.[17] The first letter of the designation was the type of ship for which the reactor was designed—"S" for submarine, "A" for aircraft carrier, "F" for frigate, "C" for cruiser, and "D" for destroyer. The number following the initial letter denoted the model of that type of plant by that designer, and the final letter designated the designer—"W" for Westinghouse, "G" for General Electric, "C" for Combustion Engineering, and "X" for unassigned projects. The form which this list of reactor plants took late in 1955 can be seen in table 4.

Table 4.   Designation Symbols for Navy Nuclear Propulsion Plants

Project	Old Symbol	New Symbol
Westinghouse		
Submarine Thermal Reactor, Mark I	STR MK I	S1W
Submarine Thermal Reactor, Mark II	STR MK II	S2W
Submarine Fleet Reactor, SSN 578, 583, SSGN 587	SFR	S3W
Submarine Fleet Reactor, SSN 579, 584	SFR	S4W
High Speed Submarine Reactor	. . . .	S5W
Large Ship Reactor, Prototype	LSR	A1W
Large Ship Reactor, Ship	LSR	A2W
Frigate	. . . .	F1W
General Electric		
Submarine Intermediate Reactor, Mark A	SIR MK A	S1G
Submarine Intermediate Reactor, Mark B	SIR MK B	S2G
Submarine Advanced Reactor, Prototype	SAR-1	S3G
Submarine Advanced Reactor, Ship	SAR-2	S4G
Combustion Engineering		
Submarine Reactor, Small	SRS	S1C
Nondesignated		
Cruiser, Guided Missile	. . . .	C1X
Task Force Escort Reactor	FER	D1X

The Development Task

The proliferation of Navy requirements in 1955 placed new demands on Rickover and his staff in Code 1500. As in the past, Bettis and Knolls would actually design and develop the hardware, but it fell to Code 1500 to translate ship specifications into design criteria for the laboratories to follow. Through the project officers and technical groups Rickover would influence the organization of the laboratories, approve the appointment of key technical personnel, direct technical activities, and assess contractor performance.

Because Rickover's system required much closer monitoring of contractors than was customary in the Navy, the rapid growth of the nuclear propulsion program posed an impressive challenge for Code 1500. In discussing possible ways of organizing the headquarters group, some of Rickover's staff assumed that Code 1500 would never be able to exercise the kind of detailed direction that had been applied in developing the Mark I (S1W) and the Mark II (S2W). This assumption, however, was not part of Rickover's thinking. Nothing about the *Nautilus* experience had suggested to him that he should abandon his system of rigorous controls. Although some of his staff believed it would be impossible, Rickover's assistants discovered that by applying the same techniques they had used in developing the S1W and S2W, they could gain a tighter hold on contractor operations than ever before. Rickover's determination and the loyal support of his staff made that possible.

The basic organization of Code 1500 and the laboratories was in terms of specific reactor projects, but there were many development activities of a general nature and of long-term significance that did not fall within the scope of a single project. While new types of propulsion plants were being designed and built, Rockwell, Mandil, Radkowsky, Marks, Grigg, Shaw, and others in the technical groups were struggling with basic problems in reactor physics, engineering, component and system design, and fabrication techniques. The whole range of fundamental studies undertaken in 1949 and described in chapter 5 continued as a permanent and vital part of the nuclear program.

The results of long-term development were often hard to measure; but in some instances, such as the development of better fuel elements, the evidence was dramatic. Rickover and Code 1500 insisted that the laboratories continue devoting substantial effort to improving the design and performance of fuel elements. The development of better materials, such as new alloys of zirconium, improved mechanical design, and the use of burnable poisons all

helped to improve corrosion resistance and increase core life. The *Nautilus'*
first core, which cost $4 million to build, powered the ship for 62,000 miles.
The second core lasted for 90,000 miles. The third core, costing about $3
million, achieved 140,000 miles. Savings were realized not only in lower fabricating costs but also in longer periods of operation between overhauls.[18]
Under Code 1500's direction, the laboratories made similar design improvements in many reactor components.

Changing Direction at Knolls

The new requirements for nuclear-powered submarines had little direct impact on the engineers and scientists employed by General Electric at the
Knolls Atomic Power Laboratory. In the summer of 1955 Code 1500 was
still struggling with the Knolls staff to bring the Mark A prototype (now
called the S1G) into full operation at West Milton. Now more than two years
behind the Mark I prototype (S1W), the sodium-cooled plant was not yet a
fully reliable system.

Part of the long development time reflected the laboratory's problems in
adjusting to the hard-headed engineering approach which Rickover's staff
demanded. As explained in chapter 3, General Electric had established
Knolls as a laboratory for general research in the nuclear sciences. The laboratory's early work on the power-breeder had been in the hands of scientists
rather than engineers, and the company had made a conscious effort to
broaden the laboratory's competence beyond reactor development. Although
Knolls had made some progress toward becoming an effective reactor engineering organization under the Navy project, Rickover and his staff were still
largely dissatisfied with the Knolls operation. Even after five years of argument with Rickover, General Electric had not permitted Knolls to become
a single-purpose facility devoted exclusively to Rickover's projects. From
the company's perspective Rickover was trying to replace a highly talented
and diversified scientific research team with well-qualified engineering specialists. Rickover insisted that Knolls would never be able to build reliable
reactors until the company accepted this kind of transformation.[19]

Another reason for the protracted development of the S1G was the inherent difficulty in handling both sodium and water in the steam plant. The
slightest leak from the primary system containing sodium to the secondary
steam system could lead to disaster. To reduce this possibility to a minimum,
the engineers at Knolls had designed double-walled tubes for the steam gen-

erators. A third fluid, mercury, was placed between the walls of the tubes to serve as a leak detector. The presence of any mercury in the sodium or steam in the mercury would indicate a leak in the steam generators. The design and testing of the steam generators was but one example of the special precautions required in the S1G plant. Similar complications were involved in designing and fabricating valves and pumps.

As a result of these difficulties the S1G at West Milton was not ready to operate until the spring of 1955, more than two years after the initial criticality of the S1W. In many respects the plant operated well. In June the S1G virtually duplicated the S1W's "trans-Atlantic voyage" by running at full power for more than two thousand equivalent miles. A few weeks later Chairman Strauss and other members of the Commission went to West Milton to witness the first commercial distribution of nuclear power from a small generator coupled to the S1G.[20] With the Shippingport reactor still eighteen months from operation, the S1G attained a moment of glory in the international race for civilian nuclear power.

In the realistic world of naval propulsion, however, there were growing reservations about the reliability of the S1G. During July traces of mercury in the sodium indicated a leak in a steam generator. Before the end of the summer a leak had appeared in each of the superheaters. The troubles with the superheaters could be avoided simply by bypassing those units in the steam plant and accepting the consequent loss of power, but the steam generator leak was more serious. Without high integrity in this component the plant would not approach the reliability required in a submarine. By this time the *Seawolf* had been launched at Groton, and installation of the reactor and propulsion machinery had started. Knolls began an intensive effort to determine the cause of the leaks and to obtain better components.[21]

The tight schedule for completion of the *Seawolf* made it impossible to substitute new equipment in the ship before the initial trials. By January 1956 Knolls had succeeded in plugging the leaks in the S1G steam generator so that the prototype could for the first time in six months operate on both steam loops. There was no assurance, however, that similar difficulties would not be encountered in the *Seawolf*. In fact during preliminary low-power runs in June and July the S2G plant performed well. Not until a full-power run on August 19, 1956, did a failure occur in the steam plant. A leak of sodium-potassium alloy, now used as the third fluid in the steam generator, had aggravated stress corrosion in the system, causing two cracks in steam piping and a leak in a superheater. Once again Code 1500 decided to bypass the superheater while Knolls began extensive tests on the steam generator. There was for a few weeks a real possibility that the *Seawolf* would not be able to go to sea without a long delay for extensive repairs.[22]

Although makeshift repairs permitted the *Seawolf* to complete her initial sea trials on reduced power in February 1957, Rickover had already decided to abandon the sodium-cooled reactor. Early in November 1956 he informed the Commission that he was taking steps toward replacing the reactor in the *Seawolf* with a water-cooled plant similar to that in the *Nautilus*. The leaks in the *Seawolf* steam plant were an important factor in this decision but even more persuasive were the inherent limitations in sodium-cooled systems. In Rickover's words they were "expensive to build, complex to operate, susceptible to prolonged shut down as a result of even minor malfunctions, and difficult and time-consuming to repair." If the water-cooled plant in the *Nautilus* had failed, solving the steam generator problems in the *Seawolf* would have been imperative, but the success of the *Nautilus* made that effort unnecessary. The *Seawolf* would be operated with its original plant until the new one was ready for installation, and the S1G during that period would provide technical back-up for the ship.[23] Eventually the new S3G prototype for the radar picket submarine and the D1G prototype for a destroyer would replace the S1G at West Milton.

During these same years progress had been slow on the submarine advanced reactor, or S3G as it was now called. In the two years since the project had been established at Knolls in 1953, development had not proceeded much beyond the paper stage. There were reasons for this slow pace. Knolls was investigating several coolants, and it took time to shift personnel from other projects; but Rickover and his group were becoming impatient. Early in 1955, when the Navy fixed the power plant specifications, Rickover discovered that Knolls had not assembled enough reliable engineering data to indicate whether the plant would meet the Navy's needs. Rickover considered Knolls's performance so bad that he questioned the laboratory's ability to assemble a technical organization strong enough to build the reactor. When high-level attention in General Electric failed to improve the situation at Knolls during the spring, Rickover and his staff seriously considered transferring the project to Bettis. Leighton, who was Rickover's project officer for the S3G, thought Bettis could probably have built the plant without a prototype and in less time and at less cost than could Knolls.[24] On the other hand, he thought the project was the best way of bringing the laboratory into water-reactor technology.

There was in fact some argument for canceling the project altogether in the summer of 1955. On the basis of the most recent design studies, the ship characteristics board in August had favored a single-reactor plant or twin reactors of much less power than those planned for the radar picket submarine. Some members of the board argued that the Navy could save $25 million on the project by sacrificing two knots of speed with a smaller propulsion plant. Revisions in the design characteristics by the Bureau of Ships during June called for a much larger and more expensive ship. Another way of stating the problem in Rickover's realistic terms was that neither Code 1500 nor Knolls had been able to find any obvious way to achieve the bold objectives the Navy had set for the plant: less weight per shaft horsepower, greater plant reliability, less complicated reactor control, and longer reactor core life than had been obtained in the *Nautilus* plant. In Rickover's view the design Knolls had developed so far showed no significant differences from the S1W except for a new type of fuel element which had not yet been tested.[25] Rickover might have terminated the project had it not been for two long-range considerations: the need to develop a competitor to Westinghouse in water-reactor technology and his conviction that the two-reactor S3G plant would prove the ultimate solution to the Navy's needs for a missile-launching submarine.

Lacking enthusiastic support from the Navy on one hand and facing a discouraging situation at Knolls on the other, Rickover's organization tried to keep the S3G project alive. Fortunately the Commission was supporting the work at Knolls, and Rickover did not bother the Commissioners with his problems. Gradually in the fall of 1955 the S3G project began to gain momentum. Rickover approved Knolls's basic design of the reactor in October. A subcontract with Electric Boat made it possible to start the design of the prototype hull, and procurement contracts were placed for the major components. The pace of development, however, was never fast. The S3G prototype did not go into operation until August 1958, the same month that the hull of the radar picket submarine *Triton* was launched at Groton.[26] By that time Bettis was well advanced in building the new S5W reactor, and the *Skipjack,* the first attack submarine to be powered by that reactor, had already been launched.

Knolls's performance was outstanding in certain areas of research, but the laboratory's lack of experience with water-cooled plants and the difficulties of transforming the laboratory into an efficient engineering center delayed completion of the new reactor design almost beyond its point of usefulness.

By the time the *Triton* put to sea in 1959, the fast-moving technology of radar detection systems had eliminated the need for the radar picket submarine. The S5W plant, already in multiple production at Bettis, had preempted any hopes at Knolls that the S4G would become the standard propulsion system of the nuclear submarine fleet.

Knolls and the Issue of Control

Behind these technical troubles at Knolls there lurked the controversy between Rickover and General Electric over the role Code 1500 would play in controlling laboratory activities. The situation had not changed significantly when Van Tassel replaced Milton as general manager at Knolls in 1952, and Rickover became increasingly concerned when a reorganization of General Electric in April 1953 placed Knolls under Francis K. McCune, the general manager of the company's new atomic products division. McCune's job was to develop commercial power reactors independent of the Navy and Knolls. The dispute between Rickover and McCune was largely over the question of what was a sufficient number of engineers to be recruited for Knolls, but just below the surface lay the familiar question of how much control Code 1500 would exercise over activities at Knolls.[27]

This fundamental issue came into focus in 1955, about the time Rickover was thinking of transferring the S3G to Bettis. McCune proposed to Admiral Mumma that General Electric join with the well-known marine architectural firm of Gibbs & Cox, Incorporated, and the Bath Iron Works Corporation (a famous Maine shipyard which had built destroyers for decades) to study various applications of reactor systems to surface vessels. McCune believed that reactors, turbines, and other components of naval propulsion plants could be produced commercially. By building its own nuclear facilities at no cost to the government, the company could explore commercial applications of reactor technology without a government contract.

Rickover attacked the proposal on technical grounds. He charged that the proposal had no substance because it gave no attention to the nuclear propulsion plant but treated it as another component to be developed later. Rickover had evaluated dozens of proposals of this nature. The only result, Rickover claimed, was that he and his staff had wasted valuable time on what he called a "political" rather than a technical proposal. By that he meant to suggest that such proposals did not rest on any sound engineering idea but merely represented an attempt to get Navy contracts. Everyone, including the Navy, wanted lighter, cheaper, and more powerful nuclear power plants, but Rick-over contended that it was not possible to start with those objectives and work backward. First it was necessary to find some sound engineering principle which would make these goals achievable.[28]

This technical consideration was Rickover's primary reason for opposing the General Electric proposal, but the old question of control was still involved. In April 1956 McCune suggested that General Electric would be willing to finance one-quarter of the costs of the proposed study by the three companies if General Electric would "have complete freedom to select the personnel to perform the study and to be solely responsible for direction of the study." Rickover would never accept that condition. Furthermore, he charged that despite McCune's reassurances, General Electric was attempting to transfer the best engineers from Knolls to its commercial projects. A Navy study, free from Rickover's controls, would mean that the existing Navy projects at Knolls would suffer.[29]

Despite Rickover's objection, Mumma authorized a Navy contract with the three companies. After a year of work the companies concluded in July 1957 that a gas-cooled reactor offered the best hope for building a propulsion plant light enough and small enough for installation in a large destroyer or frigate. After years of study Code 1500 had already rejected the gas-cooled reactor as unsuitable for naval use, and Rickover's group had no trouble demolishing the claim of the three companies that the reactor they proposed would make possible a substantial reduction in the weight of the propulsion plant.[30]

Even so, McCune and William Francis Gibbs did not give up their idea. They enjoyed the encouragement if not the open support of Mumma, Burke, and two former Chiefs of Naval Operations, and they did not hesitate to carry their cause to Chairman Strauss, the Secretary of the Navy, and the Secretary of Defense. Despite Rickover's convincing arguments that the project lacked technical substance, he and his staff did not succeed in killing the proposal until late in 1958, more than three years after it had first been presented to Mumma.[31]

If, as Rickover claimed, the purpose was nothing more than an attempt to get a government contract, why were McCune and his associates able to keep it alive for so long? No one answer seems to fit all the groups involved. For General Electric, the proposal was important because it challenged Rickover's attempt to establish firm control over Knolls. Rickover had already been successful at Bettis. If he should succeed at Knolls, he might be able to establish a new type of relationship between the Navy and its contractors, one which would run directly counter to the "customer policy" of General Electric. Pursuing the issue was obviously worth the effort, both for Rickover and for the company.

For Mumma, Burke, and others in the Navy, the issue was whether the Navy could rely on Rickover's judgment alone to determine the technical feasibility of new ideas. The technical competence of Code 1500 was unquestionably strong, even outstanding, but was it wise to let one technical group decide what path the Navy would follow in developing nuclear propulsion? From this perspective the three-company proposal did not seem unreasonable even if, as Rickover claimed, there was no clear-cut technical advantage in sight. Although Rickover's technical judgment in this case seemed correct, the absolute certainty with which he asserted his opinion did not help to convince others that Code 1500 was open-minded on the subject of new reactor designs. It was tempting to conclude that Rickover was simply trying to establish a monopoly to keep himself in power. For others in the Department of Defense who were less well acquainted with Rickover than were Mumma and Burke, Rickover's impatience with new proposals and his unrestrained hostility to them suggested that there might be good reasons for investigating the possibilities of a lighter and cheaper reactor. For Rickover, who refused to look beyond the technical aspects of a proposal, such ideas were at best annoying and wasteful diversions and at worst deliberate attempts to drive him and his system from the Navy.

Multiple Projects at Bettis

By 1955 Bettis was well into the difficult transition from single to multiple development projects. Originally Weaver had organized the laboratory along functional lines, but the reorganization which Rickover had initiated in September 1954 had replaced the functional divisions with four projects. The STR (submarine thermal reactor) project was responsible for operation of the Mark I prototype (S1W) in Idaho and for providing technical support for the *Nautilus* plant (S2W). The PWR (pressurized-water reactor) project, described in chapter 8, for the moment overshadowed all other activities as a massive attempt to design and develop components for the Shippingport reactor. The first priority for the Navy was the SFR (submarine fleet reactor) project, to design the propulsion plant for what was expected to be the first fleet of nuclear submarines. The fourth project, still in the early design state, was the large ship reactor (LSR).[32]

Early in the year, before the performance of the *Nautilus* upset the Navy's plans for the submarine fleet, the new small reactor under development at Bettis was to be the backbone of the new submarine force. Less powerful than the S2W plant in the *Nautilus,* the submarine fleet reactor was also more compact and therefore compatible with the smaller, more maneuverable submarine the Navy had requested in 1953. A smaller reactor meant a higher concentration of energy within the plant and intensified problems of thermal and radiation effects. But Code 1500 was also insisting that the new plant be a significant advance over the S2W. The specifications which Rickover's group prepared for the Bureau of Ships called for a control system much simpler than that on the *Nautilus.* The plant would have to operate under extreme casualty conditions, a capability requiring design features which conflicted with other specifications. The plant had to be rugged and resistant to shock. It had to be designed to use a minimum amount of fissionable material. As a general requirement, the design had to be susceptible to multiple production techniques if the reactor was to be the power unit for a fleet of submarines.[33] Most important of all, Code 1500 and Bettis would have to develop the reactor without the aid of a land prototype.

Development studies at Bettis since 1952 had enabled Code 1500 to establish the general configuration of the reactor and such parameters as power output and the operating temperature and pressure of the water-coolant. In January 1955 the SFR project under Alexander Squire, who had directed the design of the first zirconium production plant at Bettis, began detailed studies of the reactor core. First the SFR design group would study a zero-power critical assembly of the core at Bettis. Later a complete reactor core would be tested in the S1W in Idaho. Some of the mechanical equipment would be set up and tested at Bettis, but there was no time to build a complete prototype of the propulsion plant as had been done for the *Nautilus* and *Seawolf.*

In the absence of a prototype, the design engineers had to rely on facilities at Bettis, Idaho, and Groton. Electric Boat, which would build the first submarine in the new class, constructed a full-scale wooden mock-up of the reactor compartment at Groton. The mock-up proved invaluable in working out the arrangement of components, piping, and controls, particularly because there was some uncertainty about the type of steam generators to use. The steam generator design would profoundly affect the arrangement of the primary system and of the shielding around the reactor compartment. When Rickover could not bring his staff into agreement on the better of the two most promising designs, he decided to use both, at least in the first few submarines in the class.[34] Thus the reactor plants were designated S3W and S4W, the only difference between them being the design and arrangement of the steam generators and reactor compartments.

Despite the significant advance in reactor technology represented by the S3W/S4W plant, Bettis was successful in bypassing the prototype and moving directly into final design and procurement. Most of the major components, such as the pressure vessel and main coolant pumps for the first ship, were ordered in the spring of 1955. Electric Boat laid the keel of the *Skate* (SSN 578) on July 21. Three more attack submarines—the *Swordfish, Sargo,* and *Seadragon*—and one guided missile submarine, the *Halibut,* would use the same propulsion plant. But even before the *Skate* was launched, the special trials of the *Nautilus* had demonstrated the inadequacy of this small propulsion plant for high-speed submarines. By the fall of 1955 Code 1500 would turn to a still newer design.

Because the Navy itself had not yet clarified its needs for nuclear-powered surface ships, Bettis had only begun to focus on specific propulsion systems for this purpose in 1955. All that the engineers in the LSR project at Bettis knew was that they were to design a very large reactor for use in a surface ship such as an aircraft carrier or a cruiser. Rickover's group in Washington had established the specifications and design objectives for a land-based prototype to be built at the Idaho site. Like all other naval plants developed by Bettis, the large ship reactor was to be of the pressurized-water type. Because of its size and the fact that each ship would require several reactors, special care would be necessary to design a reactor that would be extremely economical in using nuclear fuel.[35]

Rickover and his staff had resolved most of these uncertainties before the end of 1955. The land-based prototype, now called the A1W, would consist of two reactors driving one shaft of an aircraft carrier. The reactors would have different types of cores, and different materials would be used in the steam plants with different types of steam generators; but like the submarine prototypes, the A1W would be a practical, operating propulsion plant built to ship specifications. The A1W project would also provide basic design data for other surface ship propulsion systems—initially for the F1W, which was expected to use the A1W core in a somewhat larger reactor in a frigate, and later in the C1W plant, which was at that time expected to employ four reactors of the A1W type in a guided-missile cruiser.[36]

Reactors for the Submarine Fleet

Although the submarine fleet reactor, large ship reactor, and Shippingport projects dominated activities at Bettis during most of 1955, Code 1500 soon gave the highest priority to an attack submarine larger and more powerful than the *Skate* class. The Navy needed the new submarines to take advantage of the striking intelligence coming from the *Nautilus* trials. The new S5W plant, and not the S3W/S4W, was to be the submarine fleet reactor for the Navy.

Bettis did its best to take in stride the sudden shift in emphasis from the S3W to the S5W during the autumn of 1955. The Navy requirement reached Bettis on September 20. By this time Weaver had moved up to become the Westinghouse vice-president in charge of atomic power development, and Simpson had replaced him as the director of Bettis. In October, after setting the power specifications for the new plant, Code 1500 approved the creation of the new S5W project under Douglas C. Spencer, and the S5W design group soon won Code 1500's approval of most of the plant characteristics. Within another month Code 1500 and Bettis had fixed the principal features of the reactor, including the size of the core, the number of control rods, the size of the pressure vessel, the type of refueling system, and all the thermal and hydraulic parameters.[37]

Compared with all previous projects at Bettis, the speed with which Code 1500 could make these decisions was extraordinary. The accomplishments reflected not only the exigencies of the situation, but also the ability and experience which the Bettis engineers and scientists had assembled since 1948. Rickover persuaded Bettis to staff the project initially with men who had an intimate knowledge of the *Nautilus* and *Skate* reactors. Not only could they select a design which would most likely lead to a reliable and practical power plant; they could also take advantage of their experience in designing the S3W. Although there were improvements in design, the hydraulic and nuclear characteristics were similar enough to the *Nautilus* plant to make unnecessary the building of a test core, much less a land-based prototype. The reactor core, including the fuel assemblies and the control rods, represented the most radical departure from the S3W design, and Bettis could accomplish all of this work with a critical assembly at the laboratory.

In monitoring core design, Mandil followed Rickover's instructions to concentrate not on novelty but on practicality and simplicity, and particularly on those features which would be amenable to mass production. Although the new reactor would resemble the S2W in some respects, Rickover's group asked Bettis to avoid those features of the *Nautilus* plant which required painstaking alignment, special adjustment, or reworking of components during assembly. There was a concerted effort to simplify the refueling process and to provide easier and more flexible access to the fuel assemblies. The circular cross-section of the *Albacore* hull provided more space in the reactor compartment of the new submarine than in either the *Nautilus* or the *Skate*—an important advantage in attaining these objectives.[38]

Under these favorable circumstances, development of the S5W plant moved swiftly during 1956 and 1957 despite the usual setbacks and delays. Strikes and problems in fabrication postponed the delivery of the pressure vessel and steam generators for the *Skipjack,* the first submarine of the new type to be constructed at Electric Boat. But there was nothing exceptional in these difficulties. The real problems still lay in the adequacy of design and the performance of components. Such matters as shielding design posed major questions which could be resolved only with the development of new computer codes by both Bettis and Knolls under the guidance of Radkowsky and Brodsky from Code 1500. In May 1957 Rickover authorized Bettis to begin fabricating the first S5W core, and early in 1958 the laboratory could report that most items for the first propulsion plant were not far behind schedule. By that time, however, the Navy had imposed new requirements for a larger number of the new attack submarines on an accelerated schedule. Even before the first S5W plant was in operation, Bettis was faced with the task of moving into multiple production.[39]

The Elements of
Multiple Production

Almost from the beginning of the S5W project, there was the possibility that the new plant would become the submarine fleet reactor of the Navy. This requirement, plus existing commitments at Bettis, would force the laboratory to grapple with all the elements of multiple production. Five S3W/S4W plants for submarines already under construction, a modified S2W replacement plant for the *Seawolf,* and as many as ten plants of the A1W type for the Navy's first surface ships in addition to the six new S5W plants would commit Bettis to building at least twenty nuclear propulsion plants for the Navy, and this barely three years after the first prototype went into operation.

Although Rickover and his staff had been thinking about the problems of multiple production since the fall of 1955, the specific measures Bettis should adopt were not at all clear in the spring of 1956. If Bettis had followed the project type of organization which Weaver had established in 1954, the S5W project group would have been responsible for everything related to the plant, including procurement as well as design and development. That had proved a practical approach on the S1W, but it was no longer feasible in 1956. The task of finding qualified suppliers, developing production specifications, supervising fabrication, and inspecting the components for six submarine plants would have left the engineers and scientists in the S5W project with little time to design and develop the initial plant. Rickover, who warned Weaver and Simpson of the dangers of diluting engineering talents at Bettis, was also concerned that Bettis was growing too large to be effective as a laboratory. In the year since June 1955 employment at the Bettis site had climbed almost nine hundred positions to a new high of more than 2,800 persons.

From the conventional Navy perspective, there was no reason to burden Bettis with this additional responsibility. Once the laboratory had developed the reactor plant, the Navy would have been willing to take over the negotiation and administration of contracts with suppliers. But Rickover had no intention of letting these tasks fall to the Navy. He was convinced he could not meet his commitments if he had to move at the ponderous pace which resulted from following regular procedures. He was equally convinced that the Navy's methods of contract administration would never produce equipment of the quality required for a nuclear propulsion plant. Because all of the components to be procured were for Navy ships, Rickover could not rely on the Commission for help. The only solution was to call on Westinghouse.

Both Bettis and Westinghouse had earlier attempted to enlarge their capabilities for producing and procuring components of nuclear propulsion plants. In September 1953 Westinghouse had created an atomic equipment department, an organization completely independent of Bettis, which would produce nonnuclear components of reactor plants on a commercial basis. Housed temporarily at McKeesport, near Bettis, the commercial department moved to a new plant at Cheswick, Pennsylvania, late in 1954. For the Navy project, however, a separate facility was needed. At Rickover's insistence, Weaver had agreed late in 1953 to set up a special procurement office as a Bettis sub-division. In January 1954 a group consisting of fifty power plant engineers and seventeen administrative personnel under Squire's direction moved into office space at Large, Pennsylvania, some fifteen miles south of Bettis. The new group began making plans to take over all procurement responsibilities for nonnuclear components required at Bettis, but the effort was short-lived. The reorganization of all Bettis activities on a project basis in September 1954 wiped out the mission of the subdivision.[40]

In creating a new procurement organization in 1956 Rickover followed some of these earlier patterns. The division of responsibility between Bettis and the new organization would be essentially that which had existed between the subdivision at Large and other parts of Bettis in 1954. Bettis would be responsible for procuring all components for prototypes or for the first propulsion plant of each type. The new organization would handle all aspects of procurement for all components except reactor cores for all successive plants of that type. Unlike the Large subdivision, however, the new organization would be a completely separate department of Westinghouse. To avoid generating questions or opposition in the Navy, Rickover wanted to give the new department an innocuous or meaningless name. He liked the name "plant apparatus department," or better yet, simply "PAD."

From the beginning PAD had an existence almost completely separate from Bettis but closely tied to Rickover and his Washington office. William L. Borden, the general manager of the new department, had not come up through the executive channels of either Westinghouse or Bettis. A lawyer rather than an engineer, he had served as executive director of the Joint Committee on Atomic Energy during the chairmanship of Senator Brien McMahon. In this position Borden had become one of Rickover's most trusted allies on Capitol Hill, and Rickover had recommended him for the position of special assistant to Weaver at Bettis when the Republicans took control of Congress and the Joint Committee in 1953. When Borden set up the first PAD organization in an old Westinghouse plant in Pittsburgh during the summer of 1956, he took with him only six engineers from Bettis. The initial composition of PAD reflected the conviction that Borden and his associates would be concerned almost exclusively with the legal, contractual, and administrative aspects of procurement.[41]

Within a matter of weeks, however, it became painfully evident that the original conception of PAD's function was inadequate. PAD could not be limited to administrative activities but would have to be proficient in both component and plant engineering. Even experienced manufacturers of power plant equipment did not understand the special problems of fabricating components for nuclear plants. PAD needed experienced engineers to explain specifications, to revise quality control procedures in vendors' plants, and to introduce changes in specifications when they appeared necessary. Changes in component specifications would lead ultimately to changes in the design of the reactor plant itself even though theoretically the following plants were to be identical to the prototype or initial design created at Bettis. Recognizing these broader requirements, Borden began an intensive effort during the summer of 1956 to recruit additional engineers for PAD. Because Bettis was already short of engineers, Borden had to recruit mostly from other sources. The newly hired engineers were then sent to training courses at Bettis, where they set about mastering the thousands of drawings and specifications for the S5W plant.[42]

Translating Bettis's original designs into plans and specifications for suppliers was never an easy task. Initially the PAD engineers had to work without any specific experience with nuclear plants, and to meet Rickover's schedule they often had to draft specifications before Bettis had settled on the final design. Usually the performance characteristics of components such as steam generators or pressurizers could be taken from Bettis's specifications. Then the PAD engineers prepared arrangement diagrams and master drawings which showed the external configurations, mountings, and connections necessary to make the component a part of the plant. In the early years, when the PAD engineers were relatively inexperienced, details of the internal design of components were often left to the supplier, a temporary expedient which did not achieve uniformity in design or complete standardization of performance. In time, as the PAD engineers became more expert in design details and fabricating techniques, uniform specifications and standards became the rule.[43]

In addition to the engineering, contractual, financial, and administrative functions performed in Pittsburgh, PAD had critical responsibilities in the vendors' plants. Very early in their work the PAD engineers in the field discovered that even old-line equipment manufacturers had little idea of how to plan and organize production of the kind of sophisticated equipment needed for nuclear propulsion plants. There was an initial inability to believe that the elaborate specifications for tolerances, integrity, and purity of materials had to be met literally. The precision and quality which PAD was demanding were simply unheard of in the heavy equipment industry. Only the experience of failure could teach most suppliers that elaborate planning of production processes, special training for technicians, meticulous adherence to prescribed steps in processing, and constant checks on quality control were necessary to achieve an acceptable product. Manufacturers of power equipment were accustomed to having customer representatives regularly visit their plants, but never with the frequency and intensity of PAD field engineers. Some spent months at a time in vendors' plants helping to get production started and then spurring the fabricator to get production up to specifications. Once a major supplier was ready for something approaching regular production, PAD assigned a resident engineer to the plant, mainly to follow quality control. PAD supervisors in turn kept close check on resident engineers even to the point of recording how many welds the engineer actually witnessed during the fabrication process. Not satisfied with a detailed inspection of every step in fabricating each component, PAD instituted the practice of reinspecting the entire unit when it was completely assembled and ready for shipment. The initial reaction of vendors was that reinspection was unreasonable, but the system did reveal faults that had not been detected earlier. The resulting system of quality control surpassed in extent and rigor any previously used in Navy procurement.

As the number of nuclear ships increased, so did the burdens on PAD. By the end of 1958 PAD was handling Navy orders with 400 suppliers, of which fifty-five had contracts of $100,000 or more and twenty-one had contracts ranging from $1 million to $15 million. Borden assigned resident engineers to the plants of all vendors with orders over $1 million; on the average there was one quality control representative for each $2 million of equipment under contract.[44]

Borden, still struggling to build technical competence in PAD, accepted Rickover's demands for a major reorganization. The original structure built around the two major projects, the S5W and the A2W, was replaced by one along component lines. Thus there were five main subdivisions—for reactor equipment, heat exchangers, pumps, instrumentation, and valves and auxiliary equipment. Quality control was centered in a separate PAD office with responsibilities for all kinds of equipment. The new organization made it possible for each subdivision manager to take a personal part in each technical decision within his area of responsibility. It also flattened the organizational structure by eliminating one layer of technical supervision between Borden and the men on the job and thereby helped to improve communications.

Rickover also gave Borden some technical support by arranging to have two experienced engineers assigned to PAD as Borden's assistants. The first of these was Marshall E. Turnbaugh, Rickover's senior technical representative at Bettis for several years, now retired from the Navy. The second was Squire, who as a former project manager and director of the special procurement office at Large, was well qualified to serve as assistant general manager for reactor plant components.

By any standards PAD's mission was difficult to accomplish. In the first place, the kind of organization Rickover was trying to create required a level of technical competence which would come only with years of experience on the job. The work required engineers of real ability, but more than that: men who would find the pursuit of hidden flaws in equipment an exciting challenge rather than a tedious and unreasonable chore. Ultimately the engineers at PAD would have to reflect Rickover's conviction that they would have to master machines, a task which demanded nothing less than perfection. In this sense PAD would never reach the goal Rickover held before it. Despite all the care and ingenuity employed, equipment would continue to fail; bad welds and substandard materials would continue to slip by the most elaborate echelons of inspectors. But if perfection escaped PAD, improvement did not. The organization did in time deliver equipment which on the whole met the reliability standards required for nuclear submarines. In October 1959 Rickover extended the PAD system to General Electric, where he created a similar organization with an equally cryptic name—the "machinery apparatus operation," or "MAO." As the number of nuclear ships increased, PAD and MAO became essential parts of the multiple-production system.

There was always a tendency among uninitiated contractors to complain that the extreme specifications imposed on nuclear equipment were unreasonable or, even worse, unnecessary. Yet reactor engineers outside the Navy project also encountered this misconception. The exceptional difficulty in finding fabricators who initially could or would meet nuclear specifications was a common obstacle in the nuclear industry long after Rickover established PAD and MAO. Despite the best efforts of Rickover and other engineers on other projects, the failure to meet equipment specifications continued to harass reactor projects into the 1970s. Rickover's relentless attack on the problem did not solve it entirely, but he helped to dramatize both for his own contractors and others the critical importance of quality control in nuclear technology.

Commercial Zirconium

In the long term the mission of PAD and MAO was to develop commercial producers of materials and components for nuclear propulsion plants. In the case of such conventional items as steam generators, valves, and pumps, it was reasonable to assume that old-line manufacturers of power plant and naval propulsion equipment would in a short time be able to meet the exacting standards imposed by nuclear technology. As we have seen, this task took much longer than most engineers at PAD expected. As for the more exotic materials and precision devices to be placed within the reactor itself, no simple extrapolation of conventional technology was available. The commercial production of such materials and equipment was clearly more remote. The Commission, through the Bettis and Knolls laboratories, would have to undertake initial production with the hope that commercial sources could be developed in several years. In this process Rickover and his associates provided the primary impetus.

One example of the move toward commercialization was the production of zirconium. As we saw in chapter 5, Geiger had organized production on a large scale in 1950 even before all aspects of the process had been explored. In order to meet the schedule for the S1W prototype, Rickover had authorized Westinghouse to construct a crystal bar plant at Bettis even while the Oak Ridge laboratory and other research contractors were still investigating methods of producing this unfamiliar metal. The production chain as it existed in early 1951 was complex, clearly based on expediency, and closely tied to government installations. The Commission bought crude zirconium tetrachloride from a single supplier at a relatively high cost. Oak Ridge processed this material in its pilot plant to remove the hafnium and convert the material to an oxide. The U. S. Bureau of Mines plant at Albany, Oregon, chlorinated the material and reduced it to zirconium sponge. Because the sponge was not believed pure enough for reactor use, it was shipped to the Foote Mineral Company and to Bettis, where the metal was refined by the crystal bar process. This clumsy, expensive production chain was the source of all the zirconium for the fuel elements and internals of the first S1W core.[45]

The production effort and continuing research sponsored by the Commission led to substantial advances in both the process and the product by 1952. The quality of zirconium sponge was so improved that Rickover decided it was possible to eliminate the tedious and expensive crystal bar process. Furthermore, the study of zirconium alloys resulted in the development of a new material called zircaloy-2, which was far superior to the original material used in the first core.

With the elimination of the crystal bar process, Code 1500 supported Geiger's efforts to develop a commercial source of supply for low-hafnium zirconium sponge. Geiger late in 1951 made a special effort to solicit proposals from a large number of companies. In 1952 he negotiated a five-year contract with the Carborundum Metals Corporation to produce 150,000 pounds of zirconium sponge per year in a plant which the company built at Akron, New York. The plant reached routine production in 1954 and eventually attained a production rate of 200,000 pounds per year at a price of about $13.10 per pound. By 1955 Carborundum Metals was able to meet all immediate requirements and the Bureau of Mines plant was shut down.

The Navy requirement for six S5W plants and several A1W units was in large part responsible for Geiger's efforts to increase the zirconium supply substantially in 1956. Declassification of most information on zirconium production processes in 1955 made it easier to interest new commercial suppliers. The Commission's solicitation resulted in three new five-year contracts. The National Distillers and Chemical Company agreed to produce one million pounds of sponge per year at a cost of $4.50 per pound in a new plant at Ashtabula, Ohio. The Columbia National Corporation built a plant at Pensacola, Florida, to produce 700,000 pounds per year at a price between $6.50 and $7.50 per pound. Carborundum Metals also constructed a new plant, this one at Parkersburg, West Virginia, to produce 500,000 pounds per year at a price between $7.50 and $8.00 per pound.

To meet the increasing demand for zirconium Geiger also arranged to reactivate the government plant at Albany, Oregon, and to award a contract to the Wah Chang Corporation to operate the plant. About a year after resuming production in 1957, Wah Chang built its own plant on the outskirts of Albany and continued to sell zirconium to the Commission without a long-term contract. During the same years the Commission negotiated a contract with the Toyo Zirconium Company of Tokyo to produce four hundred thousand pounds of sponge under a commodity barter agreement.

By the time the five-year contracts expired in 1963, zirconium had become a commercial product. In fact, delays in the construction of commercial nuclear power plants had left the industry with so much surplus capacity that zirconium sponge prices became highly competitive, falling in some cases below $4 per pound. This figure compared to a cost of several hundred dollars per pound for inferior material before 1950. Several companies dropped out of the zirconium business, but successors to Wah Chang and Carborundum Metals continued to provide a reliable supply of high-grade sponge at competitive prices for the nuclear Navy and the American nuclear power industry.

Reactor Cores

The rapidly increasing demand for zirconium was a measure of the much larger effort which Code 1500 had asked Bettis to undertake to develop and manufacture reactor cores, those complex assemblies of zirconium-clad uranium fuel elements, zirconium and stainless-steel support members, and hafnium control rods, in which the nuclear reaction took place. Both the literal and technological heart of the propulsion plant, the core was the one component of the nuclear submarine for which there could be no alternate system or back-up. If the core failed at sea, the submarine would no longer be capable of military operations. Cores also required the highest degree of sophistication in engineering design, the most extensive application of novel techniques and technical skill in manufacture, and the greatest care in inspection and testing of any part of the plant. For this reason no segment of the propulsion plant was less amenable to industrial production on a commercial basis.

Core design and fabrication commanded a large share of the engineering talent and facilities at Bettis from the beginning of the project. The F Building, a part of the original plant which Westinghouse constructed at Bettis, housed the equipment used to produce the core for the S1W prototype. As the need for more extensive facilities became apparent, Bettis made plans for three new structures: the critical experiment or CX Building, a new fuel fabrication building called "G," and an assembly and test building designated as "AT." These new facilities produced the first *Nautilus* core and provided for preliminary studies of core assemblies for the next generation of reactors. By 1954 these new reactors had become projects which would later be identified as the S3W, the A1W, and the Shippingport reactor.[46]

The introduction of multiple projects had implications for all activities at Bettis, but the impact on core work was particularly complex. In addition to designing several cores simultaneously, Bettis would also have to handle both development and manufacturing functions. There were several possible ways of organizing the work on cores, depending in part on the degree of similarity between the various core designs. If the cores were not too dissimilar, there would be obvious economies in using the same equipment for several types. A single development or manufacturing group, however, would bypass the advantages of specialization and concentrated attention which the project approach offered. A Bettis study which attempted to weigh the relative merits of several types of organizations indicated that some combination of project activities would ultimately be desirable; but until new types of cores were ready for manufacturing, Bettis would rely on the project approach to core development as it did for most other activities in the laboratory.[47]

The sharp upturn in Navy requirements in 1955 removed any thought which might still have existed in Code 1500 that Bettis could manufacture all the cores required for the nuclear fleet. Code 1500 suggested that Bettis concentrate on highly developmental cores and farm out the production of other cores which would follow designs already established at the laboratory for other ships in the *Skate* class, the second Shippingport core, and later cores for the S1W.[48] The next step was for Code 1500 to decide that Bettis would build the first core of each type and that subsequent cores would be built under contracts with commercial suppliers.

Finding commercial suppliers would be more difficult in the case of cores than it was for reactor plant machinery. Surveying the prospects in late 1955, Mandil could not find more than four or five companies with any hope of qualifying for this work. Any successful fabricator would have to have a trained staff of at least sixty engineers, scientists, machinists, and inspectors organized around a nucleus of personnel experienced in designing, developing, manufacturing, and testing fuel elements. The company would need at least 15,000 square feet of manufacturing space devoted solely to fabrication, inspection, and testing. Of this space, at least 2,500 square feet would be used exclusively for melting, rolling, and machining fuel alloy. The plant would also require special security and health facilities for handling enriched uranium. Machine tools would have to be specially designed, and the fabricator would have to obtain such equipment as inert-gas arc furnaces and welding boxes, acid etching baths, and special inspection devices.[49]

While Mandil concentrated on technical qualifications for commercial core fabricators, Rickover added a few requirements of his own. In a conference at Bettis in August 1956 he described some of the capabilities he expected core manufacturers to possess. First, the contractor would have to have experience in actual fabrication, at least in trial lots, of the type of fuel elements involved. Second, the contractor would have to possess, or have on order, the special fabrication equipment required. Third, he would have to have begun

recruiting the necessary technical personnel.[50] These were not unusual requirements, but they did illustrate Rickover's determination to get down to
realities in terms of actual equipment in place and qualified personnel to operate it.

Two further demands which Rickover imposed were less common. The
first was that all contracts were to be awarded on a lump-sum or fixed-price
basis as the result of competitive bidding. The usual practice in the Navy was
to use cost-plus-fixed-fee contracts for all new types of development until
fabrication techniques and specifications had become well enough established
to warrant routine procurement on a fixed-price basis. In the middle 1950s
virtually all procurement of reactor components outside the Navy project was
under cost-plus contracts. That practice normally would have applied to anything as novel and complex as submarine reactor cores, but not in Rickover's
organization. For both the S3W and S5W projects, commercial fixed-price
contracts were awarded long before Bettis had refined all the fabrication steps
and perfected specifications. But Rickover had no intention of using the cost-
plus system even at this early stage of core fabrication. The fixed-price competitive contract would surely save money and help to give Rickover more
nuclear ships for the limited funds available. Even more important, this kind
of contract would encourage the hard-headed competitive situation needed
to create a commercial core industry in the United States.[51]

Rickover's second demand stemmed in part from the same consideration.
Code 1500 required that every core manufacturer do all work in company
plants without any reliance on government-owned facilities or financial assistance from the government. This restriction, seldom invoked in ordinary
defense procurement, saved millions of dollars in government investment.
Not only did the rule force the core fabricators to stand on their own feet
without a government crutch, it also gave Code 1500 a strong position in
subsequent negotiations with the contractors. If on a later procurement an
established fabricator did not submit a competitive proposal, Code 1500
could take the work elsewhere without having to face the question of dispos-
ing of an expensive government plant which had been built to meet the needs
of a particular contractor.

In funding the core contracts Rickover also instituted an unusual arrangement. He insisted that the contracts be negotiated and administered by the
Commission even though the funds would come from the Navy. Because the
cores were to be used in Navy ships, the Navy was the logical source of
funds; but Rickover maintained that the highly developmental nature of the

work justified close ties with the Commission. This argument was valid in the sense that Bettis and Knolls were Commission laboratories, and the use of Commission contracts would enable the laboratories to maintain closer controls over activities in the commercial plants. From another perspective (one never mentioned in official correspondence) the arrangement helped Rickover to keep core fabrication entirely within his own area of responsibility. Like the activities under PAD and MAO, core fabrication would not fall within the purview of the Navy procurement system where, Rickover feared, these critical items would not receive the special attention they deserved.[52]

By early 1957 Mandil had established the organizational pattern for core production and procurement which Bettis would continue to follow for more than a decade. Code 1500 would work with the nuclear core department on the fabrication of the first core of each type and the outside procurement of successive cores for ship propulsion plants. The new arrangement accomplished for production and procurement the kind of consolidation of functions Code 1500 had urged at Bettis in 1954. Core design and development would continue to be the responsibility of the individual Bettis projects. Supplementing this work at Bettis would be the commercial fabricators, which would produce most of the cores for the nuclear fleet. Geiger negotiated the initial procurement, involving four S3W cores, through three contractors: the Babcock & Wilcox Company, which had a fabrication plant at Lynchburg, Virginia; Combustion Engineering, Inc., with a plant at Windsor, Connecticut; and the atomic fuel department of Westinghouse, which had plants at Cheswick and Blairsville, Pennsylvania. In 1957, when S5W procurement went into high gear, Code 1500 added to the original three contractors two more, both of which had experience in fuel element fabrication. Metals and Controls Corporation had produced fuel alloys for critical experiments at Bettis and was fabricating fuel elements for the *Seawolf* plant (S2G) and the submarine advanced reactor (S3G). A division of the Olin-Mathieson Chemical Corporation (later to become part of the United Nuclear Corporation) had also been manufacturing S3G fuel elements and experimental assemblies for Bettis. All five companies began producing S5W cores in 1957 and continued to provide cores for nuclear ships under successive contracts well into the 1960s.[53]

The core procurement system established at Bettis in 1956 had the superficial aspects of a commercial enterprise. Although Bettis itself was still involved in development, most of the cores for naval ships were coming from the plants of commercial fabricators under fixed-price contracts. The num-

Table 5.   Nuclear Core Production for Reactors Designed by
Bettis, 1951–66

Contractor	Reactor Type					
	S1W	S3W	S5W	A1W	PWR	Total
Bettis	6	1	1	2	2	12
Babcock & Wilcox	3	3	31	10	. . .	47
Westinghouse, Atomic Fuel Department	. . .	5	8	4	3	20
Combustion Engineering	1	4	2	. . .	. . .	7
Metals & Controls	. . .	. . .	26	11	. . .	37
United Nuclear	. . .	2	42	4	. . .	48
TOTALS	10	15	110	31	5	171

Table 6.   Actual Core Manufacturing Costs for Reactors Designed by
Bettis, 1955–66 (thousands of dollars)

Fiscal Year	AEC Contracts	Work at Bettis	Total
1955	0	3,854	3,854
1956	0	4,622	4,622
1957	4,042	16,899	20,941
1958	7,340	25,008	32,348
1959	45,373	12,691	58,064
1960	58,634	6,202	64,836
1961	42,753	7,927	50,680
1962	32,318	12,042	44,360
1963	30,951	11,972	42,923
1964	21,468	16,109	37,577
1965	24,209	5,246	29,455
1966	28,908	2,662	31,570
TOTALS	295,996	125,234	421,230

ber of cores produced would quickly reach an annual rate undreamed of a
few years earlier. W. Kenneth Davis and others in the Commission took the
situation to mean that the Navy could take over core procurement for its
ships. When Rickover showed some reluctance to move in that direction,
Davis demanded an immediate plan for taking the Commission out of Navy
core procurement.[54]

Rickover opposed the idea on the strong conviction that core production
was in reality far from being a stable commercial activity. He contended that
both core development and manufacturing were part of a new and rapidly
changing technology. In his opinion the commercial core fabricators still had

very little practical experience and needed a large amount of technical guidance and supervision which only the Commission could provide through its laboratories. The Navy, Rickover argued, had no trained personnel to perform the many diverse and highly technical functions of supervision and inspection. Furthermore, building such an organization in the Navy would simply duplicate the capabilities which the Commission was only then beginning to acquire.[55]

Subsequent experience seemed to justify Rickover's contention. All the commercial contractors found that the processes copied from Bettis could go out of control, and they did not always understand them well enough to know why. An elaborate system of management appraisal, engineering inspection, and quality control similar to that exercised at PAD was necessary to produce satisfactory cores for the fleet. Rickover, Geiger, and Mandil recognized, however, that in the long term their purpose was not simply to obtain the cores the Navy needed; they were also attempting to establish in the United States a commercial core manufacturing capacity that would be able to supply the large number of nuclear power reactors expected to be built in the 1960s. By insisting on the principles laid down for the initial core procurement contracts in 1955, Code 1500 succeeded in making this larger and lasting contribution to the American nuclear industry.[56]

The New Dimension

Similar patterns of production and procurement emerged at Knolls during these same years. Although the laboratory operated by General Electric never approached the quantity production attained by Bettis during the last half of the decade, Code 1500 did convince Knolls to divest itself of the production function as the number of reactors developed by the laboratory began to increase. First Code 1500 insisted on farming out the procurement of nonnuclear components and then on establishing MAO, a separate component of General Electric. Next came the transfer of core manufacturing from Knolls to the commercial suppliers.

By 1960, then, Rickover had largely completed the vertical extension of his original project structure. He now had some control over all phases of reactor design and manufacture from the raw materials to the finished propulsion plants which would be installed in Navy ships. In the Commission this span of authority meant that Rickover was virtually independent of the division of reactor development, the Commission's laboratories, and all the

Commission's production activities except for the uranium 235 used in fuel elements. Bettis and Knolls, although still technically Commission laboratories, worked almost entirely on naval reactors. Except for a very small amount of fundamental research, none of the Commission's other laboratories had any part in the Navy project. In March 1958 Rickover erected another barrier between the naval reactors branch and the rest of the Commission. He succeeded in reorganizing the Pittsburgh and Schenectady offices so that they were concerned exclusively with naval reactors and reported directly to him.[57] Rickover was punctilious about forwarding official Commission correspondence through Davis in the division of reactor development, to the general manager, and then to the Commission, but seldom did any of these officials raise serious objections to Rickover's proposals.

In terms of the Navy, Rickover had been even more successful in isolating reactor procurement from the conventional bureaucratic process. By insisting upon the Commission's special responsibilities for the development and control of atomic energy, Rickover was able to keep this portion of his work completely outside the Navy. To be sure, the Navy was guaranteed the benefits of nuclear power, but the manufacture and procurement of nuclear propulsion units was not an integral function of the Bureau of Ships.

In terms of American industry the vertical extension of the project system gave Rickover absolute control over the standards of production and the specifications of quality. For the first time, manufacturers of power equipment and metal fabricators were learning what it meant to produce equipment for nuclear plants. They were also discovering how rigidly a government contract could be administered in the hands of a conscientious and determined public official.

In all respects the new dimensions of the naval reactor project encompassed unusual forms of organization, a striking degree of independence for those in the project, and unprecedented standards of industrial practice. Rickover had taken advantage of a new technology to create a new administrative instrument for pursuing it.

Building the Nuclear Fleet: Horizontal Extension of the Navy Project

The development and manufacture of nuclear propulsion plants for the fleet were activities closely allied to the Commission's responsibilities. Nuclear reactors, even in the late 1950s, were still largely the Commission's domain. As described in chapter 9, Rickover used this distinction to keep reactor and core manufacturing out of the Navy's procurement system. In this sense he had extended his original project organization vertically to include not only the development of nuclear propulsion systems but also the procurement of materials, the fabrication of components, and the virtual mass production of propulsion plants.

When it came to building the submarines and ships in which these reactors were to be installed, however, Rickover could neither hope nor wish to create a separate organization independent of the Navy. Ship design and construction involved a wide spectrum of specialized skills and techniques which Rickover's organization could master with little advantage. At the same time Rickover considered it imperative to maintain control over the design and construction of those parts of nuclear ships which contained the reactor and the propulsion machinery. In his opinion nuclear propulsion was not yet a conventional technology which the established codes in the Bureau of Ships could handle in a routine manner. The only hope for creating a fleet of nuclear ships seemed to lie in maintaining the same standards and discipline which Rickover and his associates had established in building the *Nautilus*.

During the development and construction of the *Nautilus,* Rickover had relatively clear responsibilities. He had full control over the design and development of the propulsion system. He had some influence on the design of the submarine, particularly on those features related to the propulsion system. He imposed his own standards on propulsion equipment and had full control over tests and sea trials of the propulsion plant. The senior officers and engineering department of the ship were selected and trained according to his standards. But the Bureau of Ships, operating through various codes, had general responsibility for the completion of the ship.

As the prospects of large numbers of nuclear ships began to emerge in 1955, there was some uncertainty in the Bureau of Ships about what Rickover's role would be. Admiral Mumma, who believed Rickover had passed the point of his greatest usefulness to the Navy, had no desire to help Rickover extend his influence beyond Code 1500 to other parts of the bureau. Rickover, for his part, was just as determined to exercise his responsibilities as a Commission official to see that the Navy's nuclear propulsion plants were properly constructed and operated.

Rickover would have to find ways of asserting the controls he thought necessary through established channels in the Navy. This he could accomplish,
not so much by extracting formal delegations of authority from the chief of
the bureau, but by attempting to apply his influence in less formal but equally
effective ways. A strong challenge to this effort was the *Polaris* program in
which Admiral Burke, as Chief of Naval Operations, gave the Special Projects Office under Admiral Raborn full authority for developing and building
the *Polaris* submarine. Here, in effect, the tables were turned, and Rickover,
as director of just one of the technical codes in the bureau, was attempting
to maintain his leverage in a project which the Navy placed on an even
tighter schedule than the *Nautilus* had faced.

If the activities described in chapter 9 could be called an extension of the
project in the vertical direction, those covered in this chapter could be considered horizontal extensions of the original project which produced the *Nautilus* and *Seawolf*. In the latter case the purpose was not so much to create
an independent and self-sufficient organization which was proficient in building propulsion plants, but rather to find ways of extending the technical competence and discipline of the nuclear project horizontally into a wide variety
of Navy activities. In this sense, the responsibilities of Code 1500 in the Bureau of Ships became essentially permanent, and the original conception of
a discrete project, organized to accomplish a specific task within a given
period of time, virtually disappeared. How this new type of organization
evolved during the years from 1955 through 1962 is the subject of this and
the following chapter.

Building the *Skate* Class

On July 21, 1955, the nuclear submarine *Seawolf* slid down the ways into
the Thames River at Groton, Connecticut. As the tugs nudged the ship to
the pier, Electric Boat shipyard workers laid the keel of the *Skate* (SSN-
578), the first of a new class of attack submarines using the S3W/S4W propulsion plant. Although the two events were timed to give maximum publicity to the company and to the nuclear submarines, they pointed up a
significant fact: that Electric Boat was the only American yard which could
build nuclear ships. That fact was hardly surprising only six months after
the first sea trials of the *Nautilus,* but it suggested an important step the Navy
would have to take in building a fleet of nuclear ships.

Once the Navy had decided to build a class of nuclear submarines at more than one yard, it was possible to apply the lead-and-follow system which had been common Navy practice for decades. In essence the lead yard designed and usually built the first ship. The follow yards used the design and experience of the lead yard to construct the "follow ships." The arrangement avoided a vast amount of duplication, particularly in producing the thousands of drawings required to design a modern ship. On the other hand, since no two yards had precisely the same layout or the same equipment, the follow yard could not blindly accept the design drawings and procedures of the lead yard. For the *Skate*-class submarines, Electric Boat would be the lead yard and Portsmouth and Mare Island the follow yards.

The selection of Electric Boat as the lead yard for the *Skate* class was an obvious choice. No other yard, private or Navy, had yet constructed a nuclear ship. Electric Boat had designed and built both the *Nautilus* and the *Seawolf,* and the company had been working for months with Bettis on preliminary designs and mock-ups of both the S3W and S4W versions of the *Skate.* Any other yard would need months to reach the competence Electric Boat had already attained for designing the new class. The company's familiarity with the design was especially important because the *Skate* class was to be built without the benefit of a land-based prototype.

There was a danger, however, of placing too great a load on Electric Boat. Admiral Mumma began worrying about this possibility after he became chief of the Bureau of Ships in April 1955. In addition to the two *Skate* designs, Electric Boat was also starting work with Knolls on the S3G plant for the radar-picket submarine. On top of this was the design of the new *Skipjack* class which Admiral Burke had approved soon after he became Chief of Naval Operations. The Navy would have to rely on Electric Boat for the *Skipjack* (SSN-585) as well.

As a way of lightening the load on Electric Boat, Mumma favored transferring to Portsmouth the design as well as the construction of the *Swordfish* (SSN-579), which would use the S4W plant, while Electric Boat completed the *Skate* with the S3W plant. The assignment made sense because later Portsmouth would also build the *Seadragon* (SSN-584), which would be an S4W ship. Rickover, bristling at the suggestion, pointed out that Leggett had raised the same issue when the need for alternate designs for the *Skate* class had first become apparent and had decided to leave the design for both versions at Electric Boat. Since then the yard had proceeded to build mock-ups of the two plants and prepare detailed plans. Furthermore, in Rickover's opinion, Electric Boat was still the only yard qualified to design nuclear

26. Preparations at 7:30 A.M., July 21, 1955, for a dual ceremony at Electric Boat. Keel laying of the *Skate* (SSN-578) is to take place in the building way at the center, and the *Seawolf* (SSN-575) is to be launched at the right.

ships. Transferring design work to Portsmouth would only delay completion of the *Skate* class and increase the already heavy demands on Code 1500 and Bettis. After reading Rickover's arguments Mumma dropped the idea.[1]

The differences between the two men probably went deeper than this particular issue. Mumma might well have seen assigning the *Swordfish* design to Portsmouth as a way of breaking the hold of Electric Boat on this activity. His action could also have been part of an effort to bring nuclear propulsion back into the bureau's fold. Implicit in Rickover's opposition was his determination that he would be the one who would decide when a yard was ready to build a nuclear ship.

The Navy Yards

Bringing Navy yards into the nuclear program was a natural development. Portsmouth and Mare Island had been among the several constructors of submarines during World War II and in the lean years after the conflict had, along with Electric Boat, remained as the only builders of conventional submarines in the country.

Both Rickover and Shugg recognized that Electric Boat had to play a key role in preparing the Navy yards for nuclear work. The problem was to make sure that the training interfered as little as possible with construction at Groton. In June 1954 Rickover had gone to Portsmouth to work out the relations between the two yards. The Navy installation agreed to send a group of twenty-nine men to Groton for about a year. Some were to be supervisors, but most were to be mechanical and electrical specialists, pipe fitters, outside machinists, and electricians. All had "waterfront" jobs with no front office responsibilities. Recruiting engineers knowledgeable in steam propulsion was up to the shipyard, but Rickover promised help in getting men who had some experience in this area as well as in the nuclear field.[2]

A similar group from Mare Island arrived at Groton in April, 1955. The training was heavily technical. Men spent their time listening to lectures on such matters as shielding and radiation control and visiting special facilities such as electrical and welding shops. When possible they were given practical training by helping Electric Boat personnel in reviewing specifications, drawings, and procedures, inspecting components arriving in the yard, and participating in tests of plant systems on board the ships under construction.

Every week each man reported his activities back to the home yard and to Code 1500 in Washington. Code 1500 watched for any signs that instruction was drifting from the strictly practical to the theoretical.[3]

Seeing that the yards had the necessary equipment was another part of the preparation. Special facilities approaching clinical standards of cleanliness were needed for fabricating and assembling components of the propulsion plant. Equally unfamiliar in the shipyards was the variety of X-ray machines and other laboratory equipment used to check the quality of materials and assembly procedures. The unseen but lurking presence of radiation required new safety procedures and emergency measures. In almost every department the shipyards needed separate and often elaborate facilities to handle and to store nuclear components. At one time there was some thought of preparing a general manual on the nuclear facilities required in a shipyard, but Rickover and his assistants decided it would be more practical for each yard to prepare its own manuals setting forth the procedures and specifications to be followed. Panoff arranged to have Code 1500 personnel give lectures which the yards recorded and later transcribed. This material in some cases became the basis for yard manuals. Once Code 1500 had approved the manuals, the yards were expected to follow them to the letter.[4]

Once the Navy yards actually began doing nuclear work, Rickover had to face the question of how best to control their activities. His tactic was to extend his authority over all work related to the propulsion system. He set up a distinct organization, the nuclear power division, at both Portsmouth and Mare Island. The director of the division, called the nuclear power superintendent, was an engineering duty officer who had served in Code 1500 and had been thoroughly indoctrinated into its operations. He was to be responsible for all work related to the reactor, steam plant, auxiliaries, and controls. Through the shipyard organization he would direct plant installation, quality control, plant testing, and inspection.[5]

Although the technical problems were the same whether the yard was private or Navy, the nuclear power superintendent had a more difficult task in some respects than did his counterpart at Electric Boat. For one thing, the Navy yards suffered from the inertia and complacency of all mature bureaucracies. For another, the nuclear power superintendent was not a "customer" representative as were the Code 1500 officers assigned to Electric Boat, but rather just another naval officer of relatively junior grade in a large Navy installation. To make certain his representatives were not swallowed up in the yard organization, Rickover at various times gave some of his most experienced men these assignments—Commanders Marshall E. Turnbaugh and John J. Hinchey at Portsmouth and Commanders Edwin E. Kintner and David T. Leighton at Mare Island. But even these veterans ran into trouble in the Navy yards. Hinchey voiced a common complaint when he charged that his division got short shrift from the yard. The existing departments in his opinion were too willing to assume that they could take nuclear work in their stride even though their performance hardly justified such confidence.[6] Placed between the pressures from Rickover and the entrenched structure of the Navy yard organization, the nuclear power superintendent often found himself in a position that could be detrimental to his professional career. Eventually Rickover was to use some civilians for the job, but that change occurred gradually.

Private Yards

The lead-and-follow pattern used for the *Skate* class was an adequate response to the Navy's relatively modest requirements for nuclear ships early in 1955, but the three-yard arrangement could never answer the demands imposed after Admiral Burke became Chief of Naval Operations in August. The six submarines in the *Skipjack* class, to say nothing of later classes, would outstrip the combined resources of Electric Boat, Portsmouth, and Mare Island. It was a foregone conclusion that the additional capacity would come from private yards.

Within the shipbuilding industry there was no lack of interest in nuclear propulsion. In fact, the industry was eager for new business. The end of the Korean War and the decline in stockpiling of strategic materials had cut heavily into shipping activity in the United States. Although the industry had enough back orders and repair work to keep busy for a time, the long-term outlook was somber. Because almost 80 percent of American commercial tonnage had been built during World War II, the average age of the merchant marine fleet was only ten years. Rapidly rising costs in the shipbuilding industry caused operators to delay ship replacement as long as possible. The yards, caught between a declining market on one side and higher wages and tighter government fire and safety specifications on the other, looked to the Maritime Administration and the Navy for help.

To the Navy perhaps the most important private shipyards were those which could build large combat ships. Only three companies—all on the East Coast—fell into this category: the Bethlehem Steel Company at Quincy,

Massachusetts; the New York Shipbuilding Corporation at Camden, New Jersey; and the Newport News Shipbuilding and Dry Dock Company at Newport News, Virginia. To two of these yards the Navy had awarded a *Forrestal*-class carrier. As an added measure of support the Eisenhower administration had assigned every project in the Navy's 1954 shipbuilding program to private industry and had continued the policy of negotiating contracts in a way that would ensure stable employment and reasonable incentive in the private yards. By early 1954, 64 percent of a total of 121,000 employees in the private yards were dependent on Navy work.[7] With some reason the shipbuilding industry was interested in plans for a nuclear Navy.

Two of these private companies approached the Navy even before the *Nautilus* went to sea. In September 1954 Daniel B. Strohmeier, vice-president in charge of the shipbuilding division of the Bethlehem Steel Company told Rickover he was willing to begin studies at the company's own expense on the application of nuclear power to Navy vessels. A month later, William E. Blewett, president of the Newport News Shipbuilding and Dry Dock Company, offered to build a submarine yard on the James River and to train personnel at no cost to the government. This was the second attempt by Newport News to build nuclear ships. In 1952 the company had agreed to work with Westinghouse in designing the carrier reactor prototype and had been heavily involved in this effort by the time the Eisenhower administration canceled the carrier project in the spring of 1953. Plans for the 1956 shipbuilding program, with its total of five nuclear submarines, gave added impetus to private yards. In May 1955 Monro B. Lanier, president of the Ingalls Shipbuilding Corporation, proposed to undertake the construction of conventional and nuclear submarines at Pascagoula, Mississippi.[8]

Rickover was strongly disposed toward the private yards. Since 1954 he had been encouraging them to study nuclear technology and train their personnel so that they would be qualified to accept contracts when, as Rickover confidently predicted, the Navy fully grasped the significance of nuclear propulsion. In the fall of that year he had encouraged the shipbuilding division of the Bethlehem Steel Company to submit a proposal to study the possibility of adapting the A1W reactor plant to a guided-missile cruiser. Bethlehem, in Rickover's mind, would provide some healthy competition for Newport News, which was developing the A1W prototype at the Idaho site with Bettis. Newport News in turn had expressed a lively interest in building submarines as well as aircraft carriers. Rickover urged Mumma to bring Newport News in quickly so that the company would be qualified to build one of the *Skip-*

27. The Electric Boat yard at Groton, Connecticut, in 1958. Machine shops and building sheds are along the waterfront. A submarine under construction can be seen in the nearest building shed.

jack submarines. Again Rickover argued on the basis of the company's experience, its interest in nuclear propulsion, and the need to create some competition in the submarine field for Electric Boat.[9]

Mumma had some reservations about concentrating nuclear work at Newport News. The company was already involved in the A1W project and would probably build the first nuclear-powered carrier when such a ship could be authorized. A more attractive possibility in terms of geographical dispersion of shipyards was the Ingalls yard at Pascagoula. Mumma quickly accepted the Ingalls' proposal that the company begin to learn the fundamentals of nuclear shipbuilding at its own expense.[10] Rickover helped by providing training courses for Ingalls' personnel while Bethlehem and Newport News continued their studies of surface ship applications.

Rickover's preference for private shipbuilders rested largely on his conviction that only the private yards had the flexibility necessary to recruit competent engineers and craftsmen. The Navy yards, tied to the Civil Service system and hamstrung by the political influence of labor unions, could not on short notice acquire the necessary talent for nuclear shipbuilding. It was also easier to impose the high standards of nuclear shipbuilding on the private yards than on the Navy yards. In the spring of 1956 Rickover urged Mumma to assign the six *Skipjack* submarines and a guided missile cruiser in the 1957 shipbuilding program to the six yards which were either building or preparing to build nuclear ships: Electric Boat, Portsmouth, Mare Island, Newport News, Bethlehem (Quincy), and Ingalls (Pascagoula). He specifically recommended that no additional Navy yards be brought into submarine construction at that time and that most nuclear construction be placed in private yards.[11]

The trend toward private shipbuilding in the Navy was but the latest chapter in the long history of competition between government and private yards. In the late nineteenth century when the nation began building the "New Navy," it had turned to industry. Not completely satisfied with the efforts of private shipbuilders, Congress in 1902 had authorized the construction of one battleship in a Navy yard, and other government yards later received orders for large ships. The mixture of private and government yards was an attempt to lower costs and to quicken the building pace through competition. For similar reasons, Congress often debated during these years the merit of establishing a government-owned armor manufacturing plant. To a certain extent, Portsmouth's entry into submarine construction in 1917 followed a similar pattern. Some Navy officers had been dissatisfied with the vessels built by private industry and hoped through Navy rivalry to force better designs.[12]

The primary role of private industry in Navy construction was never seriously challenged, and the policy of the Eisenhower administration in nuclear shipbuilding marked no new departure. Of the seven submarines originally in the *Skipjack* class, five were built in private yards and only two in Navy yards. One of these two was the *Thresher,* in which the Navy incorporated so many changes that it was eventually considered the lead ship of a new class. Although Portsmouth built the *Thresher,* most of the submarines in the class came from private yards. Industrial yards appeared to have a monopoly of nuclear surface ships. Newport News was already working with Bettis on the development of the A1W prototype for an aircraft carrier and Bethlehem Steel had a contract for the guided missile cruiser *Long Beach.*[13]

Builders of Two Nuclear Submarine Classes

Skate class	Builder
Skate (SSN-578)	Electric Boat
Swordfish (SSN-579)	Portsmouth
Sargo (SSN-583)	Mare Island
Seadragon (SSN-584)	Portsmouth

Skipjack class	Builder
Skipjack (SSN-585)	Electric Boat
Scamp (SSN-588)	Mare Island
Scorpion (SSN-589)	Electric Boat
Sculpin (SSN-590)	Ingalls
Shark (SSN-591)	Newport News
Snook (SSN-592)	Ingalls
Thresher (SSN-593)	Portsmouth

Birth of *Polaris*

This pattern of ship construction presented in graphic terms the Navy's determination to build a fleet of nuclear submarines and at least a few nuclear-propelled surface ships. Code 1500 was preparing for the added responsibilities this expansion would bring, but new forces already at work would create even larger requirements before the initial expansion could be realized. At least two factors explained the swift movement of events. The first was the deterioration of United States relations with the Soviet Union as the tensions of the Cold War heightened. The second was the rapid development of military technology, particularly in missile propulsion systems and in the design of nuclear weapons.

As related in chapter 9, Admiral Burke had quickly capitalized on earlier Navy studies of missile systems to establish the Special Projects Office in the Bureau of Ordnance within a few months after he became Chief of Naval Operations in the summer of 1955. The Special Projects Office under Admiral Raborn was to develop a missile launching system for a surface ship which would use the new intermediate-range ballistic missile being developed in a joint Army-Navy project at the Army's Redstone Arsenal in Alabama. The only engine capable of transporting the 1,600-pound nuclear warhead 1,500 miles was a huge device developed by North American Aviation. Almost incredible was the idea of trying to stabilize a missile six stories high in a true vertical position on the deck of a surface ship operating at sea under all conditions and depending on large quantities of liquid fuel.[14]

Only the threat of operational Soviet missiles and President Eisenhower's decision to meet the Soviet challenge in kind made such an undertaking seem worthwhile. On December 2, 1955, Burke heard the president tell the National Security Council with some fervor that the United States had to have a reliable missile system quickly, even if he had to run the project himself. Even before the council meeting, Burke had concluded that only the service which first developed a satisfactory launching system would be able to count on having a long-range missile capability. In his opinion the Navy could spare no effort on the fleet ballistic missile.[15]

Although the Navy was forced by the limitations of technology to give its first attention to a surface launch system, there was full acceptance of the fact that a submarine would have advantages over a surface ship. A submarine would be less vulnerable to enemy attack and furnish a more stable launching platform than a surface ship could provide. The only hope for a submarine missile seemed to be in using a solid propellant in a much smaller missile than the *Redstone* design, one which might reasonably be expected to fit within the hull diameter of a submarine. The Navy had been investigating solid propellants for use in weapons since 1942. Raborn obtained permission early in 1956 to investigate solid propellants and then had contractors draw up plans for a solid-fueled *Jupiter,* as the joint Army-Navy missile was now called.[16]

At the same time Raborn arranged with the Bureau of Ships to have the preliminary design section (Code 420) determine the optimum missile characteristics for a nuclear-powered submarine.[17] Under unwritten orders from Admiral Burke, Raborn and Mumma excluded Rickover from all the preliminary studies. The three officers believed that Rickover's participation at this stage would lead to domination of the new project by Code 1500 and threaten the close cooperation which Raborn had established with the Bureau of Ships. The intention was to bring in Code 1500 at a later date when the principal features of the design had been fixed and then to restrict Code 1500 activities to the propulsion plant. By June 1956 members of Code 420 were working full-time in the Special Projects Office on studies for a missile-launching submarine. This effort, however, was at best a speculative venture because the surface-launched *Jupiter* was still the most promising for immediate development.

The prospects for a submarine launching system began to improve, however, during the summer of 1956. At a meeting of the Undersea Warfare Committee at Woods Hole, Massachusetts, Edward Teller, the distinguished nuclear physicist and weapon expert, raised the possibility of developing a much lighter warhead for the missile by the time the system was expected to be operational in 1963. By September the Commision's weapon laboratories had confirmed the possibility of developing a warhead weighing only 600 pounds with a yield similar to that of the 1,600-pound warhead which the *Jupiter* would use.[18]

This information, plus some encouraging progress in developing a solid propellant, led Raborn and his Special Projects staff to propose giving first priority to a submarine-launched, small, solid-fueled missile which they called *Polaris*. Because the technology of solid fuels was still far behind that of fluid systems, Raborn agreed to continue some work on the *Jupiter;* but before the end of 1956, the Navy dropped out of the *Jupiter* project and began developing a solid-fueled *Polaris* missile 28 feet in height, 60 inches in diameter, 15 tons in weight, and with a planned range of 1,500 nautical miles. The first *Polaris* submarine was to be ready for trials in 1963 and for fleet assignment in 1965. This decision brought a new urgency to submarine design. The design personnel from the Bureau of Ships served on the special task group which considered literally hundreds of possible submarine and missile combinations. By March 1957 the group had developed the basic parameters of the system.

Rickover's first exposure to the *Polaris* design came on April 16, 1957, when Code 1500 received a copy of the first description of the submarine which Special Projects had submitted to the ship characteristics board. The proposal called for a single-screw submarine 350 feet long with a beam of 32 feet and a submerged displacement of 6,500 tons. The sixteen vertical missile tubes were to be designed so that a missile could be serviced in a loaded tube while the submarine was submerged and the missile fired either while the submarine was fully submerged or fully surfaced.[19]

As for the power plant, the proposal stated only that the submarine would have a single screw and specified its submerged speed. There was little question that the S5W propulsion plant would meet these requirements, but Rickover was worried about an additional statement in the proposal that the ship should be capable of safe and efficient operation under the north polar ice pack. In independent studies of a missile-launching submarine, Rickover and his staff had already concluded that such a capability would be highly desirable, if not essential. But operation under the ice was inconceivable to Rickover unless the submarine had twin reactors and propellers. Two reactors seemed essential to provide reliable power in an area where surfacing was not always possible in an emergency, and twin propellers appeared mandatory for maneuvering in the close quarters and high winds encountered in lakes in the ice pack.[20]

Rickover succeeded in incorporating these ideas in the formal comments by the Bureau of Ships. Either the design should be changed to a twin-reactor, twin-propeller system, or the under-ice capability should be eliminated from the ship characteristics. The trouble with the first alternative was the lack of a proved twin-propulsion system. Such a plant (the S4G) was being built in the *Triton* at Groton, but that ship would not go to sea until September 1958. Under the heavy pressure to develop the *Polaris* submarine, there was no real choice but to delete the under-ice requirement. Admiral Burke formally approved the ship characteristics on June 17, 1957. The *Polaris* submarine now had the highest priority of any project in the Navy.[21]

Response to the Soviet Challenge

As each week passed, the pace of events quickened. Raborn already had his organization exploring ways of accelerating the development of *Polaris,* but no one in the Navy could have predicted the new incentives which the Soviet Union was soon to provide. The first was the Soviet announcement in late August 1957 that Russian scientists had succeeded in launching an intercontinental ballistic missile (ICBM). Five weeks later, on October 4, came the shocking news that the Soviet Union had placed a satellite in orbit around the earth. The appearance of *Sputnik II* within another month demonstrated the depth of the Soviet capability.

Burke's reaction to this succession of Soviet triumphs illustrated the complexity of the Navy's position in the autumn of 1957. The Navy was in a strong position with *Polaris*. Burke's realization of the importance of missiles, Raborn's ingenuity in getting the project in motion, and Rickover's accomplishments in providing a suitable platform for the missile system all gave the Navy a head start in meeting the demands which the Russian accomplishments would produce. In Congress, Senator Henry M. Jackson's subcommittee on military applications of the Joint Committee on Atomic Energy took the Russian ICBM announcement in August as an invitation to spur the development of both land-based and *Polaris* missile systems.[22]

Burke could welcome this kind of support, as could Rickover and Raborn, but as Chief of Naval Operations, Burke spontaneously voiced a note of caution which reflected the broader interests of the Navy. In suggesting the kind of reply the Navy might make to an inquiry from Senator Jackson, Burke stressed the danger of an over-commitment to *Polaris*. Missiles were vital to national defense, but so were the ships and aircraft of the Navy. He noted how many times in recent years the Navy had answered calls for action with a fleet still consisting largely of remnants from World War II. The Navy in Burke's opinion had to avoid investing a large portion of its procurement funds in any one project, even one as promising as *Polaris*.[23]

Polaris was, after all, more than a hypothetical threat to building a fleet of nuclear ships. The top priority for *Polaris* would certainly delay the completion of nuclear attack submarines and might forestall altogether the construction of nuclear-powered surface ships. With some careful budgetary planning the Navy had succeeded in including the nation's first nuclear-powered aircraft carrier in the 1958 shipbuilding program. But even before the keel of the *Enterprise* could be laid, officials in the Department of Defense and even President Eisenhower himself had raised questions about the wisdom of building such ships. At a meeting of the National Security Council on July 25, 1957, Eisenhower asked whether the services were being sufficiently cost-conscious in procuring new equipment. Were the advantages of a nuclear carrier worth the 50-percent increase in costs over conventional carriers? In view of the administration's determination to hold the 1959 defense budget to $38 billion, the president's questions were more than rhetorical.[24] Perhaps the *Enterprise* was safe enough, but Burke knew he would have to fight hard for a second carrier proposed for the 1959 shipbuilding program.

The administration's economy drive posed a direct threat to the Navy's plans for a nuclear-powered frigate or large destroyer (DLGN). After the meeting with the president, Burke made a note to check with Strauss whether the Commission would continue to support research and development on a destroyer prototype reactor called the D1G, which Knolls was planning to build inside the test sphere at West Milton. Within a few weeks Burke learned from Rickover that the Commission, under heavy pressure for budget cuts, was planning to drop the D1G.[25]

At that moment the D1G was at a critical stage. The project had started in 1956 when Code 1500 began to explore possibilities for a propulsion plant small enough and light enough to fit into a frigate. Theoretically the most promising system appeared to be one using an organic material as the heat-transfer medium, largely because organics would not become highly radio-active and thus would require relatively little shielding. In March 1956 Rick-over had asked Knolls to study the feasibility of a naval organic reactor. After a year's study Code 1500 and Knolls had concluded that an organic reactor had no particular advantages over the pressurized-water type and did present some additional difficulties. Instead they had decided to develop a pressur-ized-water reactor, and in September 1957 the laboratory was just getting organized to design and build the D1G.[26]

To assure continued Commission support of the project, the Navy relied as always on Rickover. In a meeting with Kenneth E. Fields, the general manager, and with Roddis on September 27, Rickover learned that the D1G was being eliminated from the Commission's 1959 budget because the Com-mission had been told that a gas-cooled reactor would be much more prom-ising for this application than a water-cooled plant. Rickover recognized this as part of a proposal (described in Chapter 9) by a group of companies to establish an independent naval propulsion project. Another division of Gen-eral Electric (not associated with Knolls) and the naval architectural firm of Gibbs & Cox proposed to design a nuclear-powered destroyer which the Bath Iron Works would build. Approaching the Navy through Admiral Mumma, these companies were still hoping to set up a nuclear surface ship project independent of Code 1500. In this instance Rickover had only to point to an analysis of the gas-cooled reactor design by the Commission's reactor experts which indicated that this approach was not promising. He also reminded the Commission officials that the Navy had an urgent need for the nuclear-powered frigate and that the Navy was determined to keep the destroyer project in the 1959 shipbuilding program. An exchange of let-ters between the Navy and the Commission reestablished the D1G project at Knolls. Now all Rickover had to do was to keep it in the Commission's 1959 budget.[27]

By this time the fate of the nuclear fleet rested largely in the hands of Admiral Burke. Only he was in a position to defend the Navy's interests before the higher echelons of the administration in the crisis atmosphere which pervaded Washington after the Russians launched their first two *Sputniks*. Burke quickly discerned the deep concern reflected in the discussions of the National Security Council on November 7, 1957. The meeting had been called to discuss the Gaither Report, the work of a committee appointed by President Eisenhower to assess the national security; but the discussion soon turned to the impact of *Sputnik*. Several officials were concerned about the rapid development of the Soviet economy and the apparent willingness of the Soviet Union to invest a very large share of its gross national product in military technology. *Sputnik* seemed to pose a direct threat to the technological superiority of the United States.[28]

The specifics of the defense situation were ominous. The security council and the president learned that the United States had no real defense against high-altitude or low-altitude bombers or against missiles. The only protection the nation had was the Strategic Air Command, which was itself vulnerable to Soviet attack. There were obvious ways of lessening this vulnerability and strengthening the air command, but among all the possible measures for improving the national defense, *Polaris* seemed one of the most valuable. Jerome B. Wiesner, a member of the Army's science advisory committee, told the council that the existing plans to have six submarines capable of launching *Polaris* missiles by 1965 should be tripled to eighteen. In response to suggestions of this type, Secretary of the Navy Thomas S. Gates the following week presented a plan for accelerating *Polaris*. By increasing obligational authority in the 1958 and 1959 budgets by $389 million, the Navy could have three operable *Polaris* submarines instead of one by the end of 1960.[29]

The *Sputnik* crisis and the attractive features of *Polaris* gave the Navy an argument for increasing the size of its nuclear submarine fleet, but there were dangers involved as well. The other services were just as prepared to take advantage of the situation, as Burke discovered at a meeting of the Joint Chiefs of Staff on November 16, 1957. All the chiefs had proposals for a supplemental request to the Congress on the 1958 budget—General Maxwell D. Taylor of the Army for more troops and equipment in Korea, General Thomas D. White of the Air Force for more bombers and land-based missiles. Burke found General White critical of the Navy's proposal to accelerate *Polaris* development and especially to start a second nuclear-powered aircraft carrier in the 1959 program. White opposed the carrier altogether and wanted to hold *Polaris* down to the one submarine the Navy had proposed earlier in the year. Burke would not yield, and finally at a second meeting on November 17 won Army support for two *Polaris* submarines in the 1959 program and acceleration of *Polaris* development beginning in the 1958 budget.[30]

The next day Burke learned the price the Navy would have to pay for these additions. Secretary of Defense Neil H. McElroy told him that the president still had reservations about authorizing a second nuclear carrier, especially because of its cost and vulnerability to submarine and air attack. Burke knew that the chances of saving the carrier were slim, but he had been careful not to bargain it away in the Joint Chiefs' meeting. Now he was prepared to make a deal. The proposal was to postpone the second carrier from the 1959 to the 1960 program in exchange for two additional submarines, one *Polaris* and one attack, in the 1959 program. McElroy was not even able to get a firm commitment from the president for the carrier in 1960, but he did gain Eisenhower's consent to the additional submarines.[31]

This momentary success gave Burke no room for complacency. He thought *Polaris* was probably a better missile system, certainly smaller, and perhaps less costly than any other being developed; but it had a severe disadvantage: even on the most optimistic schedule it would not be ready before the autumn of 1959 and could not be installed on the first submarine before January 1961. The Air Force planned to have the *Thor* missile operational by 1960. Burke predicted a big drive in the Defense Department to reduce funds for *Polaris* in favor of land-based missiles. To avoid this tactic Burke ordered Raborn to squeeze every drop of time out of the *Polaris* schedule. This Raborn was prepared to do, even to the extent of reducing the range of the missiles to be installed in the first few submarines from 1,500 to 1,200 miles and accepting, also on an interim basis, 600-pound nuclear warheads with somewhat less yield than had been specified.[32]

Although Raborn as the director of the Special Projects Office had over-all responsibility for the submarine as well as the missile, Burke personally verified the plans of the Bureau of Ships in a meeting with Mumma and Rickover on November 26, 1957. Burke's concern was that shipbuilding was vital not only to *Polaris* but to other parts of the fleet, and he was still being careful to maintain a balance in fleet composition. Mumma had no hesitation in assuring Burke that the first *Polaris* submarine would be ready by October 1960, but Burke was now hoping the Navy could catch up with *Thor* by having the first *Polaris* submarine ready by late 1959 or early 1960. Mumma could not promise that, but Rickover was confident the reactor plant would be ready ahead of schedule.[33]

The Emerging Nuclear Fleet

By the end of 1957 the Navy's requirements for nuclear ships were beginning to take on the dimensions of a large shipbuilding program. First, there was the task of completing the ships and submarines authorized in earlier years. These included the four attack submarines in the *Skate* class, of which only the *Skate* itself was ready for fleet service. Although the mission of the *Triton* (SSRN-586) as a radar-picket submarine had been largely superseded by the rapid development of high-flying radar-equipped aircraft, Rickover had convinced Burke that the ship should be completed as designed to provide a prompt evaluation of the twin-reactor, twin-screw submarine. Still more than ten months from completion, the *Triton* would face competition for shipyard personnel at Groton as *Polaris* construction gained momentum in 1958.

Of all the nuclear submarines under construction at the end of 1957 none was of greater interest to the Navy than the *Skipjack,* also on the ways at Groton. Not only was the *Skipjack* the lead ship for the new class of fast attack submarines using the S5W propulsion plant; the ship was also, as a result of *Polaris* acceleration, to be the heart of at least the first three *Polaris* submarines. In fact, to make the 1960 completion date, construction of the *Scorpion* (SSN-589), which had been laid down at Groton in November 1957, would stop in January 1958 so that a missile section of 130 feet in length could be installed between the bow and stern sections. The converted ship would be renamed the *George Washington* (SSBN-598), the first *Polaris* submarine. The *Skipjack,* then, would test the design of the S5W reactor plants in the first *Polaris* ships as well as in five other fast attack submarines in that class.[34]

Of critical importance at a later date would be the submarine *Thresher* (SSN-593), to be laid down at Portsmouth in the spring of 1958. Mumma gave the highest priority to construction of the *Thresher* because it would provide the first test of certain nonnuclear machinery modifications and hull improvements which would be used not only in follow ships of the *Thresher* class but also in two new types of submarines. The first included three submarines of the *Permit* class, which would be designed originally to carry the *Regulus* missile. Later, when *Polaris* made *Regulus* obsolete, the three ships

28. Commander James B. Osborn and Rear Admiral William F. Raborn inspect a missile hatch aboard the *Polaris* submarine *George Washington* (SSBN-598).

29. The aircraft carrier *Enterprise* (CVAN-65) after launching on September 24, 1960. The ship is being towed to the dock where she will spend another year being completed. In the background is the yard of the Newport News Shipbuilding and Dry Dock Company. *U.S. Navy*

28

29

would be completed as attack submarines. The second group was the *Ethan Allen* class of *Polaris* submarines, the first class of that type to be designed from the keel up for *Polaris*. Other projects of lesser urgency but still a part of the shipbuilding program were the guided-missile submarine *Halibut* (SSGN-587), the small hunter-killer submarine, *Tullibee* (SSN-597), and the conversion of the *Seawolf* to a pressurized-water reactor plant.

Only two nuclear-powered surface ships had been authorized by December 1957: the guided-missile cruiser *Long Beach* (CGN-9) and the aircraft carrier *Enterprise* (CVAN-65). Both ships would depend upon reactor plants being developed in the A1W prototype which had been under construction at the Idaho site since the spring of 1956; the first reactor in the prototype would not achieve criticality until October 1958. As for the ships themselves, Newport News had just been awarded the contract to build the *Enterprise* at the end of 1957 and would lay down the ship in January 1958. Bethlehem Steel laid the keel for the *Long Beach* at Quincy, Massachusetts, on December 2, 1957.

Future shipbuilding requirements projected in the annual shipbuilding programs and other Navy plans showed that the construction of nuclear ships was only beginning. The original 1958 shipbuilding program, established early in 1957, called only for the three guided-missile submarines in the *Permit* class, the *Tullibee,* and the *Enterprise*. The 1958 supplemental program, which Burke negotiated with the Joint Chiefs of Staff and the administration in November 1957, added the first three *Polaris* submarines. The 1959 shipbuilding program, presented to Congress in February 1958, included five attack submarines of the *Thresher* class, two more *Polaris* submarines of the *George Washington* class, the first four *Polaris* ships of the *Ethan Allen* class, and the nuclear frigate *Bainbridge* (DLGN-25).[35]

Even more startling were the Navy's long-range plans for nuclear ship construction. Early in December 1957 Rickover learned that the Navy was expecting to build thirty-nine *Polaris* submarines in the five years beginning in July 1959. In addition plans called for twenty-six attack submarines, two aircraft carriers, four frigates, three cruisers, and the *Seawolf* conversion. As it turned out, not all these surface ships would be authorized, but the impact of these plans on shipbuilding facilities would nevertheless be severe.[36]

Impact on Shipbuilding

In total perspective the number of nuclear ships under construction began to move up sharply toward the end of the decade. As illustrated in chart 7, the number almost doubled from 1957 to 1958 and continued to show a steady increase to a peak of thirty-seven in 1961.

Accommodating this many ships required a substantial expansion of shipyard capacity. Still concerned about the difficulties of training new yards to meet nuclear standards, Rickover did not want the Bureau of Ships to disperse ship construction much beyond the yards already selected. During the late 1950s the bureau added only one new yard, the New York Shipbuilding Corporation at Camden, New Jersey. Rickover had no choice but to accept the bureau's decision to permit New York Shipbuilding to build attack submarines. Always suspecting the worst, Rickover suggested that the company's low bid for the contract probably reflected a lack of understanding of the amount of inspection, quality control, and engineering which nuclear shipbuilding required. Until the company had proved its ability, Rickover wanted to limit the yard to one ship. Despite this warning, the bureau awarded New York Shipbuilding two attack submarines in March 1959. Only after the most severe difficulties was the yard able to complete the submarines before the end of 1964. Ultimately the yard was shut down.[37]

The Pattern for Production

The magnitude of the nuclear shipbuilding effort by 1959 far exceeded Navy expectations of a few years earlier. The nuclear project now involved six shipyards, two Commission laboratories, three prime development contractors, two procurement organizations, four land-based prototypes, six reactor core contractors, and hundreds of vendors and suppliers. Within the Navy, Code 1500 had only a small part in the shipbuilding process in terms of the total numbers of people involved. A dozen or more codes within the Bureau of Ships, several divisions of the office of Chief of Naval Operations, a number of other technical bureaus, and various organizations within the Department of Defense, such as the Defense Contracts Administration Services, all had a part to play in the shipbuilding process. In fact the broad scope of operations and the complex interrelation of organizations had greatly enlarged the original structure of the nuclear project. All these organizations could and did cooperate to a greater or lesser degree in producing nuclear ships, but many of them, particularly within the Navy and the defense establishment, did so within the context of their regular functions and not as a part of a special project.

Rickover and his associates in Code 1500 knew too much about the Navy and its methods to try to change the system in any far-reaching way. Never presuming to take responsibility in areas beyond the technical competence of Code 1500, Rickover preferred to use—and perhaps on occasion to abuse—the capabilities of existing organizations; only as a last resort did he ignore or by-pass them. At the same time, the tradition of Code 1500 would never permit a passive acceptance of the status quo. Code 1500 still proclaimed nuclear technology to be something unique, demanding the kind of special care and attention which the routine activities of the Navy did not seem to require. And if that appeal did not work, Rickover could always impose his authority as a Commission official acting under the mandate of the Atomic Energy Act.

Although the sum total of activities required to build a nuclear ship encompassed a wide variety of industrial and government organizations in many parts of the nation, the shipbuilding function naturally centered about the shipyards. Here the hundreds of thousands of intricate parts were brought together to create a fighting ship. As we have seen, the expansion of shipyard facilities took place entirely in the private yards. No additional Navy yards followed Portsmouth and Mare Island in constructing new nuclear ships after 1956, and these yards produced only a small percentage of the nuclear ships constructed after 1959.

Supervising the Private Yards

The pattern of supervision in the private yards stemmed directly from the initial experience at Electric Boat. There at Groton, Rickover had made his first impression on the shipbuilding industry. Through Carleton Shugg and other officials at Electric Boat he had imposed a new concept of technological development to a degree unprecedented in the yard's experience. It was no exaggeration to say that Rickover changed the perspective, the standards, and the quality of shipbuilding at Electric Boat. And yet Code 1500 had no formal responsibility beyond the propulsion system itself. In follow ships, the jurisdiction of Code 1500 would not extend beyond the reactor plant. An engineering officer called the supervisor of shipbuilding represented the Navy at the private yards. He monitored construction of the hull and the assembly, inspection, and testing of the entire ship outside the reactor compartment.[38]

On the *Nautilus* and other lead ships at Groton Code 1500 exercised a special influence which grew directly out of the project system. The exceptional feature of these first ships was obviously that they contained nuclear

Chart 7.   The number of nuclear ships under construction increased sharply from two (the *Nautilus* and the *Seawolf*) in 1955 to thirty-seven in 1961. By 1962 more than half the total were *Polaris* submarines.

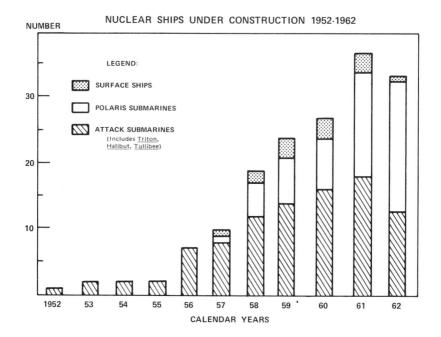

NUCLEAR SHIPS UNDER CONSTRUCTION 1952-1962

NUMBER

LEGEND:

SURFACE SHIPS

POLARIS SUBMARINES

ATTACK SUBMARINES
(Includes Triton, Halibut, Tullibee)

CALENDAR YEARS

Chart 8.   The Pittsburgh and Schenectady offices now reported
directly to the Naval Reactors Branch. A land prototype
was started at Windsor, Connecticut, and three more
shipyards were added.

THE NAVY NUCLEAR PROPULSION PROJECT IN JULY 1958

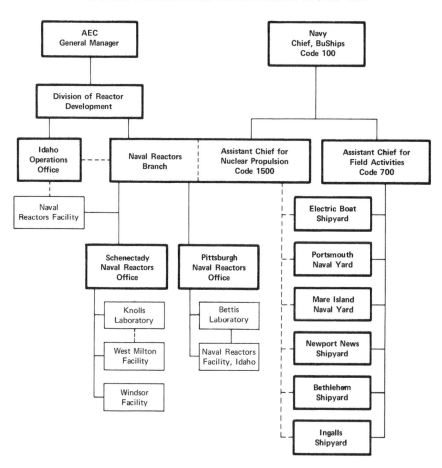

propulsion plants. In that sense Rickover's personal representatives at Groton could be expected to possess an influence equaling if not surpassing that of the supervisor of shipbuilding. During those critical early years from 1952 through 1956, Rickover was careful to assign only his most experienced and aggressive officers as his Groton representatives. The first two officers to hold that position—Commanders Samuel W. W. Shor and Arthur E. Francis— had both completed the graduate course in nuclear engineering at the Massachusetts Institute of Technology and had worked in Code 1500. Both had the tenacity and drive to ferret out the information Rickover wanted, and they were usually capable of convincing the contractor that they spoke for Rickover.[39] Rickover often found it necessary to convey his wishes in person, but even at a distance he could set the style and pace of activities in the yard. Others could request; Rickover could demand and get immediate action.

As the number of nuclear ships under construction increased, there was a natural tendency within the Navy and the shipyards to take nuclear propulsion for granted and to concentrate attention on the thousands of other details which did not involve the propulsion plant. This tendency was especially strong when the yards began building *Polaris* submarines, for which, on Burke's orders, the Special Projects Office had supreme authority. Although Rickover did not openly challenge Raborn's role, he never permitted Electric Boat or any of the other shipbuilders to overlook his special interest.

At Groton, Shor and later Francis were responsible for keeping the interests of Code 1500 in the forefront of activity. This they accomplished by sheer energy and cussedness. Both officers spent literally all their waking hours in the shipyard; the time or day made no difference. Constantly looking for the symptoms of trouble and signs of weakness, they crawled through inaccessible portions of the hulls, haunted the shops, followed foremen on the job, witnessed every critical installation no matter at what hour, asked questions, and made notes—endless notes which became the substance of daily telephone reports direct to Code 1500 in Washington. No technical detail was too small if it could be the forerunner of a significant deficiency. Never satisfied with passive observation and inspection, Rickover demanded imaginative probing and creative analysis, an untiring quest for evidence of inevitable errors and oversights.[40]

The result was that Rickover and Code 1500 often knew more about the job than the bureau personnel assigned at Groton. They did not have to wait for reports from Electric Boat to discover what was happening or not happening in the yard. Code 1500 usually heard of new developments at Groton before any one else in Washington, or even the company officials in Groton, could receive reports. The ultimate in fast reporting occurred some years later when one of Rickover's representatives in another shipyard noticed a column of smoke rising in the yard while he was making his daily telephone report to Rickover. Quickly terminating the call and placing another to the president of the company at the yard, Rickover was able to report the fire before the shipyard alarm sounded. This was an accomplishment which Rickover with obvious amusement enjoyed recounting years later. In a more serious vein, it illustrated an important source of Rickover's influence in the yards. Knowledge was a source of authority, whatever the organization charts and formal descriptions of duties might indicate.

In the other private yards the pattern of surveillance established at Groton did not apply. The situation at Electric Boat was unique in that the yard had no significant work other than building nuclear submarines. The company was thus almost wholly dependent upon Rickover and the Navy. In the other yards, particularly Newport News, Rickover did not initially have such leverage. As one of the largest shipyards in the nation, Newport News would continue to build a substantial number of conventional ships for the Navy, a significant number of commercial vessels, and a large volume of nonmarine equipment. Here Rickover's tactic was one of isolating the nuclear work from the rest of the yard. Long before Newport News began building nuclear submarines, while the company was working with Bettis on the A1W prototype, Rickover had insisted that all nuclear activities be concentrated in the atomic power and atomic installation divisions. These divisions reported to retired Rear Admiral Norborne L. Rawlings, a former engineering officer in the Bureau of Ships who was now executive vice-president of the company.

Early in 1958, when the yard started construction of the *Enterprise* and the *Shark,* its first submarine, Rickover ordered Commander Crawford, his representative at Newport News, to review the responsibilities of the atomic power division with Rawlings and other company officials. Crawford was to insist that the division be the center of all activities related to nuclear ship construction, including the technical adequacy of the entire reactor plant. The division would schedule all design, procurement, plans, construction, and testing, and would serve as liaison with other divisions of the company, between the shipyard and the Navy, and between the shipyard and all reactor plant contractors. Once again Rickover was requiring the shipbuilder to assume full responsibility for each phase of the project. This responsibility was to be focused on one man, in this case the head of the atomic power division.[41]

Not all shipbuilders were eager to accept the responsibility which Rickover thrust upon them. There was always the danger, in Rickover's mind, that the shipyards would fall back on Code 1500 for technical support, especially when they ran into difficulties. He sensed this implication in an exchange of correspondence late in 1957 with Monro B. Lanier, the vice-chairman of the board of Ingalls and principal source of the company's interest in nuclear shipbuilding. There was no requirement, Rickover observed, for the company to come to him for advice; nor did he intend to waste time by giving his opinion after the company had decided upon a course of action. Ingalls had a right to expect prompt delivery of the documents and material called for in the contract. The company would receive the same consideration that any other private shipbuilder could expect while constructing a nondevelopmental follow ship. Lanier disclaimed any intention of using Code 1500 as a crutch. What troubled him was the impression in Rickover's letter that Ingalls would be expected to meet every difficulty on its own. Lanier wrote Rickover that he had come to appreciate "more and more, as we proceed, the complexities of the problems and the difficulty we face. I am doing my utmost to develop a competent technical staff which would not have been possible within this time without your help."[42]

Rickover expected the shipyards to take responsibility, but that did not mean he would let them blunder into trouble. Early in 1959 Edward L. Teale, president of the New York Shipbuilding Corporation, expressed his reluctance to segregate the nuclear work under a project officer. Because he could see no sharp distinction between the nuclear and nonnuclear portions of the ship, Teale could not define the duties of the project officer. A visit to Portsmouth did not help. Teale discovered that much of Turnbaugh's work as nuclear power superintendent stemmed from the fact that Portsmouth as a submarine yard had never built steam-propelled ships. About 60 percent of Turnbaugh's functions were already being performed by experienced divisions at New York Ship. Despite these arguments, Teale had no success in convincing Rickover that the project system was unnecessary. In March Teale agreed to set up a separate group under one engineer who would have no other assignments and who would report directly to Teale himself. Teale assured Rickover that the project leader would remain in this position at least until the first two attack submarines were completed at New York Ship or until both the company and the Navy agreed to terminate the contract.[43]

Teale's reaction was to be expected from an official of a well-established shipyard. In the past it had been possible for a shipyard to accommodate itself to new developments without a drastic reorganization. But past techno-logical improvements in naval architecture had been evolutionary compared to the revolutionary impact of nuclear power. In insisting on the project orga-nization and an elaborate training program for shipyard personnel, Rickover and his associates were simply trying to prepare the new private yards for the unprecedented difficulties which lay ahead. Despite the warnings and even the earnest efforts by the yards to prepare themselves, none found it easy to make the transition. Newport News, by exercising an uncommon amount of ingenuity and effort, was able to make some original contributions to nuclear shipbuilding relatively soon after entering the field. Ingalls, Beth-lehem, and New York Ship all shared the common experience of early opti-mism, a growing sense of concern, and finally a desperate feeling of inade-quacy as technical difficulties mounted and schedules slipped farther behind. Rickover and Code 1500 did what they could to help, even to the extent of encouraging Bethlehem to hire Laney and New York Ship to hire Dunford after these two veteran officers retired from the Navy. These were exceptional steps, but the task was more than one man could hope to accomplish. Only the yards themselves could resolve the problems they faced, and solutions often seemed to require an unprecedented and even unreasonable amount of effort.[44]

The Shipyard Representatives

In the initial experience at Electric Boat both the local representatives and Code 1500 had a part in imposing Rickover's imprint on the company and the yard. Their constant probing of every facet of shipyard activity, their seemingly endless reports, and their daily telephone calls established a pat-tern which Rickover would use at all the private yards. But the representa-tives of Code 1500 were more than mere observers and reporters; they were also spokesmen for Rickover. For this purpose the Code 1500 representatives had to have direct access to the president of the company and the senior yard management. These conferences were not an occasion for exchanging cour-tesies over coffee, but sessions for plain talking. A constant threat to the sys-tem was the familiarity and even sympathetic understanding which developed from the frequent contacts between the Code 1500 representatives and the shipyard officials. Subtly and almost imperceptibly the representative often could find himself drifting into the company's perspective. There was also the danger that his very knowledge of the yard and its problems would lead the representative to take matters into his own hands and to attempt to settle problems without bothering Rickover with them. Rickover was constantly alert to signs that his representatives were losing their independence, and he warned them to ward off the temptation to become "a company man" or a "good guy." When Rickover received an especially perceptive report from his representative, one demonstrating a courageous show of independence in a difficult situation, he sometimes sent the letter to a few of his key staff. One such report came from a young lieutenant, William Wegner, who found himself lecturing the president and vice-chairman of the board—one old enough to be his father, the other to be his grandfather—on the failings of their yard. Rickover distributed the report so that his staff "could see how effective such meetings regularly held can be. Be sure you have something specific to say and say it clearly. The meeting should be held in *your* office; you are the *customer*."[45]

The Bettis Resident Engineer

Indispensable to the mass production of nuclear ships was the use of standardized propulsion plants—the S5W for most submarines and the A1W for the *Enterprise* and *Long Beach*. Despite some modifications such as those which made it possible to distinguish the *Thresher* class from the *Skipjack*, all of the attack and *Polaris* submarines laid down after 1956 used what was fundamentally the S5W plant. Rickover insisted that Bettis and PAD see that all power plant components for an increasing number of ships were delivered on time and in good condition. Under the circumstances Bettis had no choice but to station resident engineers at each of the yards. In the past Westinghouse and other manufacturers of large machinery had sent engineers to shipyards to oversee the installation and test of their own equipment, but these had always been temporary assignments. As the number of nuclear ships under construction at Electric Boat increased, the presence of Bettis engineers at Groton became more a matter of residence than temporary assignment. The selection of additional yards made it all the more important to have experts on the propulsion plant available in the yards, not only to supervise the delivery and installation of components but also to provide technical assistance.[46]

The Bettis engineer at the shipyard was concerned only with the reactor plant. Although the engineer was a Bettis employee, the shipbuilder learned to recognize him as part of the Rickover organization. The resident engineer

lived by the manuals and written instructions from Bettis. Whenever deviations seemed necessary, he called Bettis and confirmed the change in writing.
A large part of his job was inspecting components as they arrived at the yard.
This task kept him in close contact with both Bettis and PAD, but he also
worked with engineers from the major vendors when their products were
found defective. As the plant was completed, the Bettis engineer participated
with the shipyard and the crew in drawing up test procedures. Later, as the
fleet of nuclear ships grew in size, the Bettis engineer acquired additional responsibilities for refueling. (These duties will be described in chapter 11.)
The functions of the resident engineer were fundamentally the same whether
he was assigned to a Navy or private yard. In either case he worked closely
with the Code 1500 representative.[47]

As in all activities in which Code 1500 had an influence, the organization
of the shipyards changed with the workload and the abilities of personnel.
Simply in the interests of efficient administration, Bettis attempted to stan-
dardize the responsibilities and functions of the resident engineers at the various yards, but such efforts could never be completely successful.[48] Mounting
pressures for ever-increasing effort seemed to outstrip attempts to establish
routines. At any particular time and place, the shipyard organization normally reflected little more than the pressing needs of the moment. But, whatever might be the momentary ramifications of organization, responsibility for
activities in the yard continued to rest with the shipyard's project manager,
the Code 1500 representative, and the Bettis resident engineer. All had different duties, but all worked together on the same project. And all were ultimately answerable to Code 1500. Through these three channels—the shipyard, Code 1500, and Bettis—Rickover asserted his unmistakable influence
on shipbuilding activities. Technically his responsibilities stopped with the
nuclear power plant; actually his presence permeated the life of every yard
where nuclear ships were being built.

Quality Control— "The Never-Ending Challenge"

For decades inspection had been an integral part of the shipbuilding process;
but as the technology of weapons, communications, and propulsion systems
became more complex, especially after World War II, the Navy had found
it necessary to devise more elaborate inspection procedures and systems. By
the 1950s the Bureau of Ships had two types of organizations to inspect
equipment destined for naval vessels. One was headed by inspectors of ma-

chinery who were stationed at the few key industrial plants which manufactured propulsion machinery, including boilers, generators, and other large items. The other consisted of inspectors of material who worked from regional offices and visited factories which furnished smaller products. Within the shipyards a variety of groups performed the inspection function. In the Navy yards the planning department prepared test instructions and data for some types of work, and the production department—which did the actual construction—had an inspection division. In private yards the task was one of the many that fell to the supervisor of shipbuilding. At Groton the Navy had the usual inspection organization, but Rickover had never been confident that either the Navy or Electric Boat personnel had the technical competence to inspect nuclear components and monitor their installation in the *Nautilus*. One of Shor's duties was to keep Code 1500 informed of the latest developments on inspection systems at Groton.[49]

Code 1500 found many reasons to believe that the inspection system was inadequate, but it took one dramatic incident to bring this point home to company and Navy officials. On the night of September 16, 1954, the *Nautilus* lay alongside the pier, a few weeks from sea trials. Shor was aboard, watching the officers and men in the ship's company and Electric Boat personnel conduct tests of the steam system with steam from the dock. Shortly before midnight a small pipe burst, filling the reactor compartment with steam. The engineering officer on duty tripped the boiler safety valves and the engine room crew cut off the shore supply. In outward appearance the incident was minor. Personnel injuries were slight and of course there was no radiation hazard. Electric Boat and the Navy had been responsible for inspecting the piping in that part of the ship, but Rickover quickly took the initiative in probing the cause.[50]

Investigation revealed that the situation was more serious than first believed. Specifications had called for seamless pipe, but ordinary stanchion pipe had been installed. Even worse, there was the possibility that the same mistake had occurred in the Mark I and Mark A, since Electric Boat was responsible for constructing the same parts of those facilities. After two weeks of investigation, Shugg was forced to conclude that there was no positive way of knowing which of the installed pipe in the smaller dimensions was welded and which was seamless. The only thing to do was to rip out all the suspect pipe.[51]

The *Nautilus* incident triggered a series of events. Rickover tightened up his own procedures and called for a permanent marking stamped into the surface of pipe so that pieces installed could be readily identified. Electric Boat assumed full responsibility for the unacceptable pipe in the *Nautilus* and the two prototypes. The company also assigned an outside consultant to make an independent survey. The bureau disseminated information on the incident to all yard personnel and set up its own investigation under Captain Philip W. Snyder, commander of the Boston Naval Shipyard.[52]

Snyder completed his report in December 1954 and sent it to the bureau. The *Nautilus* incident he attributed mainly to Electric Boat for failing to control the issue of carbon steel pipe from its warehouse department; second only to the company in line for blame was the supervisor of shipbuilding at Groton, who failed to detect the substitution of the wrong piping. But much more significant was Snyder's criticism of inspection methods in private and Navy shipyards generally. He found them inferior to other types of industry, and the Navy yards worse than private yards. One section of the report cited some glaring examples of the improper use of materials, which in a few cases had resulted in fatalities in Navy yards. Snyder recommended intensive education of personnel in the shipbuilding and repair industry and establishment of quality control systems in all shipbuilding and ship repair facilities under direct control of the Bureau of Ships. There was an urgent need in Snyder's opinion to replace perfunctory checks on shipbuilders with searching surveillance of actual shipyard performance. Despite the obvious faults which Snyder's report demonstrated, the bureau did not establish a quality control engineering office until 1959, and by that time several more costly and time-consuming mistakes had occurred.[53]

The term "quality control" was intended to draw a distinction between the kind of surveillance Snyder was recommending and the traditional activity called "inspection." Whereas inspection was simply a physical check aimed at weeding out substandard items, quality control attempted to determine the step in the manufacturing process at which the defect occurred. It would then be possible to prevent the defect at its source rather than try to find all the defective parts produced. Quality control also differed from inspection in its implications for management. Inspection was usually a fairly local matter carried on in a shop or department, but quality control brought under scrutiny all the operations of a plant, or even a complex of plants. Like many management techniques the origin of quality control was hard to pinpoint. However, in the years immediately after World War II, it had become a recognized consultant field, complete with textbooks, charts, and jargon.[54]

In the production of nuclear power plants Rickover had always exercised his own version of quality control. As described in chapter 9, he insisted that the reactor contractors develop exacting specifications for nuclear equipment and then demanded that the suppliers meet them. Quality control at Groton was originally the responsibility of Electric Boat's operations department. But as the number of ships under construction increased, quality control began to slip. Rickover then decided that the company should transfer all quality control, inspection, and welding engineering activities to a separate quality control department, with the understanding that the Code 1500 representative would keep close check on the company's performance. Quality control was not something a shipyard could take in stride, particularly when the customer would have direct access to the reports of deficiencies. Electric Boat—whose entire business now depended on building nuclear submarines —had no choice but to comply with Rickover's demand. Even so, the company did not formally establish a quality control section until October 1957.[55]

At Newport News Rickover set out to make quality control a part of the shipyard function from the beginning. He did not want the yard to depend upon Navy inspectors. During the first six months of 1958 Newport News had laid the keels for three nuclear ships, the aircraft carrier *Enterprise,* the attack submarine *Shark,* and the *Polaris* submarine *Robert E. Lee.* As the hulls of these ships took shape on the bank of the James River, Newport News was planning for the inundation of components from the suppliers. In August 1958 Rickover took Rawlings, the executive vice-president, to Idaho to inspect the A1W prototype which the company was helping Bettis design and build. The immensity and complexity of the prototype with its two mammoth reactors gave Rickover a good opportunity to expound on the importance of quality control. Rawlings was reluctant to accept the idea, not because he opposed quality control but because it would further complicate his already elaborate organization. In the end Rawlings relented and agreed to find a man to head the new department.

The choice of Richard S. Broad as head of the quality control department at Newport News indicated what Rickover expected the job to be. Broad was a graduate of the Newport News apprentice school and the son of a former company official. He had a master's degree in marine engineering from the University of Michigan and had attended the Oak Ridge school of reactor technology when Newport News accepted a study contract for the first nuclear aircraft carrier back in 1949. When the Eisenhower administration canceled that contract, Broad had gone on to other jobs at Newport News which had given him practical experience in cost estimating, contracting, and procurement. Commander Crawford, who was now serving as the Code 1500 representative at Newport News, reported Broad to be intense, hard-working, intelligent, and thorough. Equally impressive to Rickover and Crawford were Broad's three years of experience on the waterfront as a machinery installation subforeman. It was a dirty, grimy job involving all the practical realities of shipbuilding.[56]

One of Broad's first assignments was to draw up a description of his responsibilities. He checked the draft with Crawford, who sent it to Code 1500 for comment. The final version established Broad as the quality inspection engineer reporting directly to Rawlings for quality control, inspection, and health physics for nuclear propulsion plants. Broad was to propose and carry out the measures necessary to assure that the propulsion plants were fabricated and installed in strict accordance with approved procedures and design. If he discovered violations he could halt the work. He was responsible for the inspection of components and materials coming into the yard, in the shops, and on the ships. He could go at any time into any part of the yard doing nuclear work and expect complete cooperation.[57]

Broad's sweeping mandate disturbed some company officials at Newport News. They objected to the new division because they believed the company already had a sound inspection system. They also feared that the transfer of quality control to a special division would destroy a sense of responsibility in the shops. Just below the surface was a current of resentment that Rickover was interfering in company affairs. By doing a good job with Rickover's support, Broad consolidated his position in the company. Opposition to the new organization slowly disappeared over the years.

The quality inspection group was organized around seven sections covering health physics, quality control, incoming components and materials, piping, machinery, electrical work, and shielding. The titles of the section and the number of people assigned to them changed as needs dictated. Above all Broad sought the best men available; he set out consciously to break the industry habit of assigning to inspection men who were not capable of doing the work they inspected. Broad wanted top-flight technical personnel who were respected by the men in the yard, but he greatly underestimated the numbers he needed. At first he believed he would require about forty people, but by early 1960 he had reached 150 and was still looking for qualified men. When he could no longer find the type of man he needed at Newport News, Broad began hiring retired Navy warrant officers and chiefs, who had years of experience in their technical specialities.[58]

The problems Broad encountered at Newport News were no different from those met by his counterparts at Electric Boat, Ingalls, Bethlehem, New York Shipbuilding, and the Navy yards. Nuclear propulsion involved shipbuilders in handling materials with which they had little experience. Errors could cause at the worst the loss of the ship and the crew; they could also cause the failure of a component located in a high radiation area where accessibility was impossible during reactor operation. Extremely high standards rigorously upheld were essential, and these in turn meant that formal written procedures were needed. Both, however, had to evolve from experience. The technical data necessary for setting standards could come from the laboratories, as did the drafts of manuals, but Code 1500 in Washington was the only group authorized to approve them.

Deficiencies were much the same from one yard to another although their frequency might vary. The most common faults were improper welds and incomplete or inadequate tagging on instruments and components. An ever-present danger was the unintentional use of conventional materials in the nuclear portions of the plant. A special problem was cleanliness. It was not easy to convince manufacturers and shipyard workers that a small metal chip or a bit of wire could irreparably damage a primary coolant system.

The quality control system also raised management issues. The imposition of a special quality control group meant that vendors or shipyard personnel had no opportunity to investigate and correct deficiencies before they were reported to senior officials or, even worse, to Code 1500 in Washington. There was little chance to explain unusual situations or to cover up foolish oversights. Sometimes the quality control system disrupted established procurement methods when a pattern of deficiencies pointed to a company which had been a regular and trusted supplier of the yard. Sometimes manufacturers challenged unfavorable reports or even withdrew from bidding on contracts for nuclear equipment; others welcomed a frank appraisal of their products as an opportunity to improve their production methods. But in either case Rickover demanded relentless attention to quality control from his own representatives and the quality inspection divisions in the yards.

Another part of the quality control system in the shipyards was the Code 1500 audit. At first an informal check, by the late 1960s the audit had become a regular procedure for auditing the quality of nuclear work. At a prearranged date several Code 1500 personnel, augmented perhaps with specialists from the laboratories or another contractor, visited one of the yards for several days. Before the visit the inspection team studied reports from the yard management, the quality control divisions, contractors, and the Code 1500 representative. For several days the team walked the yard, talking with supervisors and workers and attempting to find the underlying causes of defects. The team had its own quarters and temporary offices isolated from the rest of the yard. In some instances the team had a private telephone line which did not go through the yard switchboard. After several days of intensive work, the team members drew up a preliminary report which they discussed with the yard officials and the Code 1500 representative. The bulky reports spared no detail in citing specific discrepancies between practice and procedures. Late on the final day of the visit Rickover flew in from Washington. On the way from the airport to the yard the senior team member described some of the difficulties and suggested possible solutions. In a frank and sometimes brutal session with the yard management, Rickover probed deeper to discover the often hidden or unrealized source of the deficiencies the team had detected. Rickover recognized that errors would always happen and that individuals through ignorance or pressure to meet schedules would try to act unilaterally. His purpose was to keep the inevitable to a minimum. Quality control was, as Rickover told an audience at the National Metal Congress in 1962, "The Never-Ending Challenge."[59]

Tests and Trials

Launching a submarine was usually a colorful ceremony in which high dignitaries of the Navy, the Department of Defense, and the political world watched the wife of a distinguished citizen swing a champagne bottle against the bow of the ship. If all went well, the bottle shattered and the vessel slid down the ways—slowly at first but with gathering momentum as she hit the water. For those who were constructing nuclear submarines after 1957 the occasion was only a momentary respite in an otherwise unbroken hum of activity. As soon as the submarine could be moored to the wet dock, workmen swarmed aboard and the ship was once again enmeshed in a tangle of cable while brilliant sputtering arcs of light showed that welding had begun again. Yet launching made a difference. Although the massive hull seemed inert in the water, the changing curves of the mooring lines were reminders that the ship was afloat and well on the way to becoming part of the fleet. Before this could happen, the shipyard would have to complete the installation of the reactor plant and prepare for a grueling series of dockside tests and trials.[60]

The same shipyard workers, engineers, and Navy personnel who had scrambled over the ship on the ways now climbed over each other in the crowded compartments of the floating vessel. Shipyard employees dominated the scene in terms of numbers, but representatives of the reactor contractor and other major suppliers were always on hand when major pieces of equipment were being installed. The Code 1500 representative and other engineering officers from the bureau haunted the ship. Of growing importance as construction progressed was the "ship's force," the group of Navy officers and sailors who would eventually man the ship at sea.

The prospective commanding officer, usually a lieutenant commander or commander, arrived at the yard with a few engineering officers and enlisted men before the ship was launched. Following the practice Rickover had established in building the *Nautilus* and the *Seawolf,* the ship's force participated in the actual construction and testing of the plant. In the prenuclear submarine fleet the ship's force did not usually come aboard until the vessel was almost complete. During sea trials the ship's force only observed operation of the propulsion plant by shipyard personnel. By contrast, the officers and crew of a nuclear ship were expected to be generally qualified on propulsion plants when they arrived at the shipyard. During their year of training in the yard they learned to service and operate every piece of equipment in the propulsion plant. All had completed at least a year of intensive training at the nuclear power schools (which will be described in chapter 11). The ship's force was on hand to witness the installation of each piece of equipment in the ship, and the officers and men were expected to take a personal interest in the quality of workmanship on which their lives would depend. For this reason, the prospective commanding officer could be one of the most demanding individuals in the yard.[61]

When the reactor plant neared completion, a new organization, called the joint test group, was established to coordinate the testing of the propulsion plant. The group consisted of several senior engineering specialists under the direction of the shipyard test engineer. Other members were the Code 1500 representative, the Bettis or Knolls resident engineer, and in private yards the supervisor of shipbuilding. The prospective commanding officer attended all sessions, although he was not a member of the test group.

The first task of the joint test group was to read and discuss every paragraph in the detailed test specifications, which had been drafted by the reactor plant contractor and approved by Code 1500. The test group could not authorize deviations from the test specifications but was expected instead to devise procedures that would make it possible to perform the complicated series of tests on that particular ship. Since the circumstances surrounding construction activities were never the same for any two ships, no two test schedules were ever the same. Some adjustment in the schedule was always necessary to avoid construction bottlenecks or the temporary lack of equipment and to take the best advantage of available manpower. The joint test group made certain that each participant understood his job and that every step in the test conformed with the approved documents.[62]

For the actual tests, the ship's engineering force manned the controls aboard the vessel. In a submarine, there was not enough space below decks to accommodate all the specified testing gear. Thus the crew remained below while others monitored some of the test instruments in a shack topside. When each component of the propulsion plant had passed a rigorous series of inspections, the entire plant was tested as a whole, first by filling the primary system with cooling water and eventually raising it to operating pressures and temperatures with shore-based power. If the primary system operated properly, the reactor was loaded with fuel and, with all control rods inserted, the system was checked again with hot and cold runs. The slightest expression of uncertainty or disagreement by any member of the test group could bring the complex procedures to a halt. At each step in the tests, the procedures called for dozens of checks and measurements before the next step could begin. Depending on the performance of the plant, the preliminary tests could take days, weeks, or months. Finally the reactor was brought to criticality, and then, usually within a few days, to full power.[63]

For the commanding officer a major change in status came when the sea trials were only a few weeks away. He then became the "officer-in-charge," and although the ship still belonged to the builder, he was finally responsible for the vessel, its physical integrity, and the safety of the personnel. The submarine was operated insofar as possible as a Navy ship and more than ever the tests were those of both the crew and the vessel operating together.

The first time the prospective captain had the ship to himself was during the "fast cruise." For about four consecutive days the ship was sealed and moored "fast" to the dock. In all respects the submarine was ready for sea with her full crew, stores, and essential spares. No one except the assigned officers and crew was on board. The ship's force conducted drills and operated the equipment—including the propulsion plant—insofar as possible as if they were at sea. The captain had an opportunity to check the condition of his ship and the training of his officers and men. His only communication with the shore was by telephone, by which he reported at least daily to Rick-over. The purpose of the fast cruise was to make certain by actual operation that the ship was ready for sea trials. Failure of equipment or crew during the fast cruise was sufficient cause to postpone further trials.[64]

The readiness of the ship at the time of the fast cruise usually made it possible to begin actual sea trials a few days later. During the sea trials the ship was crowded with "riders" in addition to the full crew. The typical complement of riders on a submarine built at a private yard and containing an S5W propulsion plant included several engineers from Code 1500, the Code 1500 representative at the yard, senior company officials and engineers, the supervisor of shipbuilding, a few officers from the technical desks in the bureau, a captain from the submarine force, the Bettis resident engineer, some contractor and vendor personnel, and Rickover himself.

The presence of these high-ranking officials made the sea trial anything but a routine experience for the crew, but Rickover's presence made the greatest difference. Even by 1957 he had become something of a legendary figure in the Navy, and his arrival aboard ship visibly affected the entire crew. Quickly boarding the ship in civilian clothes Rickover customarily climbed to the bridge to observe the departure. Later Rickover inspected the propulsion plant. During most of the trial he worked in his cabin and appeared in the attack center or the maneuvering area (as the reactor control compartment was called) only during critical tests.

One of these was the submerged emergency stop, an evolution which placed a heavy strain on the propulsion plant. The signal for a crash stop came when the ship had been steaming below the surface at full speed for several hours. Usually there was a flurry of disciplined activity as the men at the control panels spun steam throttle wheels and manipulated switches so that the plant could answer the order for full speed astern. Rickover and a few of his technical group crowded into the small maneuvering area to observe the instruments or performance of the crew. Occasionally Rickover might reprimand one of the officers for improper procedures in giving commands or in crew response. Aside from the dials it was usually difficult to tell that the submarine was coming to a stop.

The sudden loss of power was another important test. The intentional tripping of an alarm shut down the reactor and left the entire ship on emergency power. By following proper procedures the well-trained crew could bring the reactor back into operation without delay. Obviously it was vital to avoid a long shutdown of the reactor plant. Rickover watched every move of the men

30. Rear Admiral Rickover and Lieutenant Dean L. Axene, executive officer of the *Nautilus,* as the ship prepares to leave the dock on sea trials in January 1955. This was the first of almost sixty trials in which Rickover personally observed during the next decade.

at the control panels, ready to criticize the slightest hesitancy or careless action. Each engine room watch performed the emergency tests under Rickover's practiced eye. Occasionally Rickover would tap the engineering officer of the watch on the shoulder and declare him "dead." That left three young enlisted men, some of whom were going to sea for the first time, to handle the critical procedure alone.[65]

The sea trial was, in Rickover's opinion, a true test and not just a simulated exercise of the ship and her crew. The emergency drills were performed on a real ship at sea, below the surface of the ocean. There was an actual, if remote, possibility that a crew error or an equipment failure could endanger the ship. Sometimes, though rarely, the unexpected occurred. During the sea trials of the *Triton* the submerged emergency reversal brought the ship close to danger. All submarines were somewhat unstable in a full-power reversal and had a tendency to "squat," or sink deeper stern-first as the ship backed down. Because the *Triton* was unusually long and narrow, this instability proved especially strong. As the ship lost her forward motion and the now-reversing screws began to bite the water, the vessel started to oscillate and sink by the stern. Quick action by the crew regained control of the ship.[66]

Moments like this sometimes exposed the strong sense of responsibility which Rickover felt for the ship and her crew. If any equipment directly affecting the safety of the ship appeared to be operating abnormally, Rickover investigated the matter himself on the spot. He alone would decide whether it was safe to proceed. Although he would never lightly dismiss any fault or malfunction, he would not tolerate an overly cautious approach, especially if it threatened continuation of the trial. He maintained that a naval vessel had to be ready to perform its mission under any circumstances and that it was the crew's responsibility to find a way to operate the ship safely even under less than perfect conditions.

On one trial there occurred a steam leak which the engineering officer believed would worsen during the full-power run. Rickover disagreed and backed up his opinion with action. He cleared the area of all personnel and sat down to watch the leak himself while the trial continued. On another trial a leak appeared to develop in a double-hatch leading to the deck. Rickover was convinced that the hatch was not leaking and that the water found between the hatches had resulted from an improper alignment of valves. When the commanding officer appeared reluctant to proceed, Rickover climbed into the space between the hatches with a flashlight while the ship submerged. Had there been a leak, there would have been no way to remove Rickover from the space between the hatches until the ship surfaced. Now more concerned than ever, the commanding officer took the ship down while Rickover proved his point.

These incidents were not foolish heroics. Rickover did not take such risks needlessly. He was trying to convey the idea that on a nuclear-powered ship, particularly a submarine, the officers and crew had to bear full responsibility for their actions. They were not involved in a simple drill, but in a deadly serious enterprise on which their lives might depend. Rickover's own presence on nearly every trial and his obviously deep personal concern about the performances of the ship and her crew instilled in the men of the nuclear fleet an attitude which a thousand hortatory letters could never have evoked.[67]

The Area of Influence

By the early 1960s Rickover had succeeded in extending his influence into many areas which were customarily administered by other codes in the Bureau of Ships. By insisting upon the unique requirements for building nuclear ships, Rickover had established his own representatives within both the Navy and private yards. Although formal authority for naval ship construction continued to rest with the Bureau of Ships, Code 1500 was able to exercise an unusual amount of leverage simply by out-working and out-maneuvering other organizations. It would have been an overstatement to maintain that Rickover actually ran the yards building nuclear ships, but it was no exaggeration that he left a lasting impression on every yard in which nuclear ships were built. In the private yards especially he acquired a practical, if not a formal, authority which was more clearly acknowledged than any other in the Navy. Rickover was perfectly willing to leave the hundreds of housekeeping details to the Navy bureaucracy, but on the really crucial issues affecting nuclear ships neither the Navy nor the private builders would challenge his actions without carefully considering the possible consequences. Thus Rickover achieved a horizontal extension of his influence which went far beyond the formal limits of his original charter.

Fleet Operation and Maintenance

In the construction of nuclear ships, Rickover's nuclear power directorate (Code 1500) had a clear role based upon its responsibilities for the propulsion plant. Through its ties with the Commission and its laboratories, Code 1500 provided the essential technical knowledge and experience which made nuclear propulsion possible in the Navy. As we saw in chapter 10, Rickover had actually extended his authority beyond the formal limits set by the Bureau of Ships and the Chief of Naval Operations. As a result no phase of nuclear ship construction from preliminary design to sea trials could escape the scrutiny of Code 1500 and its contractors.

With the completion of sea trials, however, further controls by Code 1500 might have seemed no longer necessary. At this point the nuclear ship entered the fleet and became subject to the authority of the Chief of Naval Operations and the operational commanders. Matters of deployment, performance, and discipline were commonly recognized as the proper concern of the operational rather than the technical arm of the Navy. Rickover himself accepted this distinction, but he did not intend to permit the operating forces to debase or ignore the unusually high standards he had imposed on the operation of the prototypes and the *Nautilus*. From this premise Rickover moved on to establish during the late 1950s an influence in operational matters which was unprecedented for a restricted line officer in a technical bureau.

Seeds of Conflict

In extending his authority over the fleet, Rickover was running the danger of reopening an old dispute between those officers who were engineers and those who conned and fought the ships. About two decades before the Civil War, Secretary of the Navy Abel P. Upshur established an engineer corps to operate and maintain the steam engines coming into use in the Navy. The engineers were a separate group aboard ship. They wore distinctive insignia and their duties were confined to the engine room. Command of the vessel, and of all the personnel on board, remained in the hands of the line officers. The two groups were in conflict from the beginning; the line officer was contemptuous of the greasy engineer, the engineer disdainful of those who were in command but knew nothing of machinery. In the period after the Civil War the bickering grew in intensity as machinery aboard Navy ships grew more complex and as engineers demanded more recognition. Attempts to overcome the differences were unsuccessful until 1899, when Congress enacted legislation prepared by Theodore Roosevelt when he was Assistant Secretary of the Navy. The intent of the law was to end the quarreling over rank and responsibility by doing away with the engineer corps. Provisions of the law were complicated, but the basic idea was that all officers aboard ship—with the minor exception of such specialists as doctors and chaplains—would have military responsibilities. The old distinction between the line and engineer aboard ship vanished.[1]

In certain respects the amalgamation of the engineer corps into the line worked well. There was initial difficulty in getting line officers actually into the engine room to work as engineers, and there were some painful episodes —one an engine room disaster that cost several lives. Although the line officers in time proved capable of operating shipboard machinery, the dissolution of the engineer corps meant that the Navy no longer had specialists in designing and developing machinery. The disadvantages were not immediately apparent because the legislation of 1899 allowed older members of the engineer corps to continue their specialties. When these men retired, it became difficult to replace them because the system did not give officers a chance to specialize in engineering as a profession without detriment to their chances for advancement.

The Navy had to find some way to attract and retain officers capable of designing and building ships. One solution was to permit officers once again to choose engineering as a specialty while retaining the status of a line officer and thus assuring their prospects of promotion. In 1916 Congress authorized the selection of certain line officers for engineering duty only and provided for uniform promotion standards. These men wore the gold star of the line on their sleeves and might serve at sea in their early careers, but eventually they were assigned to shore duties and could not exercise command afloat. These engineers later became part of a group designated "restricted line officers."[2]

The unrestricted line officer was by comparison a generalist. As an officer who commanded ships at sea and led them into combat, the unrestricted line officer looked upon the engineering duty officer as a valued but limited specialist. The men in the technical bureaus might design and build ships and weapon systems, but the essential elements of command rested with the unrestricted line. From this group alone could come the Chief of Naval Operations, the admiral with the highest military position in the Navy.

Against this background Rickover could expect to encounter resistance in his efforts as a restricted line officer in a technical bureau to impose regulations on fleet operations. In this sense his plan was an inversion of the accepted relationship between the restricted and unrestricted line. Some officers were convinced that Rickover was simply making a bid for personal power artfully concealed behind the allegedly exceptional dangers associated with nuclear power plants. Others, who observed the pains Rickover took in design, development, and testing, recognized his efforts as an intense concern with safety.

The Criterion of Safety

The extraordinary extension of Rickover's authority during these years rested upon the criterion of safety. There were two sources of his sense of responsibility: one personal, the other legal. Since his first days at Oak Ridge in 1946, Rickover had been aware of the dangers of radioactive materials and understood the exceptional emotional and political impact which a minor accident could have. From his experience with the fleet he knew that it would be difficult to instill in naval officers a healthy respect for nuclear power plants, and he realized that one major accident on a nuclear ship, especially if it caused damage or injury to the public, could jeopardize the use of nuclear power in the Navy. In addition, as chief of the naval reactors branch in the division of reactor development, Rickover had certain legal obligations. The Atomic Energy Act of 1946 gave the Commission responsibility for protecting the public from the hazards of atomic energy. The 1954 Act broadened these provisions and made them more specific.[3] Neither by personal inclination nor by the phrasing of the law could Rickover set aside his duty to see that the fleet operated nuclear reactors safely.

Rickover had begun to exercise his safety responsibilities as a Commission official long before the *Nautilus* was completed. He had worked closely with the Commission's advisory committee on reactor safeguards in designing, building, and operating the Mark I prototype. In July 1953 he had devised an intricate procedure which would enable the advisory committee to evaluate the potential hazards of the *Nautilus* at each of several stages—initial dockside operations, initial full-power operations dockside, initial sea trials, fleet operations, and refueling. In this way the committee could consider each safety question as the crew gained operating experience without having to commit itself to blanket approval of the entire operating plan.[4]

Well before Wilkinson and his crew were ready to use these procedures, Rickover proposed a general agreement for dividing responsibility for naval propulsion plants between the military services and the Commission. Under the agreement which the Commission accepted in February 1954, the military departments would be responsible for the safe operation of their own reactor power plants, including the establishment and enforcement of their own safety standards. The Commission's responsibilities would end at the moment the reactor plant was transferred to the military department, except that the Commission, upon request of the military department, would evaluate operating procedures, general safety standards, and security arrangements for protecting nuclear fuel. The agreement also stipulated that the department would make available to the Commission the safety and security standards it established for each type of reactor plant and all pertinent data on operations under these standards.[5]

Although the Department of Defense did not formally approve the new procedures until late in 1954, both the Commission and the Navy accepted the principles set forth, and Rickover's draft of a memorandum of understanding covering the transfer of the Mark II propulsion plant in the *Nautilus* to the Navy was based on the agreement. Bettis would prepare a reactor hazards report for the advisory committee on reactor safeguards. The Commission (that is, Rickover's naval reactors branch) would recommend a safe operating plan to the Navy. The reactor, including the core and fissionable material, would be transferred to the Navy at no cost. The Navy would become fully responsible for the reactor and its operation under the draft agreement, but the Commission, as requested by the Navy, would continue to evaluate operating procedures, safety standards, and security arrangements. Both the Commission and the Navy complied with every provision of the agreement. During the initial tests and trials of the *Nautilus,* Rickover's organization prescribed the operating procedures and limits at each step. There was no question that Rickover had full authority over matters of operational safety.[6]

In applying the same safety review procedures to the *Seawolf* in 1956, Rickover had the complete support of the Commission and its advisory committee on reactor safeguards. In fact the committee made clear that its approval of fleet operating plans for the *Seawolf* in the spring of 1957 rested primarily on the rigid safety procedures which Rickover had established for the initial testing of both the *Nautilus* and *Seawolf*. The committee urged that the same type of procedures be applied to all future nuclear ships. Going even further, the committee recommended that the Navy formulate rules which would assure the safe entry of nuclear ships into ports in heavily populated areas.[7]

The opinions of the safeguards committee carried even greater weight in

1957 after Senator Clinton P. Anderson, a senior member of the Joint Committee on Atomic Energy, introduced a bill giving the safeguards committee
statutory authority. On the Joint Committee's recommendation, the Congress
adopted this amendment to the Atomic Energy Act in September 1957. That
action, plus the Joint Committee's firm backing for Rickover, assured that
he would continue to have a strong voice in operational safety matters for the
nuclear fleet.[8]

By the time the *Skate* was nearing completion it became obvious that the
project approach to safety would no longer be adequate. The reactor safeguards committee warned the Commission of the increasing risks of operating nuclear ships in populous ports as the number of nuclear ships increased.
In response to a request from the committee for a general operating plan,
Admiral Burke in January 1958 reported to the Commission the steps the
Navy was taking to assure safe operation of the nuclear fleet. For each nuclear ship the Navy was providing a detailed operating manual. The Bureau
of Ships was also preparing technical manuals on the nature of reactor hazards aboard nuclear ships and on radiation problems associated with normal operations. The operating and technical manuals, consisting of several
bulky volumes crammed with engineering drawings and instructions, bore all
the marks of the Rickover influence.[9]

With all their technical detail, the manuals provided a useful but static
form of control. They would serve as reliable guides to crews operating on
lonely missions beneath the seas around the world. But without some form
of enforcement, the procedures set forth in the manuals could hardly be effective. Early in 1958 the Navy issued three instructions prepared in Code
1500 which provided a more dynamic form of control. The basic directive
from Admiral Burke on the operation of nuclear-powered ships incorporated
into Navy practice the safety procedures set up in the 1954 agreement and
elaborated during the trials of the *Nautilus* and *Seawolf*. Under Burke's directive the Navy would continue to submit reactor hazards reports to the
Commission on each new type of reactor and would make available to the
Commission any changes in design or data on operations which might affect
the safety of the reactor. The chief of the Bureau of Ships, in cooperation
with the chief of the Commission's naval reactors branch, was to insure that
the Navy complied with all Commission safety requirements. The bureau
(Code 1500) was to prepare manuals on reactor safety and radiation protection and supervise the preparation of operating manuals for the propulsion system of each nuclear ship. For the repair and maintenance of nuclear

ships, the bureau was to provide the necessary equipment, instructions, and technical knowledge. The instruction assigned responsibility to the Chief of Naval Personnel for selecting and training officers and crew for nuclear-powered ships and to the chief, Bureau of Medicine and Surgery, for health safeguards. Finally the instruction set forth provisions governing the entry of nuclear-powered ships into populous ports.[10]

The other two instructions spelled out procedures in the areas of personnel training and the operation and maintenance of propulsion plants, but the directive from Admiral Burke summed up the central issues which had emerged since 1954 in testing and operating the first nuclear submarines. It also recognized the three areas in which Code 1500 would continue to exercise its influence in the operating fleet in the years after 1958: the selection, training, and assignment of personnel for nuclear ships; the safe operation of nuclear propulsion plants; and port entry for nuclear ships.[11]

Personnel: The Key to Control

The directive from the Chief of Naval Operations in February 1958 reflected the strong reliance which the Commission and the reactor safeguards committee had placed upon personnel training in the safe operation of nuclear ships. Code 1500 could draft reams of technical instructions and guides, but unless the officers and crews aboard nuclear ships could understand these materials and apply them in practical situations, there was little possibility of reliable and safe operation. What many officers in the Navy failed to realize was the radical difference between World War II diesel boats and the new nuclear attack and *Polaris* submarines that would soon be entering the fleet. As one experienced nuclear submarine commander put it, the difference was as great as that between a bicycle and a modern automobile. The World War II diesel boat, like a bicycle, was slow, limited in range, and uncomfortable, but it was simple in design, easy to repair, and amenable to a display of individual dexterity and even daring by the operator. The nuclear submarine, like the automobile, was fast, capable of long-range operation, and comfortable, but it was an extremely complex and expensive vehicle requiring specialized skills and facilities for repair, and demanding caution and self-discipline more than flamboyance from the operator.

Few submarine officers really understood the technological revolution which was about to engulf them. Some of the more far-sighted among them realized that the submarine force faced a crisis, that the ships which had performed successfully in the Pacific were outmoded at the end of World War II. For this reason officers such as Beach, Grenfell, and Momsen in the office of the Chief of Naval Operations had been willing to help Rickover in 1947 get the approval of Nimitz and the Secretary of the Navy for a nuclear submarine project. These officers were interested in other types of propulsion even though they recognized that the closed-cycle approach and the improved diesels could never produce a true submarine. As time went on, the advantages of nuclear propulsion became even clearer. The closed-cycle system with its dangerous chemicals was unattractive, and the improved diesel submarines—the *Tang* class—had proved fiascos. Beach, who commanded one of them, dubbed it an "inglorious failure," and he was even more disturbed at the apparent indifference the Bureau of Ships showed to palpable defects.[12] Rickover's hard-driving attitude, with its emphasis on achieving sound technical results, offered an exhilarating contrast.

Although the submarine force was changing fast, it had a strong personality that differentiated it from other branches of the service. Life aboard any ship imposed a sense of cohesiveness, but duty aboard a submarine had a peculiar quality all its own, for it combined close teamwork with individualism. Far more than on most ships, safety on a submarine depended upon each man knowing his job exactly, trusting his fellow crew members, and being able in an emergency to handle the next man's job. At the final moment of attack, however, it was the captain alone who stood at the periscope and gave the order to fire torpedoes. In World War II the submarine accounted for most of the Japanese shipping destroyed and was second only to carrier-based aircraft in the destruction of the Japanese Navy. The price, however, was high: by the end of the war the Americans had lost almost one out of every five submarines. The submarine force had emerged from the war with its own legends, traditions, and heroes.[13]

Some of the high standards of the prewar submarine service had been diluted through rapid expansion, but the trappings remained. Submarines were assigned to COMSUBLANT or COMSUBPAC (navalese for Commander, Submarine Force, U. S. Atlantic Fleet, and Commander, Submarine Force, U. S. Pacific Fleet). Although the famous submarine victories had been won in the Pacific, COMSUBLANT carried more weight in the Navy. It ran the New London school through which every submarine officer and enlisted man had to pass. COMSUBLANT had far greater influence in the selection and assignment of officers in the submarine service than COMSUBPAC, for most often the man who sat at the assignment desk in the Bureau of Naval Personnel had served in COMSUBLANT.

Officers had to volunteer for submarines, and those lucky enough to be selected went for six months to the submarine school at New London. After successful completion of the course each man was assigned to a submarine. The green newcomer was known as "George," the name which was also applied to neophytes reporting aboard a destroyer, which like the submarine was a small ship with strong traditions. On the submarine "George" spent hours tracing every wire, pipe, valve, and fitting, and learning to master the operation of every component until he could perform the duties of every officer aboard. Usually after a year of shipboard training, he was examined by three senior officers of the force to see if he met the standards of the service. Only then did the young officer find his dolphins at the bottom of a ten-ounce glass of whiskey.[14]

Rickover was convinced that these selection and training methods were not good enough for nuclear submarines. At its best, the old system could produce competent and resourceful officers, some with the special flair that gave them, their crew, and the ship an individual personality. From Rickover's standpoint the great weakness of the system was that it was essentially self-contained and self-perpetuating. It served well when naval technology was developing slowly enough so that the ship itself could serve as the classroom, but assignments aboard diesel boats would not help to qualify men for nuclear submarines.

Rickover was certain that the use of nuclear energy demanded officers chosen as much for engineering aptitude as for leadership qualities. It required sober maturity, conservative judgment, and strict conformance with safety rules. COMSUBLANT was equally confident in the selection and training methods which had produced outstanding officers in the past. Occasionally these underlying differences flared into emotional outbursts from both sides. Some submarine officers maintained that Rickover was embittered by his failure to attain submarine command. Rickover frequently complained that most submariners were too stupid and unimaginative to see what the future held.

The Compromise

It was easy, however, to misinterpret the fundamental differences between Rickover and the officers of the operating submarines. The idea that every officer should be able to do the job of any other officer on a submarine was consistent with Rickover's contention that the captain, executive officer, and all the engineering officers had to be qualified reactor operators. The argument came over Rickover's insistence that he as a Commission official was responsible for selection and training. Another aspect, not readily apparent at first, was Rickover's larger intention of moulding the officer to the new technology. This did not mean that he was creating a separate officer class, a group of specialists in nuclear engineering, but that the officer in charge had to know his ship. It was an idea stemming directly from the Naval Personnel Act of 1899.

The selection of Eugene P. Wilkinson as prospective commanding officer for the *Nautilus* did not arouse serious dispute, although COMSUBLANT had proposed other names. Rickover wanted Wilkinson because he had been involved in design and construction activities for the Mark I plant at Argonne and Bettis and had demonstrated his competence in nuclear technology. Both his background and ability would be indispensable in bringing the *Nautilus* plant into operation. For its part, the submarine force recognized Wilkinson as an outstanding submarine officer.

This kind of accommodation was not possible in choosing the other officers for the *Nautilus*. There were no other submariners who had enough technical background to make them obvious candidates. The officers manning the submarine assignment desk in the Bureau of Naval Personnel were therefore prepared to exercise their usual prerogative by assigning officers to the *Nautilus*. At this point Rickover asked the desk to submit a list of names from which he would select those for nuclear training. This was a bold move on Rickover's part. If the submarine desk complied, he would be able to assert an unprecedented influence over the submarine force.[15]

The officers on the submarine desk recognized Rickover's challenge, but his reputation made them wary about picking up the gauntlet. The natural reaction of many submariners was that Rickover, as an engineering duty officer, had no right to intervene in the assignment process. Those who knew something about the activities of the Bureau of Naval Personnel would have to admit that Rickover, as head of the nuclear project, was entitled to express some opinion about the officers to be selected. Some officers—such as Commander James F. Calvert, who was serving on the submarine desk—saw a larger issue involved. They realized that the submarine force was no longer the effective fighting instrument it had been during World War II. The submarines built since the war had been badly engineered and training standards had declined. Further, Rickover was in an unassailable position. As long as he was responsible to the Commission for safe operation of the Navy's nuclear propulsion reactors, the submarine desk had to recognize his right to participate in the assignment process.

Both Rickover and the Navy were fortunate that Vice Admiral James L. Holloway, Jr., was Chief of Naval Personnel. Holloway's assignments at sea with the surface fleet—destroyers, cruisers, and battleships—gave him an impartial stance in the controversy. He was also thoroughly familiar with personnel and training activities. At the close of World War II he had headed the "Holloway Board," which had a profound effect on Navy policy for recruiting and educating officers. He directed the demobilization of the Navy after the war and was superintendent of the Naval Academy from 1947 to 1950. In February 1953 he became chief of the Bureau of Personnel, a position he was to hold for five years.

Holloway was well aware that the Navy could not stand still in a changing world. He also had the breadth of vision to recognize what Rickover was accomplishing. Holloway accepted the principle that the Navy had established in 1899: that all line officers aboard a ship should be eligible for command. He would not have agreed—even had it been proposed—to giving special training to engineering duty officers so that they could serve aboard ship as reactor operators. Because these officers could not have succeeded to command afloat, he would have considered such an arrangement a step backward. Consequently he accepted the view that ultimately all line officers aboard a submarine had to be nuclear trained.

By law, however, Holloway was responsible for selecting and assigning personnel. He agreed to permit Rickover to interview candidates for nuclear training and make recommendations, but he insisted on retaining the responsibility for selection. With this understanding, Calvert sent Rickover the files of apparently qualified candidates. Perhaps to establish his position clearly, Rickover rejected the first few men recommended after interviewing them, although later he agreed to accept them. The point was that Rickover had established a principle which Holloway had accepted. Now Rickover would have to make certain that submarine officers later assigned to the desk did not subvert it.

Establishing the Training Program

As Chief of Naval Personnel, Holloway had to find ways of integrating the requirements of nuclear technology into the Navy's personnel policies, training activities, and educational institutions. With Carney taking the first steps toward building a nuclear fleet, the Bureau of Naval Personnel wanted to introduce courses in nuclear physics and engineering at the Naval Academy and at the postgraduate school in Monterey, California. Far more important was the general pattern for nuclear training in the Navy. Wilkinson and the *Nautilus* crew had received much of their training from Westinghouse employees at Bettis, and the officers and crew of the *Seawolf* were gaining their education in reactor technology in large part from General Electric personnel at Knolls. Certainly the quality of training was high, but this method seemed to leave the Navy dependent upon private industry for training a vital segment of its officers and crews for the nuclear fleet.

At Carney's request, Holloway undertook a study of the place of nuclear training in the Navy and its impact on personnel policies. Holloway assigned the task to Rear Admiral Henry C. Bruton, a World War II submarine commander, lawyer, and engineer. Bruton set to work in July 1954 and over several weeks interviewed scores of naval officers and civilians, including Rickover and many of the staff in Code 1500, as well as officers at the New London submarine base.[16] From New London Bruton visited West Milton, where Commander Richard B. Laning, the commanding officer of the *Seawolf,* and his crew were being trained on the Mark A prototype. Bruton mentioned that the New London submarine school was about to establish a basic course in nuclear power. Laning immediately went to New London to investigate. He found that Rear Admiral George C. Crawford, who was serving as COMSUBLANT, had asked the school to submit its ideas for the course to the Chief of Naval Operations. Laning was relieved to discover that Commander Momsen had helped to prepare the material on the course and that Roddis and Wilkinson had discussed it. After reviewing the New London draft, Laning reported to Rickover that New London had the proper approach.[17]

Rickover saw the issue at once. From Laning's memorandum it was easy to conclude that Crawford intended to concentrate all of the Navy's nuclear submarine training at New London and to phase out assignments to the Idaho prototype. As a veteran submarine officer, Laning naturally assumed that COMSUBLANT would suprvise training at New London. If this assumption proved correct, Rickover believed he would be unable to provide the training necessary to assure the safe operation of nuclear ships.

Rickover moved quickly to circumvent Crawford. Two days after Laning had written his report Rickover sent Roddis to New London to meet with representatives from Naval Operations, the Bureau of Ships, and the Bureau of Naval Personnel. The group drafted an agreement whereby the Chief of Naval Operations would request the Bureau of Naval Personnel to take over the function, which had previously been exercised by the Commission, to train personnel for nuclear ships. The New London school would furnish the basic training in nuclear technology previously provided by the Bettis and Knolls laboratories, but the Bureau of Naval Personnel, not COMSUBLANT, would be in charge. After six months of training in the basic sciences, the officers and men would spend six months at the prototypes, either at Arco or West Milton. Training at the prototypes would be under the Commission's (that is, Rickover's) technical control with the Bureau of Naval Personnel having administrative responsibility.[18]

The COMSUBLANT proposal reached Admiral Carney's office several weeks later. As Rickover had guessed, the plan was a straight-forward attempt to integrate nuclear power training into the traditional New London approach. All the training would have taken place at New London with the use of special texts, models of components, full-scale mock-ups of some equipment, and simulators of the control panels in the maneuvering room—all provided by Bettis and Knolls. After four to six months, the officers and men would be assigned to submarines in the fleet for further training and qualification according to standard submarine force procedures.[19]

The crux of the proposal was the request for mock-ups and simulators. In the minds of COMSUBLANT officers, these devices would eliminate the need for prototype training, for supervision by Rickover, or interference by the Commission. They would make possible full integration of nuclear technology into the existing discipline of the submarine force. Thus the mock-ups and simulators epitomized the issue of control, not only in an organizational sense (which was important) but also in terms of the longer-range objectives which Rickover had in mind. If New London could impose its standards on nuclear training, there would be little hope for creating the new type of naval officer Rickover envisioned.

Rickover had taken advantage of the situation to fend off the COMSUBLANT proposal, but that plan also had obvious weaknesses. It would require $600,000 for buildings and $2,000,000 for simulators at New London. Rickover could rightly claim that the prototypes were already available to serve as training facilities at Arco and West Milton. The Bruton board had already confirmed the urgent need for prototype experience. At New London on November 1, 1954, Roddis participated in the final negotiations which established the training program under the bureau's control. New London would provide basic instruction without simulators, and the Commission would offer practical training on the prototypes.[20]

The Nuclear Power Schools

The Bureau of Naval Personnel was responsible for organizing the nuclear power school at New London. That arrangement was acceptable to Rickover, but he wanted to be certain that both the curriculum and the instructors would provide the personnel needed for the nuclear fleet. For his representatives at New London, Rickover had two prospective commanding officers: Laning, who would command the *Seawolf,* and Calvert, who would take command of the *Skate,* the third nuclear submarine to joint the fleet. Panoff and Rockwell surveyed the training courses at Bettis and Knolls so that they could evaluate the proposed curriculum. They also helped to find qualified instructors for the school.

The basic course for enlisted men would include mathematics through elementary calculus, basic physics, reactor and electrical theory, thermodynamics, nuclear plant systems, chemistry, metallurgy, and health physics. Officers in the advanced course would study mathematics through advanced calculus, nuclear physics, reactor theory and engineering, chemistry and metallurgy, servo-mechanisms and control, and nuclear plant systems. Rickover insisted that both the basic and advanced course be of high academic quality and concentrate on the fundamentals needed for training at the prototypes. To provide instructors with the proper academic qualifications, the Bureau of Naval Personnel arranged to have a branch of the Navy's Monterey postgraduate school established at New London. Several officers and chiefs from the *Nautilus* were sent to New London to assure proper attention to practical aspects of instruction.

The nuclear power school at New London opened in January 1956 with a pilot course for six officers and fourteen enlisted men. By 1958 the school was training four classes of more than one hundred enlisted men and two classes of about thirty officers each year. Almost 150 officers and more than a thousand enlisted men completed the courses by the summer of 1959. The difficulty of the subjects and the high standards on which Rickover insisted made this record a real accomplishment. At the beginning, when there was keen competition for admission, both the officer and enlisted students were of high caliber, but as some of the glamour wore off, the qualifications of entering students began to decline with a resultant increase in dropouts or failures. Attrition rates were about 3 percent for officers and more than 10 percent for enlisted men.[21]

To some, particularly to submarine officers at New London, these high attrition rates indicated that the courses were too difficult and needlessly academic. The long hours of grueling study on shore pay seemed a negative incentive for students who by dropping out could go back to the fleet on submarine pay. Furthermore, Rickover was extremely rigid in selecting students for the schools, and would take only those who volunteered. Enlisted men were also required to commit themselves to an additional service obligation upon entering the nuclear program. The school's reputation frightened off those looking for an easy billet, but it attracted at the same time those who liked competition and were trying to advance in the Navy. No matter how committed and energetic the students were, most soon found that they had never before worked so hard. A common reaction among graduates was that the school had presented the greatest challenge they had ever encountered, and to more than a few officers successful completion meant more than graduation from Annapolis.

As the size of the classes increased in the nuclear power school, the inadequacy of physical facilities at New London became apparent. The submarine school still had first call on buildings and equipment, and the classroom and laboratory spaces for nuclear power training were cramped at best. By the time the school reached its full operating capacity in 1958, the Navy was planning the vast expansion of its nuclear submarine force, including both attack and *Polaris* ships (described in chapter 10). Planning in the Bureau of Naval Personnel assumed that nine *Polaris* submarines would be completed by early 1961 and thirty-nine would be operational by 1965. Each submarine would require two nuclear-trained crews of five officers and thirty-nine enlisted men. Assuming an attrition rate of 10 percent, New London would have to enroll about 300 officers and 2,000 enlisted men each year to keep the *Polaris* and attack submarines manned in the early 1960s. For the surface fleet much larger requirements seemed likely, and no training school for this purpose had yet been established. As a temporary measure, personnel for the surface ships were being trained at the prototype facility in Idaho.[22]

The increasing demands on New London exacerbated the always troubled relations between Rickover and COMSUBLANT. Even if control over training was no longer an active issue, the nuclear power school, as a department of the larger submarine school, was naturally pervaded by the influence of

COMSUBLANT. Rickover had long been anxious to move out. He complained that New London lacked an academic atmosphere and declared that the lower standards of the submarine school had a bad effect on his students. First Rickover succeeded in establishing on the West Coast a new school at the Mare Island Naval Shipyard, which, with all of its other merits, was about as far as one could get from New London. He maintained that students of the West Coast school performed better than those at New London because they were free from distractions. Moving out of New London proved somewhat difficult, for the submarine force was reluctant to lose the nuclear department and its prestige, along with what influence COMSUBLANT could still exert. But the obvious inadequacy of the New London facilities and the availability of the World War II naval training center at Bainbridge, Maryland, were convincing factors. The first class opened at Mare Island in January 1959 and at Bainbridge in July 1962.[23]

Prototype Training

As explained in chapter 6, the use of land-based prototypes was a distinctive feature in the development of nuclear propulsion plants. Rickover's conception of the prototype as an operating facility closely resembling the shipboard plant helped to concentrate effort on the practical aspects of design and made possible the concurrent development of the propulsion plant and submarine. But the distinctive function of the prototype did not end with the completion of the ship. All the prototypes but one continued to function as an important part of the training program.

The pattern of Mark I operation at the Idaho site became typical for all subsequent prototypes. Even before initial startup in 1953 Mark I served both as a development facility and as a training device for the officers and crew of the *Nautilus*. After the submarine went to sea in early 1955, Mark I was used for further design improvements such as extending the life of nuclear cores. Simultaneous use for both training and development caused scheduling problems which the availability of a reactor simulator might have avoided, but Rickover was convinced that naval officers and men would never fully appreciate the skill and discipline needed aboard a nuclear ship until they had actually operated a reactor. In his estimation, training on a simulator had none of the realism which the prototype provided and therefore did not foster the sense of responsibility which he considered essential. Just as the prototype had been the focus of design and development, so in its new function it would be the unique element in the training effort.

As other prototypes were built, they were also used for both purposes. At Arco the two reactors in the A1W prototype were used to train crews for both the surface fleet and submarines. The abandonment of the sodium-cooled plant after the trials of the *Seawolf* made the Mark A at West Milton useless for training, but other prototypes would be available. The S3G for the *Triton* was taking shape nearby, and the D1G for the *Bainbridge* would soon replace the Mark A in the huge containment sphere. Also available for training was the S1C prototype of the *Tullibee* at Combustion Engineering's plant in Windsor, Connecticut.

Few of the officers and men could be trained on the prototype of the propulsion plant which they would later operate at sea. After all, most of the submarines in the fleet would use the Westinghouse S5W plant, which had no prototype. The heavy demand for personnel also made it difficult to coordinate training with future assignments. Even with the training units at the prototypes operating three shifts around the clock every day of the week, there was often a shortage of training spaces. Actually there was no serious disadvantage in having crews trained on prototypes different from the shipboard propulsion plants. All the reactors for ships entering the nuclear fleet were of the pressurized-water design, and training was about as useful on one prototype as another.

For many officers and most enlisted men the six months spent at the prototype were more demanding and exhausting than any they would face during their service in the Navy. Each trainee, officer or enlisted, was assigned to an eight-hour shift at the prototype, but all were expected to stay for an additional four hours of study. At the remote Arco site the trainees had to spend an additional three hours on the bus, which made for a fifteen-hour day. Many, if not most, students found this amount of time inadequate for completing the course in the allotted time, and some found it necessary to work at the prototype on their free days or to catch a few hours sleep in the ramshackle dormitory provided for this purpose. The Spartan appearance of the cafeteria, classrooms, and offices made the Arco facility look more like an industrial plant than a school. The proximity of West Milton to urban areas made that site preferable to the extreme isolation of the Arco school; but Rickover, to avoid getting into administrative tangles, refused all requests for a cafeteria at West Milton. To some extent the plain severity and drabness of the prototype facilities reflected Rickover's concern for economy. More important, however, was his desire to give the schools a lean and hard atmosphere. There was to be nothing soft, relaxing, or casual about nuclear training.

The curriculum at the prototype schools was sharply focused on the practical aspects of plant operation. Although the courses for officers and enlisted men differed in some ways, the two groups were intermingled in common aspects of training, and officer trainees received no special consideration from petty officers serving as instructors. Both the officers' and enlisted men's courses began with classroom work covering all electrical, mechanical, and nuclear systems of the plant. Under severe time pressures instructors coaxed, scolded, and shamed students into keeping on schedule. The daily check on each student's progress promptly revealed those who were falling behind. Students found deficient in fundamentals presumably learned at the nuclear power schools were quickly assigned additional courses in those subjects. As each student completed his study of a specific component or system, he was required to appear before an instructor for an oral examination during which the student would draw a diagram of the system or component and answer questions about it. Each instructor certified the student's performance on a signature card, and the administration checked to see that students obtained a reasonable number of signatures each week. Later audits by Code 1500 pointed up deficiencies in curriculum or instructions.

After a few weeks of classroom work, students were assigned to watches within the prototype hull. Instructors and students were paired so that every minute of the watch could be used for training. As students climbed through the machinery compartments their instructors followed them to check their performance, raise questions, and correct their mistakes. The emphasis was always on the "why" rather than the "how."

Although training took place in all parts of the prototype, the center of activity was usually the reactor control room called the maneuvering area. Four instructor-student pairs crowded into the small space designed for three enlisted men and the engineering officer of the watch. The strict discipline, close supervision, and realism of the operation made watches in the maneuvering area a stimulating and often a tense experience. Even though the plant was merely a hull section on land rather than a ship at sea, the controls were tied to an operating reactor and steam plant; the consequences of errors could be just as severe as if they had occurred at sea.[24]

The realism, the stern sense of purpose, the determined attitude of the instructors, the isolation and harsh physical environment—all made prototype training an unforgettable if not always pleasant experience. There was always more to learn than any student could master in six months, and every student was driven to work at capacity. In the corridors of all the nuclear power and prototype schools was the legend: "In this school the smartest work as hard as those who must struggle to pass. H. G. Rickover."

Selection of Personnel

In training officers and crews for nuclear ships nothing was more important than selecting competent men. Rickover had demonstrated his appreciation of that fact in insisting upon approving the officers chosen by the Bureau of Naval Personnel to man the *Nautilus*. Working closely with Admiral Holloway, Rickover accepted the principle that the bureau would select the candidates and be responsible for those selections, but he expected his recommendations to receive serious attention.

Code 1500 influenced the selection of enlisted men largely through naval instructions. In 1957 enlisted men were required to be high school graduates, volunteers, and qualified both physically and technically for submarine duty, to have high scores on intelligence and mechanical aptitude tests, and to have a minimum of forty months' obligated service. In the early years these qualifications made it difficult for any enlisted men except those in high rates to gain admission; and as personnel requirements increased, some reduction in standards became necessary. The main concession was accepting young sailors without submarine qualifications.[25] By keeping academic standards as high as possible, the nuclear power schools were able to transform recent high school graduates into reliable technicians.

Rickover took a direct personal part in selecting officers. Beginning with the *Nautilus* crew, he had himself examined the records of individual officers and brought them into his office for interviews which searched out the deepest motivations and traits of character. Only when he was convinced that they had the intellectual ability, the perseverance, and the motivation to qualify for nuclear service would he recommend them for nuclear power school. During the late 1950s, while the number of officers on nuclear ships was still small, Rickover had no trouble establishing a personal relationship with each officer. That relationship was seldom friendly, never intimate, but always frank and direct. If Rickover frequently expressed his dissatisfaction, the fire of his displeasure was convincing evidence of his personal concern. He demanded nothing less than the best from every officer.

By 1960 the demand for officers had grown so rapidly that Code 1500 was no longer able to rely upon the supply of qualified submariners. Top-ranking graduates from the Naval Academy, the Naval Reserve Officers' Training

Corps, and the Navy's Officer Candidate School were allowed to apply for nuclear training. This broadening of entrance requirements did not displease Rickover, for he contended that young men direct from college had fewer bad habits to unlearn than those with experience in the fleet. The new system also tended to reduce the influence of the New London submarine school because officers would now complete their nuclear training before going to New London.[26]

Even as the numbers of officer candidates rose into the hundreds, Rickover persisted in interviewing each man himself. The staff in Code 1500 collated information on the candidate's background, class standing, and academic interests. Several senior members of the staff interviewed each candidate not only on the usual personnel matters but also on technical subjects. The purpose always was to break through any surface gloss to gauge the ability and character of the man himself. A summary of each interview went to Rickover before he met the applicant.

The direction of each interview was unpredictable, for it depended upon the candidate's record and his responses to Rickover's questions. Striking at once for the jugular, Rickover attempted to get a measure of the candidate and force him to discuss his strengths and weaknesses without sham or self-deception. Where the record of performance was weak, the interview might well center upon the candidate's ability or desire to turn a new leaf and upon the level of his commitment to the nuclear power program. Rickover expected each man to back up his statements with deeds, even if it came to singing a solo or losing thirty pounds of overweight.

In time officers described their experiences in the Rickover interview in books or articles.[27] Often incredible, these descriptions were usually close to the fact. Probably every candidate found the interview an unforgettable experience. Some saw it as a turning point in their lives. For a few, unfortunately, it was a shattering event leaving scars that would last a lifetime. But for the most part the interviews were effective. They enabled Rickover and sometimes the candidate himself to establish some measure of his ability and commitment. They also laid a firm foundation for a direct, unglossed relationship between Rickover and the candidate—a special relationship that would last as long as Rickover and the officer were in the Navy.

In addition to choosing officers who would attend nuclear power school, Rickover also personally selected officers who would serve as instructors. Many of these officers came from the nuclear fleet itself. After serving about a year at sea, a young officer could volunteer to teach at a nuclear power school. If his academic standing had been high, his shipboard performance good, and his commanding officer's recommendation favorable, the Bureau of Naval Personnel nominated him for consideration by Code 1500. Rickover's staff chose instructors from the bureau list with his approval. Because Rickover had interviewed these officers before they entered nuclear power school, he did not take the time to interview them again. A second source of instructors was the Navy's Officer Candidate School. Newly commissioned officers with high academic standing from a good college or university could apply to teach basic subjects at nuclear power school. Those with likely qualifications were subjected to the interview process used for school candidates and were prepared for teaching at the nuclear power schools.

Rickover placed the highest demands of all upon those officers who were seeking command of nuclear ships. He had hand-picked and -trained Wilkinson. He had personally supervised the training of Calvert for the *Skate,* Laning for the *Seawolf,* and Commander William R. Anderson as the second captain of the *Nautilus.* Calvert and Anderson both worked for a year in Code 1500 in Washington. This assignment gave them a much more thorough and extensive exposure to nuclear technology than the average officer acquired in nuclear power school. Prototype training for these officers was crammed into eight hectic weeks at Arco.[28]

As numbers increased in the 1960s, some of the personal attention necessarily declined. But Rickover still insisted that all prospective commanding officers spend three months in a special training course in Code 1500. The course involved a rigorous restudy of all the elements of propulsion technology and safety, a series of detailed oral examinations by the senior technical staff, and an exhaustive written examination.

Because Rickover could select the officers entering the nuclear program, he attained an immense influence over the operating forces of the Navy, and particularly over the submarine fleet. He had broken the dominance of the old Atlantic submarine force with its traditions of an earlier day, and had replaced it with one of his own. Whether the ship was an attack or *Polaris* submarine, the captain, executive officer, and all the engineering officers had to have nuclear training. Because nuclear propulsion developed more slowly in the surface Navy, Rickover's influence in this area was correspondingly less; but the command of nuclear surface ships—especially carriers—was a prize to be sought. As a result, to an ever-increasing extent the route to flag rank lay through the nuclear propulsion program.[29]

Rickover's conviction that technical qualification was supremely important in a naval officer was deeply rooted in his sense of history. He was certain that American power was ebbing. In a threatening world, control of the sea—and the Navy's part in that task—was essential to the survival of the nation. From the position that he had gained over the years he could assess the Navy from several points. He interviewed the best of the young men who wanted to become officers; he studied the academic reports of enlisted men who were failing in the schools; he talked with prospective commanding officers and had them under his eye for months; he went to the sea trials of nearly every nuclear ship; he read the operation reports of vessels when they returned from patrol; he knew intimately the workings of private and Navy yards and the details of procurement; and he dealt directly with the Chief of the Bureau of Ships and the Chief of Naval Operations. No matter where he looked, Rickover saw waste and indifference, and he was convinced that the modern Navy was not meeting the demands placed upon it.

In an earlier time—between World I and World War II—machinery had been comparatively simple and there had been room in the Navy for tradition, technology, and even for a certain amount of individualism. Those days had vanished, and technology was developing at a far faster rate than the Navy's ability to absorb it. Rickover was convinced that such patch-work solutions as new management systems, leadership techniques, and elaborate reorganization were bound to fail. The answer lay in officers competent to handle the new technology. He scoffed at the old idea of the line officer as a Jack-of-all-trades and a master of none. He unmercifully ridiculed midshipmen who claimed that with a management course or two and a little experience they could run the Naval Academy, a large company such as General Electric, or even the nuclear power division. Rickover saw this kind of overconfidence as the myopia of a technician. A naval officer should be broader than this. Breadth of view, however, was meaningless without the focus of professionalism. As Rickover defined it, a professional person has a "mastery of a specific area of higher learning, and [an] ability to apply this specialized knowledge to practical problems. He applies to his work a broad base of knowledge and a habit of independent and logical thought that sees each problem in its overall setting." Too often older Navy officers buried themselves in routine, sheltered themselves under management systems, and by their examples snuffed out intellectual eagerness in a young person wishing to learn. That the Navy often reflected management vogues in the business and academic worlds was no excuse. Ultimately the businessman had his balance sheet but the naval officer's day of reckoning might come only

when the safety of his ship and his crew were at stake. To Rickover, personnel selection was an indispensable part of Navy reform.[30]

Revising the Operational Directive

The safe operation of nuclear ships depended not only upon well-trained officers and crews but also upon a workable definition of the safety responsibilities of both the Commission and the Navy. These Code 1500 had defined in
a general way in the three Navy instructions issued early in 1958. The basic
directive covering the operation of nuclear ships firmly established the command function of the operational forces but recognized the Commission's
authority in matters of operational safety. Actually the directive was mostly
a summation of the procedures Code 1500 had used to expedite the Commission's review of the design, testing, and operation of the first nuclear
submarines. Although these measures had proved neither burdensome nor
time-consuming, the directive turned out to be only a tentative compromise
accepted with considerable reservation by both agencies. On the Navy side,
the operational forces were uneasy about any regulations which would
threaten the full exercise of command. On the Commission's side, Rickover
kept raising the issue of whether the directive would permit the agency to fulfill its statutory obligations for safety.

There were many aspects of operational safety, some of them highly technical, but the heart of the matter was the extent to which technical authorities,
either in the Bureau of Ships or in the Commission, would be allowed to circumscribe the command function of the operational officers. Rickover's insistence that the Commission take whatever action was necessary to assure
the safe operation of nuclear ships was bound to cause resentment in the
Navy even though the general principle of operational safety was accepted.
Some officers in the Navy found it easy to believe that Rickover was using
the safety issue merely to extend his authority over fleet operations.

Rickover's unyielding position on the question of safety did not improve
his already strained relations with the operational arm of the Navy. Ever since
the fight over his promotion to rear admiral in 1953, the Navy had rankled
under the barbs which Rickover continued to launch with impunity under
the protective eye of the Joint Committee. By 1958 pressure was mounting
on the Commission and the Navy to promote Rickover to vice admiral.[31]
James T. Ramey, formerly a lawyer at the Commission's Chicago office and

now executive director of the Joint Committee, had drafted material for Joint Committee members to use at an appropriate time in advocating Rickover's promotion. That opportunity came in August when the Navy omitted Rickover's name from the list of those invited to attend the White House ceremony honoring the captain of the *Nautilus* for the first submerged transit of the North Pole. Ramey brought the incident to the attention of members of the Joint Committee, who used Ramey's material in floor speeches. Charging an attempt to persecute Rickover, the Joint Committee aroused so much sympathy for Rickover that the Navy dared not refuse him promotion.[32] As a vice admiral, Rickover would have more leverage on the Navy than ever.

Rickover's growing influence over safety matters, however, depended not upon his promotion but upon the technical competence of his organization and his strategic position between the Navy and the Commission. On the one hand, he could assure the Chief of Naval Operations that Navy directives were responsive to the complex technical requirements which the reactor safeguards committee and the Commission's staff imposed. On the other hand, he was in a position to assure the Commission that the Navy directives were effective. To make certain of this latter point, Rickover convinced Admiral Burke that the Navy should issue new instructions which would formalize in writing the specific procedures which the operating forces had developed to meet the general instructions issued in February 1958. These new instructions, issued in November 1958, offered even greater assurance of the safe operation of nuclear ships.[33]

Statutory Responsibilities for Safety

The procedures set forth in the Navy instructions were effective; but from a legislative or administrative perspective, the interagency arrangement did have certain disadvantages. The basic agreement and the implementing naval instructions skirted the difficult question of legal responsibility for the safe operation of nuclear ships. Although the Navy was to be held responsible for operating the propulsion plants once they had been placed in submarines or ships, the instructions recognized the Commission's continuing role in reviewing and approving operating procedures.

The interagency agreement, again for pragmatic reasons, reflected the inherent ambiguity of the Atomic Energy Act of 1954. One section of the Act (161b) authorized the Commission to establish "by rule, regulation, or order, the standards and instructions necessary to protect health or to minimize danger to life and property." This provision suggested, but did not unequivocally establish, the Commission's authority to police reactor operations anywhere in the government, including the military departments. Another section of the Act (91b) authorized the president to direct the Commission to transfer special nuclear materials or weapons to the Department of Defense "for such use as he may deem necessary in the interest of national defense." The president could also authorize the military "to manufacture, produce, or acquire" weapons and reactors for military purposes. It was possible to read the Act as giving the president authority to transfer safety responsibility as well. Both sections of the Act were subject to a variety of interpretations, including a number which appeared to be contradictory.[34]

As the number of nuclear weapons and reactors in military custody began to increase near the end of the decade, the vague provisions of the Act took on something more than an academic interest. One of the major issues which John A. McCone had to face after he became chairman of the Commission in July 1958 was a proposal from the Department of Defense to transfer a majority of the nuclear weapons in stockpile to military custody. McCone at once expressed his concern over the apparent gap between the Commission's statutory responsibilities and its actual control over nuclear weapons and reactors. He grew uneasy as he contemplated the potential hazards involved in having hundreds of nuclear weapons deployed at military installations throughout the world, scores of propulsion reactors moving about the oceans on Navy ships, and dozens of research and experimental reactors being operated by all three military services. Over such far-flung activities the Commission could hardly pretend to exercise effective control, and yet the Atomic Energy Act, at least according to some interpretations, seemed to say that the Commission had such responsibilities. McCone had no desire to shun his responsibilities, but he was determined to clarify them.[35]

Within the Department of Defense there was a similar interest in defining the Commission's authority over what were essentially military activities. In terms of weapons particularly, the idea of Commission responsibility seemed unrealistic. The vast proliferation of nuclear weapons within the military services seemed to place safety matters far beyond the Commission's effective jurisdiction. The development of intercontinental ballistic missiles and the resulting need for quick response to enemy threats or actual aggression suggested a flexibility in military operations which could hardly include the Commission's uncertain safety role. By 1959 some officials in the Department of Defense were prepared to argue that the president's authority under section 91 to transfer weapons, materials, and reactors to the military carried with it the authority to transfer safety responsibilities as well.[36]

Although Rickover did not presume to have any competence in weapon matters, he was quick to see in such arguments a direct threat to the system of joint responsibility between the Commission and the Navy which he had carefully established for the operation of nuclear ships. In a hearing aboard the submarine *Skipjack* while operating at record speed and depth on April 11, 1959, Rickover warned the Joint Committee that there was a "question in some people's minds as to whether the AEC has any responsibility at all for the safety of these ships once they have been turned over to the Navy." This statement gave Chet Holifield, the ranking member of the committee and an author of the Act, an opportunity to declare in no uncertain terms that Congress intended to establish the Commission's authority over reactor safety matters in section 161 of the Act.[37]

Despite this strong statement which Rickover elicited from the Joint Committee aboard the *Skipjack,* McCone proceeded with his intention to clarify the Act. At McCone's request, members of the Joint Committee introduced amendments which would have given the president authority to transfer safety responsibility to the Department of Defense, and McCone appeared before the committee in August 1959 to testify in support of the amendment. Although Holifield was not present, his *Skipjack* statement had obviously influenced the committee. The best McCone could get was an expression of the committee's willingness to consider a compromise under which the military would have operational responsibility but would be required to meet general safety standards and procedures established by the Commission. In ensuing staff discussions Rickover and the Commission's division of military application took a firm stand against such a proposal. Rickover also asserted his views through the Joint Committee staff with the result that agreement on the precise language of a compromise amendment proved impossible. After several more attempts the Commission concluded in December 1960 that the original amendment would clarify the law but was not really necessary. With this statement McCone ended his efforts to circumscribe the Commission's safety responsibilities.[38]

For Rickover the kind of clarification McCone was seeking would have threatened the procedures which he already had in operation. If under such an amendment the Department of Defense had obtained complete authority over military reactor operations, Rickover's procedures might have been challenged by the department. Neither would Rickover have been able to exercise his authority as a Commission official over the operation of nuclear ships. Rickover and other naval officers, the Joint Committee, and some members of the Commission and staff were convinced that the continued safe operation of nuclear ships depended upon the independent review and guidance which the Commission exercised under the existing agreements.

McCone's failure to amend the Act permitted Rickover to consolidate his position as a safety monitor of nuclear ship operation in the Navy. Through Code 1500 he rigorously enforced the procedures set up under the Navy instructions, and he was successful in obtaining approval of new instructions which governed such things as the handling of radioactive materials from nuclear ships and the disposal of radioactive wastes from naval facilities.[39] The new instructions, like the old, rested upon joint responsibility of the Commission and the Navy, and permitted Rickover to exercise his dual authority over these matters as well. In the larger context which included weapons and special nuclear materials as well as military reactors, President John F. Kennedy reaffirmed the principle of joint responsibility in a directive to the Commission and the Department of Defense in September 1961.[40]

Enforcing the Directive

The safety directives issued by the Chief of Naval Operations and the Bureau of Ships were important but hardly sufficient measures for assuring the safe operation of nuclear ships. One of Rickover's most common admonitions to his staff was that directives were not worth the paper they were written on unless they were enforced. Rickover was careful to see that his staff established adequate procedures to check on fleet compliance with safety regulations.

Inspection of the ship in port gave Code 1500 some indication of the condition of the propulsion plant, but it did not cover the performance of equipment at sea. To fill this gap Rickover carefully reviewed operational reports from the fleet. Rickover personally read all this information even when the number of nuclear ships in operation made that a formidable task. If this review revealed matters of general interest to the fleet, Code 1500 would issue a report as a technical bulletin.[41]

The information which Code 1500 gathered from operating ships in the fleet gradually enabled Rickover and his staff to discover incipient safety problems which might have escaped a less rigorous system. This effort added substantially to the effectiveness of safety surveillance. It also gave Rickover an intimate knowledge of the operation of the nuclear fleet, sometimes surpassing the information available to fleet and type commanders.[42] Again, as in the supervision of shipbuilding, knowledge alone gave Rickover a voice of authority in what were essentially operational aspects of command. Rickover was always scrupulous about acknowledging the limits of his formal responsibilities, but he did not hesitate to remind line officers of theirs and in the process to bring his influence to bear on the operating fleet in ways unprecedented in the Navy.

Refueling and Overhaul

The last phase in the cycle of naval ship operation was overhaul. By the middle of the 1950s the Navy had established a standard and relatively sophisticated procedure for overhauling ships in the fleet. Requirements varied from one type of ship to another, but most ships normally were overhauled every eighteen months. The Bureau of Ships had arranged with type and force commanders to compile lists of repairs and alterations to be completed while the ship was dry-docked. These lists, called "90-day letters" because they were required ninety days before overhaul began, set forth in some detail all the tasks to be accomplished in the sixty days usually allowed for overhaul. For modern ships with all their complex machinery, the preparation of the work list required a rather high degree of planning, scheduling, procurement management, and coordination. Because virtually all overhauls were performed in naval shipyards, the various departments in the yards had acquired experienced craftsmen and the equipment required for a standard overhaul.[43]

It took no unusual insight to understand that nuclear ships would impose completely new requirements on the Navy's overhaul system. Most obvious was the fact that many overhauls would include refueling of the reactor, a task which demanded the special skills of reactor engineers. Beyond that fact, nuclear ships were also highly specialized vessels containing the most sophisticated and complex equipment. The difficulties encountered in building the first nuclear submarines suggested the kinds of problems to be expected in overhauling nuclear ships.

Soon after the *Nautilus* put to sea in 1955, Code 1500 began planning for the first refueling. To assure the highest degree of capability in this first refueling of a nuclear ship and to minimize the time the ship would be out of service, Rickover arranged to have the work performed at the Electric Boat yard in Groton. Following usual Navy procedures, Code 1500 began assembling the work list in the spring of 1956 for the refueling which was scheduled for early 1957. In addition to replacing the reactor core, the work list included certain improvements in reactor components, a number of relatively minor modifications of equipment to remove defects revealed during trials and tests, and some inspection of equipment to determine the adequacy of design.[44]

Under close supervision by Code 1500, Electric Boat worked with Bettis in devising detailed plans for the refueling. Special equipment was ordered in advance, shipyard personnel were trained in necessary skills, and responsibilities were carefully assigned. Although Electric Boat was in charge of the refueling, Bettis was required to provide engineers who would check every step in the refueling process as it proceeded three shifts per day, six days each week. This careful preparation, plus the availability of experienced personnel, made it possible to complete the refueling and limited overhaul in fifty-seven days during the winter and spring of 1957.[45]

The first true test of complete overhaul and refueling procedures would not come until 1959, when the *Nautilus* returned from its second tour with the fleet. This time Portsmouth, not Electric Boat, would do the work, and the ship, having traveled 153,000 miles in almost four years, required extensive inspection and overhaul.[46] Although Portsmouth had been preparing for nuclear submarine overhaul since 1954, the assignment taxed the yard's capabilities to the ultimate. Even after months of training, many of the shipyard workers found themselves unprepared for the exacting manipulation of equipment in close quarters. Supervision and scheduling also proved weak. Some of Portsmouth's difficulty could be explained by the fact that virtually all the machinery in the engine room had to be opened for inspection and an unexpectedly large proportion had to be removed from the ship for reconditioning. Extensive alterations were also made in the superstructure, attack center, and ventilating system. The overhaul, begun during the spring of 1959, was not completed until almost fourteen months later.[47]

Rickover took a special interest in the lessons to be learned from the Portsmouth experience. One was that all concerned had underestimated the problems generated by the presence of radiation. Better equipment, training, and procedures would be necessary in the future. Supervisors and craftsmen in the trades customarily required for conventional submarine overhauls were found to be unprepared for complex work required on a nuclear ship. Again, as Code 1500 had discovered in building nuclear ships, there was some doubt whether the rotation of management in naval shipyards could provide the professional leadership required for overhauling nuclear ships.

Here the parallels were close to Code 1500's experience in building nuclear ships. There was by 1959 a need to expand capacity for overhauls and refueling. The Bureau of Ships again proposed an expansion which would provide these capabilities at yards which did not yet have nuclear experience. Rickover proposed that overhauls be restricted for a time to yards which had built nuclear ships and that overhauls for the first ships in a class be accomplished at the building yard. Rather than bringing in the naval shipyards at Norfolk, Virginia, and Charleston, South Carolina, Rickover succeeded in having most of the overhauls in the early 1960s assigned to Electric Boat, Newport News, and Portsmouth on the East Coast and to Mare Island on the West Coast. Not until the middle of 1962 did Code 1500 have Pearl Harbor geared up for its first overhaul and refueling of a nuclear ship. Charleston was almost ready for that task by the end of the year.[48]

By this time the refueling operation had taken on something of a common pattern at all yards doing this work. Primary responsibility was fixed in the shipyard, but Code 1500, the ship's force, and the Bettis or Knolls resident engineer were directly involved in every step of the process. As in reactor assembly and startup, any one of these representatives could halt the refueling operation at any moment if something appeared wrong.[49]

Because refueling usually occurred during a general overhaul, submarines more often than not were dry-docked. Once the vessel was in the dock, yard men linked her to the dockside with a web of power cables and hoses and erected a maze of scaffolding over various parts of the hull. Nearby was a cluster of buildings and equipment for handling the highly radioactive fuel removed from the reactor, for minimizing the spread of radioactivity from the open reactor vessel, and for keeping dirt and foreign matter out of the reactor. The most prominent of these facilities was the reactor access house which sat astride the hull directly over the reactor. A hole cut through the hull gave a clear path from the floor of the house to the reactor compartment and pressure vessel head. A retractable roof made it possible for the dockside crane to remove old fuel and bring in the new fuel. Near at hand on the dock was a building for temporary storage of the spent fuel, another for new fuel, and structures where personnel could change clothing, test equipment, and operate mock-ups of the reactor core and handling devices.

During refueling the yard had to guard against two types of nuclear accidents. The first was an inadvertent criticality of the reactor, which could result from improper manipulation of the fuel elements or control rods. There would be no explosion like that from a nuclear weapon but rather a burst of radiation and a sharp rise in temperature. Part of the fuel could melt and release fission products which could spread downwind from the submarine. To minimize this possibility the laboratories had carefully designed the tools, equipment, and procedures to be used for refueling. The second type of accident was the possible exposure of personnel to radiation during removal of the depleted fuel from the pressure vessel. Here the defense was adequate shielding around the operation and exceptionally rigorous procedures. Throughout the overhaul it was especially important to avoid the spread of low-level radioactivity.

Careful design of reactor core components and refueling tools with an eye toward safety was essential, but it could not take the place of training. Code 1500 and the laboratories guided the shipyards in qualifying technicians for critical tasks. The workers had to demonstrate on actual equipment in real or simulated conditions that they could perform the operation safely. The hardest lesson they and their supervisors had to learn was verbatim compliance with written procedures. No deviation was permitted without the express consent of Code 1500 and the laboratory. Special security measures were used to prevent unauthorized or untrained personnel from entering radiation areas.

The refueling procedures called for an exceptional amount of consultation and record-keeping. At each step in the process, even for such simple operations as removing nuts from bolts, the technicians worked with open manuals and discussed the operation until all agreed on what was to be done. If there was any doubt or disagreement, work stopped until higher authorities had resolved the question. Usually such problems could be resolved on the spot by the joint refueling group representing Code 1500, the laboratory, and the shipyard. Sometimes the work was halted for several hours or even days until Code 1500 in Washington or the laboratory provided the necessary approvals. At each stage of the operation the technicians were required to sign documents certifying that they had completed the task described in the manual. Code 1500 made constant checks to see that these documents were properly completed and that the manuals were up to date.

Special shielding and remote control devices were used to lift the fuel assemblies from the reactor into the shielded removal container. Then the roof of the reactor access house was opened, and the dockside crane moved the container to a dockside building where the fuel assembly was transferred to a shielded shipping container weighing over 100 tons and mounted on a railway car for shipment to the National Reactor Testing Station in Idaho. There at the expended core facility, completed in 1958, the fuel elements were disassembled and analyzed. Later the spent fuel was moved to the nearby chemical processing plant to recover the uranium.

As in all aspects of nuclear propulsion technology, refueling required an unusual amount of skill and reliability on the part of the operators and exacting administrative controls. With his safety mandate from the Commission, Rickover insisted upon full compliance with written procedures. His refusal to settle for anything less meant that every refueling yard developed a regimen and discipline that made it clearly distinctive from other naval installations and private shipyards.

The Nuclear Fleet at Sea

On October 6, 1962, the guided missile frigate *Bainbridge* was commissioned at Quincy, Massachusetts. Built by the shipbuilding division of the Bethlehem Steel Company, the vessel was propelled by the D2G reactor plant, which General Electric had designed and developed at Knolls. Although larger than some light cruisers built during World War II, the *Bainbridge* was the smallest surface ship in the nuclear fleet. Already at sea were the aircraft carrier *Enterprise* and the guided missile cruiser *Long Beach*. No other nuclear surface ships were under construction, but the Navy had awarded a contract to the New York Shipbuilding Company for another guided missile frigate, the *Truxtun* (DLGN-35).

In contrast the nuclear submarine fleet was flourishing with twenty-seven ships in commission. Nine were *Polaris* submarines and thirteen were high-speed attack submarines. The others included the *Nautilus* and converted *Seawolf,* still valuable fighting ships, and three one-of-a-kind submarines: the *Tullibee,* the *Triton,* and the *Halibut.* Even more striking was the number of submarines under construction. Eleven high-speed attack and nineteen *Polaris* submarines were on the ways, and more were being planned.[50] From Code 1500's perspective, the most significant fact was that all these submarines would use the S5W plant.

Nuclear propulsion in submarines had made possible a series of spectacular achievements, particularly submerged voyages of unparalleled distance and duration. In October 1958 the *Seawolf* completed a sixty-day submerged voyage of 13,761 miles. Early in 1960 the *Triton* circumnavigated the world in eighty-three days, independent of the earth's atmosphere. The most dramatic accomplishments were the submarine penetrations beneath the Arctic ice cap. During the summer of 1958 the *Nautilus* made the first transpolar voyage on a cruise from Pearl Harbor to Portland, England. A few days later the *Skate* also reached the geographic pole and in extensive patrols the following year proved the feasibility of winter operations in the Arctic. These voyages had obvious implications for defense strategy, but of more immediate significance was the departure of the *Polaris* submarine *George Washington* from the Charleston naval shipyard on November 15, 1960. Armed with sixteen *Polaris* missiles, the *George Washington* began a series of submerged vigils which would provide the United States a reliable and always ready nuclear deterrent for years to come.[51]

Though far smaller than the submarine force in 1962, the nuclear surface fleet seemed to hold great promise for the Navy of the future. In 1958 Admiral Burke had predicted a fleet of more than nine hundred ships by the 1970s. Of these the undersea force would have fifty missile and seventy-five attack submarines, all using nuclear power. The nuclear-powered surface fleet, Burke suggested, would consist of more than thirty ships: six carriers, twelve guided-missile cruisers, and eighteen guided-missile frigates. Although Burke's prediction proved optimistic in both categories by the end of 1962, he greatly overestimated the number of nuclear-powered surface ships. Only the *Enterprise, Long Beach,* and *Bainbridge* were then at sea.[52]

The advantages of nuclear surface ships were not as easy to grasp as submarine voyages under the Arctic ice or around the world. A reactor-driven surface ship might not be able to go faster than a conventionally-fueled ship of a comparable type, but it could maintain this speed for long periods, independent of fleet oilers. A fast, far-ranging nuclear task force promised enormous military advantages in responding to rapidly developing international crises in all parts of the world. The *Enterprise, Long Beach,* and *Bainbridge* gave some idea of the potential of such a task force in 1964, when the three ships circumnavigated the globe in 65 days, completely independent of logistic support.[53] During the Cuban missile crisis, and later off Viet Nam, the *Enterprise* demonstrated that nuclear propulsion dramatically improved the carrier's ability to remain on station to launch aircraft on sorties against the enemy, and to react quickly to changes in orders.[54] To Rickover, his staff, and many officers who served on these ships, nuclear power could change the role and mission of surface ships in naval warfare.

There were many experienced naval officers and officials in the military

31. The *Skate* (SSN-578) surfaced at the North Pole, March 17, 1959. The *Skate,* built by Electric Boat and driven by the S3W submarine fleet reactor, had voyaged beneath the Pole in August 1958.

32. The *Sculpin* (SSN-590) in a high-speed surface run in May 1961. The size of the bow wave indicates the power of the nuclear propulsion system. *U.S. Navy*

31

32

33. The *Triton* (SSRN-586) leaving New London on February 16, 1960, for a trip around the world. Driven by two S4G reactors, the largest submarine ever built circumnavigated the globe, submerged, in 83 days and 10 hours.

34. The *Polaris* missile submarine *Ethan Allen* (SSBN-608), photographed on the surface on August 20, 1961. The hydrodynamic shape of the hull was modified in Polaris submarines to accommodate the missile tubes aft of the sail. The *Ethan Allen* was the lead ship in the second class of *Polaris* submarines, the first to be designed from the keel up for missile launching.—*U.S. Navy*

33

35. Nuclear Surface Fleet, May 1964.
From left to right: the guided missile
cruiser *Long Beach* (CGN-9), the
aircraft carrier *Enterprise* (CVAN-65),
and the guided-missile frigate *Bainbridge*
(DLGN-25).
U.S. Navy

35

36. The nuclear-powered aircraft carrier *Enterprise* (CVAN-65) in action. The tremendous power of her eight-reactor propulsion system is dramatically demonstrated in this high-speed turn. *U.S. Navy*

establishment, however, who opposed the large-scale construction of nuclear surface ships. The question was not whether nuclear-powered ships were more effective but whether the extra cost in terms of construction and skilled personnel purchased commensurate military advantages. The interpretation of financial and performance data would spawn heated debates in the Department of Defense and in the Congress during the 1960s and 1970s. The controversy reflected a fundamental issue which had plagued the Navy since World War II: was the carrier a magnificent relic which had become a large, expensive, and vulnerable target, or was it—particularly with nuclear power —a vital ingredient in national defense? Rickover and others supported the latter thesis but they were not immediately successful. Not until June 1968 almost six and a half years after the commissioning of the *Enterprise* and after two more conventional carriers had been launched, did the Navy lay the keel of a second nuclear-powered carrier, the *Nimitz* (CVAN-68). No more nuclear-powered surface ships were laid down during the decade.[55]

Although the Navy did not reach the goals Burke had anticipated, nuclear power had made a profound impact on the fleet. Nuclear power had revolutionized submarine warfare and had offered decisive advantages for surface ships. Less dramatic but probably more significant were the new standards of ship design and construction, of crew training and qualifications, of ship operation and safety which the adaptation of nuclear power brought to the Navy. Before the end of the decade nuclear power would become an indispensable element in the Navy's bid for control of the sea.

The Measure of Accomplishment

By the end of 1962 Rickover's group had completed the essential process of technological innovation in bringing nuclear power to the fleet. Nuclear propulsion had demonstrated clear superiority over conventional systems in both submarines and surface ships. All future submarines and an increasing number of new surface ships would be nuclear-powered. Although Rickover and his staff were to continue their efforts to improve nuclear plants, later development would be less concerned with the problems of innovation and more with the scope and rapidity of adoption, matters which raised a series of questions outside the scope of this book. Nuclear power was also coming into its own in civilian applications. Although many issues involving its commercial use were yet to be resolved, these lay increasingly in legislative and regulatory fields and less in technological development.

The years spanned by this book represent a period of rapidly accelerating technological development. In addition to nuclear propulsion, the postwar decade produced jet propulsion for aircraft, the transistor, the high-speed digital computer, man-made earth satellites, inertial guidance systems, long-range ballistic missiles, and thermonuclear weapons, to mention only a few of the developments with significant military applications. Yet for every project which was successful, dozens failed to reach their objectives even when the theory on which they were based was sound. Of those projects which were completed, many suffered from huge cost overruns and repeated schedule delays. Often the final product fell far short of the performance specifications required to make it useful.

Against this background the achievements described in the preceding chapters are exceptional. Of all the military development projects started in the two decades after World War II, naval nuclear propulsion has been one of the most successful. As a technical feat, building a nuclear fleet surpassed the original development of the atomic bomb, and it was achieved without the open-ended commitments which the World War II project enjoyed.

Because so many development projects since World War II have failed to reach their goals, the management of technological innovation has aroused increasing concern.[1] In many instances the pace of innovation has outstripped attempts to manage it efficiently. As technology has become more sophisticated, innovation has required increasing numbers of specialists both in government and industry. Thus the management task has become much more difficult. It is not likely that any single technique or philosophy can be applied universally without considerable adaptation, but the Rickover experience does offer a promising approach. On that premise the following pages attempt to take some measure of Rickover's accomplishment and to suggest what made it possible.

The Accomplishment

In engineering development the most significant measure of accomplishment is the hardware produced. The difficulty of the undertaking and the amount of time and resources required are beside the point if the goal of the project is not achieved. In the Navy project Rickover and his associates clearly accomplished their initial task: to build land-based prototypes and operational submarines using two different types of propulsion systems. The Rickover team reached its goal on a self-imposed schedule which many experienced engineers considered impossible. Not only were the first two prototypes and ships constructed as planned; they also met or exceeded design specifications almost from startup. This fact alone was a rare achievement.

Even before they had completed the *Nautilus* and the *Seawolf* Rickover and his staff had begun to develop new types of nuclear propulsion plants for both surface ships and submarines. Scarcely had this work started when the Commission gave Rickover the responsibility for developing the nation's first full-scale central-station power plant using nuclear energy. Despite a lack of familiarity with the power industry or large-scale construction projects, Rickover's group succeeded in building the Shippingport plant on schedule. Once again the absence of start-up difficulties was virtually unprecedented outside the Navy project. In the meantime Rickover's organization had expanded the developmental and fabrication capabilities of the project for multiple production of a variety of nuclear propulsion plants. This effort, beginning in the fall of 1954, made possible the thirty nuclear-powered ships which had joined the fleet by the end of 1962.[2]

Evaluation and Comparison

Producing a fleet of combat ships, each fully operational and driven by a completely new type of propulsion system, was a striking accomplishment in itself. Even more impressive was the speed and economy with which these propulsion plants were built. The Mark I prototype was completed just four years after Rickover's group began designing it with Argonne and Bettis. Work on the *Nautilus* began before Mark I went into operation. Concurrent development (explained in chapter 6) meant that the ship could go to sea only two years later. Within seven more years, when twenty-nine more ships were operational, it was fair to say that nuclear propulsion had been widely adopted in the fleet.

It is not easy to find instances of technological innovation which are similar enough to the Navy project in circumstances and objectives to make a comparison worthwhile. Perhaps the best example for this purpose is the development of jet engines for military aircraft. Just as nuclear power revolutionized submarine warfare, so did jet propulsion change the nature of air power, and both of these innovations were accomplished in the years after World War II.

The first American interest in jet propulsion came in 1922 but a discouraging evaluation of the idea delayed engineering studies until 1938, when modest government efforts at Wright Field in Ohio met a similar fate. Most of the early development of jet propulsion took place in Europe, and the Germans had jet aircraft in service during the closing days of World War II. However, American efforts to build jet aircraft did not begin until the summer of 1941, when a turbojet engine already tested in England was brought to the United States to be incorporated in an American airframe. Even then the first operational aircraft, thirteen Lockheed Shooting Stars, were not completed until September 1944, and none flew in combat during World War II. The Navy did not have jet fighters until 1947 and the first Air Force bombers were not operational until 1948.[3]

The comparison can be easily overdrawn, but it does suggest the scope of Rickover's accomplishment. It took Rickover less than two years to convince the Commission and the Navy to undertake development of the technological innovation which science had shown was theoretically possible. Pressing war needs and the lack of an advocate as insistent as Rickover lengthened this step in jet propulsion development to nineteen years. Because a major advantage of the jet engine was its simplicity and small number of moving parts, the time from initial development to the prototype was very short, less than a year, but almost seven years elapsed between the first test of a jet engine in England and the completion of the first operational jet fighters in the United States. Rickover had the *Nautilus* operating in six. So successful was the *Nautilus* that the Navy at once adopted nuclear propulsion for general application in submarines. In jet propulsion at least a decade elapsed between initial development and general adoption of the technology.

Comparing the development time for the first submarine plants with other reactor projects is difficult because few were completed during the middle

1950s. This fact itself is significant: the Mark I and the Mark A were pioneering ventures. The four years of research and development that went into the Mark I prototype cost $47 million. The cost of all the equipment on the site, including the reactor, was $24 million.

Just a few miles from the Mark I site the Commission completed the materials testing reactor (MTR) in 1952. Like the Mark I, the MTR used water as the moderator and coolant, but there the similarity ended. The MTR, designed to produce a large quantity of neutrons for testing reactor materials, did not generate useful power. Nor was it necessary for the MTR to have the ruggedness or compactness of the Mark I. The MTR could use concrete shielding and could be modified for research purposes. Although the use of beryllium as a reactor material caused some difficulties, the MTR did not require the extensive development needed in Mark I. The $18 million dollar MTR facility cost only three-fourths as much as the Mark I, but research and development costs were only a third as much.[4] Even more striking was the fact that the much more expensive development of Mark I took only four years while the MTR required six. It is true that the MTR was plagued by administrative uncertainties and delays within the Commission, but Rickover's relative freedom from such problems resulted more from careful planning than from luck.

The success of a technological innovation, however, can never be judged fairly in terms of the schedule and costs of that one project alone. It is also necessary to consider the impact of that development on the total resources available to the parent organization. For a time after World War II there was a tendency among American political leaders to overlook this consideration. The oversight was encouraged by the belief that, given enough money, any project in technological innovation could reach its goal. President Eisenhower's virtually open-ended commitment to *Polaris* in 1957 and President Kennedy's decision to put an American on the moon in the 1960s reflected that kind of thinking.

By the end of that decade, however, many Americans had come to the realization that the nation's resources were not limitless. It is one thing to develop a machine with unlimited funds; it is something else to accomplish the same thing within budget constraints. In this respect the Navy nuclear project provides a useful example. During the years covered by this volume, the project never received open-ended funding or overriding priorities from the Navy, although Commission support was generous. The early prototype and submarine projects were all funded within the regular Navy and Commission budgets. The total construction costs for the thirty nuclear ships in operation by the end of 1962 were slightly more than $3 billion. The cost of all the machinery plants (of which the reactors were only a part) was just over $500 million. We can better appreciate the modest impact of the project on the Navy's resources when we consider that the *total* cost of these thirty propulsion plants was just one-fifth of the total funds expended for Navy shipbuilding and conversion for just *one year,* fiscal year 1962.[5]

Impact on Nuclear Technology

The thirty nuclear-powered ships and the Shippingport power station were only the most obvious manifestation of Rickover's accomplishment. Far more important in the long run were the contributions of the naval propulsion project to the development of nuclear technology.

Reactor engineering was in its infancy in 1946. The invitation which brought Rickover, the Navy group, and a score of engineers from industry to Oak Ridge was a frank recognition of that fact. The Oak Ridge project was intended to demonstrate the feasibility of nuclear power and to train engineers. Rickover acknowledged the primitive state of the art and the need for engineers, but he did not accept the conclusion that functional power reactors would come only in the remote future. He perceived that the nuclear sciences had already provided the essential understanding of the physical phenomena necessary to develop useful reactors. The evidence for this opinion was by no means conclusive in 1946, and Rickover's perception of the situation must stand as an almost intuitive act of great consequence.

Rickover and his associates embarked upon a quest to develop nuclear energy, not as a scientific curiosity but as a practical source of power for ship propulsion. Many of the difficulties they encountered in 1947 and 1948 stemmed from a contrary view—that more scientific data were needed before sound reactor design could begin. This opinion, held by many Commission officials and by many scientists in the laboratories, explained the small size and modest status of the Commission's reactor development branch in 1947 and 1948. During those years the Commission was giving its highest priorities to the production of fissionable materials and weapons.

Rickover's emphasis upon engineering explained his selection of Westinghouse rather than Argonne as the principal development contractor for the water-cooled reactor. It lay at the center of his dispute with General Electric over the management of Knolls and the direction of research on the sodium-cooled Mark A plant. Engineering considerations alone led him first to investigate and then reject the gas-cooled reactor for submarine propulsion. Although the Mark B plant successfully drove the *Seawolf,* Rickover abruptly terminated work on sodium-cooled systems when water-cooled reactors proved superior for ship propulsion.

Rickover thus concentrated technological development on water-cooled reactors. Now having both Westinghouse and Argonne as part of his organization, Rickover and his associates could begin converting scientific knowledge into technical specifications. Insisting always on practical engineering, Rickover drove Bettis and Argonne to accomplish the essential first steps in the creation of a technology: the collection of data on materials, the design and testing of components, and initial studies of operating systems. The results of this process can hardly be exaggerated. It led to the production of important materials like zirconium. It produced a dozen handbooks which documented the fundamentals of the new technology. It provided proven designs of essential components for water-cooled plants. This work would influence nuclear technology for decades.

Without the striking success of the Mark I, the *Nautilus,* and the Shippingport power station, water-reactor technology might not have dominated reactor development in the United States in the following decade. Shippingport demonstrated in a way a thousand paper studies never could have that nuclear power was an engineering reality rather than a scientific dream. The performance of Shippingport launched the development of civilian nuclear power in the United States and ultimately in other countries—a process which provided the industrialization of the technology. Hitherto nuclear power development had been a government monopoly. Now with the example of Shippingport before them, leaders of American industry could take practical steps to enter the nuclear field.

Just as much of that technology came directly from the naval propulsion project, so did the laying of a broad technical base in industry depend in large measure upon the techniques devised in building the nuclear fleet. The expansion of hardware production beyond Bettis and Knolls gave hundreds of fabricators and vendors their first experience in producing equipment for nuclear plants. This expansion had three important effects.

The first effect was to help create the nuclear equipment industry upon which the later rapid expansion of nuclear power plant construction depended. The same contractors who produced fuel elements, core assemblies, pressure vessels, and pumps for the nuclear fleet were prepared to fill similar orders for commercial power plants.

The second effect was to set new and unprecedented standards of precision and quality in the fabrication and assembly of nuclear equipment. The difficulties which Bettis, Knolls, PAD, and MAO experienced in obtaining suitable components for the fleet made manufacturers and vendors realize that the new standards were not an expression of unreasonable perfectionism but important to the safe and reliable operation of nuclear power plants. Industry learned the lesson slowly, but by the end of the 1960s the specifications which had seemed fantastic in the 1950s were being accepted as standards. Although this trend toward higher standards was by no means unique to the nuclear industry, it grew in this case directly out of the naval propulsion project.

The third effect of the expansion of the propulsion project was to provide the technical manpower base for the nuclear industry in the United States. The thousands of engineers and technicians trained at the Oak Ridge reactor school, in the Bettis and Knolls laboratories, in hundreds of vendors' plants, and in the nuclear power schools provided a ready supply of qualified, experienced talent to meet rapidly growing industrial requirements. Without this source of trained manpower, it seems unlikely that the nuclear industry could have grown as rapidly as it did in the 1960s.

Underlying all these accomplishments was Rickover's passion for safety and reliability in nuclear technology. A constant theme in this book, this concern for safety colored every aspect of both the Navy and industrial projects. The effects of this concern are difficult to measure because they are largely negative—the absence of widespread failure or malfunction of water reactor systems and the truly incredible record of safe operation of these plants, both military and civilian. A prudent concern with safety had been evident in the Manhattan project during World War II, but in the limited context of that effort, enforcement was relatively simple. The Commission similarly exercised great care in safety matters relating to plutonium production reactors and power reactor experiments. But the difficult task of transferring this concern to a rapidly growing military and civilian technology was in large measure accomplished by Rickover and the naval reactors branch. Without that influence, it is hard to imagine what the state of nuclear technology would be today.

It is clear, however, that the influence of the Navy project has been more than simply to accelerate the development of nuclear technology. It has also encouraged development in directions it might not have otherwise taken. Some have claimed that this same influence held back the development of new, more imaginative reactor systems which would be potentially more economical in using fissionable material. It is too early to determine the validity of that charge, but at this moment it is hard to see how concentration on more "advanced" reactor types, such as molten-salt or organic-moderated reactors, could have resulted in the large number of nuclear plants in operation and under construction today. The question of whether a new reactor system is capable of practical development is, after all, largely a matter of judgment, and Rickover's decisions were vindicated by the trend of reactor development during the next decade.

The Approach

Given the exceptional accomplishment of the naval propulsion project, what explains its success? The preceding chapters in this book contain scores of examples of the techniques used in specific situations, but like much of engineering development, they do not readily lend themselves to generalizations. In writing this book, we have followed many paths in attempting to summarize what is distinctive and useful in what we might call the Rickover approach.

Some generalizations we can draw are useful but not very distinctive. We may, for example, point out Rickover's insistence upon keeping development in the engineering rather than the scientific context. We can cite his passion for detail, his insistence upon the highest standards of quality, his preoccupation with the practical performance of equipment. Such concerns were a vital part of the Rickover approach, and they are often overlooked by managers of technical projects. But in the end they are only the elements of good engineering. It would be more to the point to say that Rickover assembled and trained a group of talented men who were able to apply the best principles of engineering in a very effective way.

It is also easy to draw generalizations which are distinctive but not very useful. The most obvious conclusion of this type is that Rickover as an individual made the difference between an ordinary development project and one which was truly exceptional. Few technical managers in our times have been willing or able to devote all their waking hours, six or seven days a week, to their jobs. Few would try to exercise the degree of control which Rickover maintained over all facets of a broad and complex project. Few would have the courage to challenge an institution as powerful and tradition-bound as the United States Navy and then carry on the fight for a generation. Even if some project leaders were sufficiently motivated to attempt such a feat, many of them would lack Rickover's intuitive skill as an engineer and administrator, qualities which have always been essential to keeping him in command of the project. Arguing the case on Rickover's unique qualities, however, has disadvantages. Uniqueness is hard to prove at best, and even if it can be demonstrated, the assertion is not very helpful. If Rickover was really unique, what then can we conclude from studying the project except that others should try to imitate him?

Important as engineering techniques and Rickover's superior personal qualities have been to the success of the project, there is an underlying principle which does have some meaning for the problems this nation faces in technological development. Put in oversimplified terms, the principle is "persons, not organizations." Many pages in this book demonstrate that this idea was more than a cliché. The first twenty years of Rickover's naval career were marked by his intense personal involvement in his assignments. These experiences strengthened his determination to retain personal control over the far-flung activities of the electrical section during World War II. While the rest of the Bureau of Ships surrendered much of its responsibility for technical design to shipyards and field installations, Rickover accepted the dispersal of design and development activities while retaining firm control over contractor and field activities in Washington.

Rickover's experiences in the electrical section served as the model for the nuclear project beginning with the Oak Ridge assignment in 1946. Avoiding commitments to organization, Rickover concentrated upon the engineering data revealed in the wartime research effort. Then he saw to it that he and his men assimilated these data in concise and accurate reports. Thus the collection of data served not only to build a base for technology but also to train men in the management of technological innovation.

Rickover's approach reached full maturity in the nuclear project. In its early years he sacrificed immediate results by concentrating on training. He sparked the formation of the graduate program in nuclear training at the Massachusetts Institute of Technology and the reactor school at Oak Ridge. Dozens of on-the-job training courses in his Washington office helped prepare a team of engineers who would be technically competent to oversee the design and development of nuclear propulsion plants. Here again the emphasis was on individuals, not systems. Rickover personally selected each engineer for his staff on the basis of the man's technical ability and personally observed his progress in training. There was no distinction between civilians and military officers. Military rank and professional standing meant nothing, technical and administrative ability everything in a project that lacked most of the organizational characteristics of a government bureaucracy.

Although the group did not exhibit many of the conventional aspects of a government project, it had a form of organization and administrative process of its own. From the beginning each member of the staff had definite responsibilities and was held personally accountable for every aspect of that responsibility even when it overlapped assignments to others (as it usually did). The creative process of design took place in the discussions involving Rickover and his senior staff—those spontaneous, probing, challenging, and usually argumentative sessions in which the validity of ideas was tested. Here each participant, including Rickover, stood on his own feet and depended upon his own knowledge, skill, and wit to advocate what he believed was right in a technical sense. Only the technically qualified took part in these discussions; administrative personnel were excluded. Intensely personal in terms of responsibility and participation, the sessions were almost devoid of personalities in that they centered on the merits of ideas and not on the institutionalized authority of those who presented them.

This application of a sort of Socratic method to the process of technological innovation provided a stimulus and a challenge for all who were involved. The method placed the stress on the unknown, the undecided, and the unresolved. It laid every assumption open to question. But most of all, it made truth and reality the supreme criteria for engineering design. In this process, Rickover functioned as the teacher and protagonist, and the validity of ideas was the only measure of merit.

In the initial project to develop the propulsion plants for the *Nautilus* and *Seawolf,* Rickover saw his relationship to Argonne, Bettis, and Knolls in this same personal context. He refused to deal with a faceless corporate entity; instead he held the laboratory director or the company president personally responsible for all activities under his authority. Like a tight-fisted customer in a country store, Rickover considered every dollar his own and demanded full value for them. He insisted that his own staff personally follow each contractor activity in detail. He frequently inspected the work of each major contractor himself and took up his differences at whatever level was required to resolve them.

In his own organization Rickover could demand full accountability from each of his staff; among the contractors he had to depend upon his leverage as the customer. At Bettis he was largely successful in imposing his principle of full personal responsibility. At Knolls he had only limited success after many years of argument. At Argonne the relationship was terminated before this issue was resolved. But in every case Rickover put the relationship in the personal context. Organization, reputation, or system did not determine the quality of a laboratory or shipbuilder. Quality was the algebraic sum of the talents of the laboratory director or company president and each member of his staff.

Rickover's determination to act in terms of persons rather than systems was one source of his troubles with the Navy, but it largely explained the success of his relationships with the Commission and the Congress. In building the *Nautilus* essentially on schedule and as planned, Rickover convinced most of the Commissioners, if not all key members of the staff, of his technical competence and administrative ability. With Commissioner Murray's strong backing, Rickover won the opportunity to build the Shippingport plant, and with its success he emerged as the Commission's most reliable producer of operating reactors. Some Commissioners sided with the staff in opposing what they considered Rickover's high-handed methods in gaining Commission support for the projects he advocated. Some bridled at Rickover's refusal to accept the technical opinions of his superiors if he believed them wrong. But most of the Commissioners could not discount Rickover's ability, his consistency, or his effectiveness. Rickover kept his facts straight and presented them with great persuasion. He seldom bothered the Commission with his problems; and when he did, he was precise about what he needed. Often tangled in a jumble of administrative and technical snarls, the Commission was usually relieved to have one less program to worry about.

Rickover's mastery of personal relationships was the key to his success with the Congress and the Joint Committee. Congress, it has been said, is a collection of individuals. Congressional committees commonly reflect the personality of their chairman, and the legislative process depends as much upon relationships between individual leaders as it does on formal procedures. Because he also was an individualist, Rickover had little trouble finding a common ground of understanding with members of Congress. The relationship was founded on mutual trust between individuals rather than on the transitory economic or political interests of legislators.

Unlike many other government officials, Rickover did not use the bureaucracy to shield himself from responsibility. Rather he presented himself as a distinctive and colorful personality, whom individual Congressmen could come to identify with nuclear propulsion. He had enough confidence in the technical competence of his own organization to speak frankly and openly to members of the Joint Committee about his successes and failures. The *Nautilus* and the Shippingport plant made a lasting impression on the committee, but so did Rickover's abrupt decision to cancel all research on sodium-cooled reactors after the *Seawolf* experience. His proprietary attitude toward government funds and his insistence upon a fair return for the government's dollar won Rickover strong support in the appropriations committees just as his success in building reactors earned the confidence of the Joint Committee. To say that Rickover was adept in the common tactics of capitalizing on the interests of individual Congressmen is to miss the point. Rickover could be as good at that game as any seasoned bureaucrat, but the source of his strength in the long run was his integrity and technical honesty. He refused to give assurances that he could not back with sound technology; he refused to promise what he could not deliver. His unwillingness to jeopardize his reputation for short-term advantages paid off handsomely in the end. Not only could he count on Congressional support for his projects; he also had in Congress an indispensable ally in his efforts to reform the Navy.

Rickover's approach was not easy to apply even in the early days when the project was small. Only stern self-discipline on the part of Rickover and his staff and a seemingly endless succession of weeks without days and days without hours made it possible to approach the standards Rickover demanded. Even then, the system was not always successful. As the "Quaker meetings" at Bettis revealed, the very intensity of the effort sometimes defeated the purpose it was intended to accomplish. An approach to management based on personal integrity and responsibility inevitably produced conflicts on the personal level—the sort of enervating, emotional clashes which conventional bureaucratic systems were designed to avoid.

All these problems were troublesome enough, but the difficulties grew with the size of the project. Rickover's approach had survived the building of the *Nautilus,* but how could it endure the demands of designing and building a nuclear fleet? Some of Rickover's staff assumed that a new approach would be necessary. The Bureau of Ships had long since given up the idea that the kind of technical management Rickover advocated was any longer practical for the highly sophisticated, diversified process of innovation in modern technology. The bureau, which in World War II had already decentralized much of the design functions to the field, now further fragmented the management functions by adopting the project system. Rickover, without giving the matter a second thought, pursued his original approach. Somehow he and his staff met the challenge, perhaps not always in the way they would have wished but at least well enough to accomplish the results which have captured our attention.

The Navy's growing reliance on the project system in the late 1950s did not in any sense mean an acceptance of Rickover's approach to technical management. In some respects the *Polaris* organization resembled the nuclear project. Rear Admiral William F. Raborn's decision to build a strong technical organization in his Washington headquarters may have been based on Rickover's success in controlling his contractors. But the sharply contrasting management styles of the two officers illustrated how important the personality of the leader could be in determining the character of a project. Raborn was a product of the Navy's officer system; Rickover fought that system throughout his career. Raborn was not a technical specialist but a seagoing line officer; Rickover was a specialist in engineering and had spent most of his career in the Bureau of Ships or in engineering duty. Admiral Burke had selected Raborn because he knew how to get along with people. Admiral Mills had sent Rickover to Oak Ridge because he would get the facts on nuclear engineering.

As a project manager, Raborn concentrated on organizational and administrative problems, leaving the engineering to his technical director. Rickover gave almost all his attention to engineering and scorned administrative activities not associated with technical problems. Raborn, who has been described as "the charismatic leader, the instinctive salesman," gave more attention to the *Polaris* image than to the realities of technology. Raborn was a master of using psychological techniques and publicity to build a feeling of competence and success.[6] Rickover focused on his technical objectives and paid less attention to publicity or organizational image.

In dealing with contractors, Raborn depended upon inspiration; Rickover, on challenge. Raborn treated contractors as members of the team, established personal ties, and used evangelistic speeches to win their support. Rickover demanded personal responsibility from his contractors but kept his relationships strictly formal. He forbade his staff to have social contacts with contractors. He was not above threatening or shaming his contractors into adequate performance.

Perhaps the sharpest distinction between the two was in the use of management systems. Rickover avoided them all, preferring to rely on his personal evaluation of a vast array of direct reporting. Raborn made a conscious effort to devise new management systems which would inspire results and build an image of managerial competence. The ultimate in this respect was the Program Evaluation Review Technique, known as PERT, a highly complex and expensive management system, which was widely adopted in government projects but produced few concrete results. According to Harvey

Sapolsky, an analyst who has studied the *Polaris* project in detail, PERT did more to give Raborn an international reputation for progressive management than did *Polaris*. Yet PERT was never accepted as a valid management device by either the *Polaris* contractors or by Raborn's staff. In the end they tolerated it, according to Sapolsky, simply because it helped Raborn sell and defend the program. Sapolsky concluded that "PERT did not build PO-LARIS, but it was extremely helpful for those who did build the weapon system to have many people believe that it did."[7] To Rickover PERT was the perfect example of the sham of management systems.

The contrasts between *Polaris* and the nuclear propulsion projects demonstrated the truism that there is no single path to technological innovation. Rickover himself denied that his success was based on any specific management methods or organization. In hearing after Congressional hearing, he proclaimed that it was the man, not the organization, that made the difference. "The key point is to assign complete responsibility for a project to a man, not to an organization. It must be understood at the top level that the man is responsible as an individual for the project. The project must be his full-time job and he must have it from beginning to end; it cannot be administered by rotating management." Furthermore, Rickover argued that unusual, even unique, methods rather than routine procedures were the essence of the project system.[8]

What made the Navy project work, Rickover argued, was the specific combination of the individual talents which were necessary to accomplish the mission. This conviction explained Rickover's opposition to the rotation of personnel either in the Navy or within contractor organizations. He held that it took years to train a man to be proficient in the peculiar kinds of technical and management problems faced in the Navy project. The idea of rotating officers after a three-year tour in Code 1500 was in Rickover's estimation the height of folly. Virtually all his senior staff agreed that the Navy's rotation system no longer made possible adequate control of technological development. Rickover convinced the Bureau of Naval Personnel to permit engineering officers to remain in Code 1500 beyond the normal tour of duty, but to do so the officer had to sign a statement recognizing that the extension would jeopardize his chances of promotion. Some officers were even willing to sacrifice their careers as naval officers by resigning or accepting early retirement in order to continue as civilian employees in the nuclear project.[9]

As Rickover and his staff accumulated years of service, they not only gained technical proficiency but the practical advantages of seniority. It was hard for an officer on a short-term assignment under the rotation system to dispute Rickover and his staff when they could muster arguments based on years of experience to counter a "new" proposal. The advantage Rickover enjoyed is suggested by the fact that from 1947 through 1962, while he was serving as the only head of the naval propulsion project, seven men served as Secretary of Defense, nine as Secretary of the Navy, seven as Chief of Naval Operations, six as Chief, Bureau of Ships, seven as Chief of Naval Personnel, five as Chairman of the Atomic Energy Commission, six as general manager, and three as director of reactor development.

Ultimately Rickover's circumvention of the rotation system, and his insistence upon technical and administrative competence especially tailored to the mission at hand, transformed the World War II conception of a project as a short-term emergency effort into a more or less permanent staff of highly specialized experts. In doing this Rickover suggested the impossibility of substituting management systems or new types of organizations for technical or administrative competence in the project manager. The point he was making was perhaps as old as human history, but it was an important one to reiterate in a day when computer technology and sophisticated systems of program analysis threatened to obscure the importance of the manager's ability.[10]

What then, in the final analysis, is the lesson of the Rickover experience? It seems clear that Rickover demonstrated the effectiveness of a highly personalized approach to technological innovation—one which was more common in the late nineteenth century than in rapidly changing, highly sophisticated technology of the late twentieth century. Though effective, the approach is incredibly difficult to apply. Its demands on the project director are so overwhelming that most would not attempt to use it. Some observers would argue that only a leader with Rickover's rare qualities could hope to use his approach satisfactorily. Others would say it is unique to Rickover himself. Yet it seems to us that the problem lies in the application and not in the fundamental validity of the approach itself. Perhaps we have become too much impressed with the complexity and sophistication of our own technology to believe that the homely virtues of intellectual integrity, technical honesty, sound analysis, and courageous decisions still have a place in managing the development of technology. Perhaps we need to remember, as Rickover has reminded us, that technology is not a self-generating, self-determining force, but an instrument which the individual can and must wield responsibly.

Appendix 1

Table of Organization

U. S. Atomic Energy Commission

Commissioners

David E. Lilienthal, chairman	Nov. 1946–Feb. 1950
Robert F. Bacher	Nov. 1946–May 1949
Sumner T. Pike	Oct. 1946–Dec. 1951
William W. Waymack	Nov. 1946–Dec. 1948
Lewis L. Strauss	Nov. 1946–April 1950
Henry D. Smyth	May 1949–Sept. 1954
Gordon E. Dean	May 1949–July 1950
Gordon E. Dean, chairman	July 1950–June 1953
Thomas E. Murray	May 1950–June 1957
T. Keith Glennan	Oct. 1950–Nov. 1952
Eugene M. Zuckert	Feb. 1952–June 1954
Lewis L. Strauss, chairman	July 1953–June 1958
Joseph Campbell	July 1953–Nov. 1954
John von Neumann	March 1955–Feb. 1957
Harold S. Vance	Oct. 1955–Aug. 1959
Willard F. Libby	June 1956–June 1959
John S. Graham	Sept. 1957–June 1962
John F. Floberg	Oct. 1957–June 1960
John A. McCone, chairman	July 1958–Jan. 1961
John H. Williams	Aug. 1959–June 1960
Robert E. Wilson	March 1960–Jan. 1964
Loren K. Olson	June 1960–June 1962
Glenn T. Seaborg, chairman	March 1961–Aug. 1971
Leland J. Haworth	April 1961–June 1963

Statutory Committees
Chairman, Joint Committee on Atomic Energy

Brien McMahon	79th Congress 1946
Bourke B. Hickenlooper	80th Congress 1947–1948
Brien McMahon	81st Congress 1949–1950
Brien McMahon	82nd Congress 1951–1952
Carl T. Durham	82nd Congress 1952
W. Sterling Cole	83rd Congress 1953–1954
Clinton P. Anderson	84th Congress 1955–1956
Carl T. Durham	85th Congress 1957–1958

Clinton P. Anderson   86th Congress 1959–1960
Chet Holifield   87th Congress 1961–1962

Chairman, General Advisory
Committee

                               Took Office

J. Robert Oppenheimer   Dec. 1946
Isidor I. Rabi   Oct. 1952
Warren C. Johnson   Oct. 1956
Kenneth S. Pitzer   Oct. 1960
Manson Benedict   March 1962

Chariman, Military Liaison
Committee

Lt. Gen. Lewis H. Brereton, USAF   Aug. 1946
Donald F. Carpenter   April 1948
William Webster   Sept. 1948
Robert LeBaron   Oct. 1949
Herbert B. Loper   Aug. 1954
Gerald W. Johnson   July 1961

The Commission Staff
General Manager

                               Took Office

Carroll L. Wilson   Dec. 1946
Marion W. Boyer   Nov. 1950
Kenneth D. Nichols   Nov. 1953
Kenneth E. Fields   May 1955
Paul F. Foster   July 1958
Alvin R. Luedecke   Dec. 1958

Deputy General Manager

Carleton Shugg   Sept. 1948
Walter J. Williams   Feb. 1951
Richard W. Cook   Oct. 1954
Paul F. Foster   Dec. 1958
Robert E. Hollingsworth   Aug. 1959

Director, Division of Reactor Development

Lawrence R. Hafstad	Feb. 1949
W. Kenneth Davis	Feb. 1954
Frank K. Pittman	Oct. 1958

Manager, Chicago Operations Office

Alfonso Tammaro	Aug. 1947
John J. Flaherty	April 1954
Kenneth A. Dunbar	Nov. 1957

Manager, Pittsburgh Area Office (later Naval Reactors Office)

Lawton D. Geiger	Dec. 1948

Manager, Schenectady Operations Office (later Naval Reactors Office)

James C. Stewart	May 1949
Jon D. Anderson	Nov. 1950
Stanley W. Nitzman	Oct. 1959

Manager, Idaho Operations Office

Leonard E. Johnston	April 1949
Allan C. Johnson	May 1954
Hugo N. Eskildson	Jan. 1962

Commission Laboratories
Argonne National Laboratory

	Took Office
Walter H. Zinn, Director	July 1946
Norman H. Hilberry, Director	Feb. 1957

Bettis Atomic Power Laboratory

Charles H. Weaver, general manager	Oct. 1948
John W. Simpson, general manager	July 1955
Philip N. Ross, general manager	July 1959

Knolls Atomic Power Laboratory

Kenneth H. Kingdon, project head	May 1946
technical manager	June 1950
William H. Milton, general manager	June 1950
Karl R. Van Tassel, general manager	June 1952
Frederick E. Crever, general manager	Dec. 1955
Bascom H. Caldwell, Jr., general manager	Feb. 1959
Kenneth A. Kesselring, general manager	Jan. 1962

The Department of Defense

Secretary of Defense

	Took Office
James V. Forrestal	Sept. 1947
Louis A. Johnson	March 1949
George C. Marshall	Sept. 1950
Robert A. Lovett	Sept. 1951
Charles E. Wilson	Jan. 1953
Neil H. McElroy	Aug. 1957
Thomas S. Gates	Jan. 1960
Robert S. McNamara	Jan. 1961

Secretary of the Navy

	Took Office
Charles Edison	Jan. 1940
Frank Knox	July 1940
James V. Forrestal	May 1944
John L. Sullivan	Sept. 1957
Francis P. Matthews	May 1949
Dan A. Kimball	July 1951
Robert B. Anderson	Feb. 1953
Charles S. Thomas	May 1954
Thomas S. Gates	April 1957
William B. Franke	June 1959
John B. Connally	Jan. 1961
Fred Korth	Jan. 1962

Chief of Naval Operations

	Took Office
Adm. Harold R. Stark	Aug. 1939
FAdm. Ernest J. King	March 1942
FAdm. Chester W. Nimitz	Dec. 1945
Adm. Louis E. Denfeld	Dec. 1947
Adm. Forrest P. Sherman	Nov. 1949
Adm. William M. Fechteler	Aug. 1951
Adm. Robert B. Carney	Aug. 1953
Adm. Arleigh A. Burke	Aug. 1955
Adm. George W. Anderson, Jr.	Aug. 1961

Chief, Bureau of Ships

	Took Office
RAdm. Samuel M. Robinson	June 1940
RAdm. Alexander H. Van Keuren	Feb. 1942
VAdm. Edward L. Cochrane	Nov. 1942
VAdm. Earle W. Mills	Nov. 1946
RAdm. David H. Clark	March 1949
RAdm. Homer N. Wallin	Feb. 1951
RAdm. Wilson D. Leggett	Aug. 1953
RAdm. Albert G. Mumma	April 1955
RAdm. Ralph K. James	April 1959

Nuclear Power Program

(The following list includes naval officers and civilian employees of the Navy and the Commission. Only senior staff whose names appear in the text are listed.)

Joseph H. Barker, Jr.	Dec. 1952–April 1958
Willis C. Barnes	June 1954–June 1964
Edward J. Bauser	Sept. 1952–Aug. 1958
James C. Cochran	Aug. 1953–Sept. 1955
John W. Crawford, Jr.	Aug. 1950–July 1963
Raymond H. Dick	June 1946–Jan. 1953
Robert W. Dickinson	Aug. 1953–April 1956
James M. Dunford	June 1946–Jan. 1961
Arthur E. Francis	July 1951–
Jack C. Grigg	Sept. 1952–
John J. Hinchey	July 1951–Dec. 1968

Theodore J. Iltis	Jan. 1951–Oct. 1965
Archie P. Kelley	Dec. 1948–Nov. 1955
Edwin E. Kintner	Aug. 1950–Nov. 1966
Jack A. Kyger	June 1948–Nov. 1954
Robert V. Laney	Dec. 1948–Aug. 1959
Vincent A. Lascara	April 1953–Sept. 1960
David T. Leighton	Aug. 1953–
Miles A. Libbey	June 1946–Jan. 1950
I. Harry Mandil	Nov. 1949–Aug. 1964
Howard K. Marks	Jan. 1950–June 1972
Sherman Naymark	April 1949–Feb. 1954
Robert Panoff	May 1950–Aug. 1964
Alvin Radkowsky	Sept. 1948–June 1972
Hyman G. Rickover	June 1946–
Theodore Rockwell	Nov. 1949–Aug. 1964
Louis H. Roddis, Jr.	July 1946–April 1955
Milton Shaw	June 1950–Sept. 1961
Samuel W. W. Shor	Feb. 1952–April 1958
Robert F. Sweek	July 1952–Oct. 1955
Marshall E. Turnbaugh	Sept. 1948–Sept. 1959
William Wegner	Aug. 1956–

Appendix 2

Construction of the Nuclear Navy, 1952–1962

No.	Name	Construction Program (Fiscal Year)	Builder	Keel Laid	Launched	Commissioned	Reactor
			Nuclear Attack Submarines				
SSN							
571	*Nautilus*	1952	Electric Boat	6/14/52	1/21/54	9/30/54	S2W
575	*Seawolf*	1953	Electric Boat	9/15/53	7/21/55	3/30/57	S2G (later S2Wa)
	Skate Class						
578	*Skate*	1955	Electric Boat	7/21/55	5/16/57	12/23/57	S3W
579	*Swordfish*	1955	Portsmouth	1/25/56	8/27/57	9/15/58	S4W
583	*Sargo*	1956	Mare Island	2/21/56	10/10/57	10/1/58	S3W
584	*Seadragon*	1956	Portsmouth	6/20/56	8/16/58	12/5/59	S4W
	Skipjack Class						
585	*Skipjack*	1956	Electric Boat	5/29/56	5/26/58	4/15/59	S5W
588	*Scamp*	1957	Mare Island	1/23/59	10/8/60	6/5/61	S5W
589	*Scorpion**	1957	Electric Boat	8/20/58	12/19/59	7/29/60	S5W
590	*Sculpin*	1957	Ingalls	2/3/58	3/31/60	6/1/61	S5W
591	*Shark*	1957	Newport News	2/24/58	3/16/60	2/9/61	S5W
592	*Snook*	1957	Ingalls	4/7/58	10/31/60	10/24/61	S5W
	Thresher Class						
593	*Thresher**	1957	Portsmouth	5/28/58	7/9/60	8/3/61	S5W
594	*Permit*	1958	Mare Island	7/16/59	7/1/61	5/29/62	S5W
595	*Plunger*	1958	Mare Island	3/2/60	12/9/61	11/21/62	S5W
596	*Barb*	1958	Ingalls	11/9/59	2/12/62	8/24/63	S5W
603	*Pollack*	1959	N.Y. Shipbuilding	3/14/60	3/17/62	5/26/64	S5W
604	*Haddo*	1959	N.Y. Shipbuilding	9/9/60	8/18/62	12/16/64	S5W
605	*Jack*	1959	Portsmouth	9/16/60	4/24/63	3/31/67	S5W
606	*Tinosa*	1959	Portsmouth	11/24/59	12/9/61	10/17/64	S5W
607	*Dace*	1959	Ingalls	6/6/60	8/18/62	4/4/64	S5W

No.	Name	Construction Program (Fiscal Year)	Builder	Keel Laid	Launched	Commissioned	Reactor
612	*Guardfish*	1960	N. Y. Shipbuilding	2/28/61	5/15/65	12/20/66	S5W
613	*Flasher*	1960	Electric Boat	4/14/61	6/22/63	7/22/66	S5W
614	*Greenling*	1960	Electric Boat	8/15/61	4/4/64	11/3/67	S5W
615	*Gato*	1960	Ingalls	12/15/61	5/14/64	1/25/68	S5W
621	*Haddock*	1961	Ingalls	4/24/61	5/21/66	12/22/67	S5W

Radar Picket Submarine

SSRN							
586	*Triton*	1956	Electric Boat	5/29/56	8/19/58	11/10/59	S4G (2 reactors)

Regulus Guided Missile Submarine

SSGN							
587	*Halibut*	1956	Mare Island	4/11/57	1/9/59	1/4/60	S3W

Hunter-Killer Submarine

SSN							
597	*Tullibee*	1958	Electric Boat	5/26/58	4/27/60	11/9/60	S2C

Fleet Ballistic Missile Submarines

George Washington Class							
SSBN							
598	*George Washington*	1958	Electric Boat	11/1/57	6/9/59	12/30/59	S5W
599	*Patrick Henry*	1958	Electric Boat	5/27/58	9/22/59	4/9/60	S5W
600	*Theodore Roosevelt*	1958	Mare Island	5/20/58	10/3/59	2/13/61	S5W
601	*Robert E. Lee*	1959	Newport News	8/25/58	12/18/59	9/16/60	S5W
602	*Abraham Lincoln*	1959	Portsmouth	11/1/58	5/14/60	3/11/61	S5W
Ethan Allen Class							
608	*Ethan Allen*	1959	Electric Boat	9/14/59	11/22/60	8/8/61	S5W
609	*Sam Houston*	1959	Newport News	12/28/59	2/2/61	3/6/62	S5W
610	*Thomas A. Edison*	1959	Electric Boat	3/15/60	6/15/61	3/10/62	S5W
611	*John Marshall*	1959	Newport News	4/4/60	7/15/61	5/21/62	S5W
618	*Thomas Jefferson*	1961	Newport News	2/3/61	2/24/62	1/4/63	S5W

No.	Name	Construction Program (Fiscal Year)	Builder	Keel Laid	Launched	Commissioned	Reactor
	Lafayette Class						
616	*Lafayette*	1961	Electric Boat	1/17/61	5/8/62	4/23/63	S5W
617	*Alexander Hamilton*	1961	Electric Boat	6/26/61	8/18/62	6/27/63	S5W
619	*Andrew Jackson*	1961	Mare Island	4/26/61	9/15/62	7/3/63	S5W
620	*John Adams*	1961	Portsmouth	5/19/61	1/12/63	5/12/64	S5W
622	*James Monroe*	1961	Newport News	7/31/61	8/4/62	12/7/63	S5W
623	*Nathan Hale*	1961	Electric Boat	10/2/61	1/12/63	11/23/63	S5W
624	*Woodrow Wilson*	1961	Mare Island	9/13/61	2/22/63	12/27/63	S5W
625	*Henry Clay*	1961	Newport News	10/23/61	11/30/62	2/20/64	S5W
626	*Daniel Webster*	1961	Electric Boat	12/28/61	4/27/63	4/9/64	S5W
627	*James Madison*	1962	Newport News	3/5/62	3/15/63	7/28/64	S5W
628	*Tecumseh*	1962	Electric Boat	6/1/62	6/22/63	5/29/64	S5W
629	*Daniel Boone*	1962	Mare Island	2/6/62	6/22/63	4/23/64	S5W
630	*John C. Calhoun*	1962	Newport News	6/4/62	6/22/63	9/15/64	S5W
631	*Ulysses S. Grant*	1962	Electric Boat	8/18/62	11/2/63	7/17/64	S5W
632	*Von Steuben*	1962	Newport News	9/4/62	10/18/63	9/30/64	S5W
633	*Casimir Pulaski*	1962	Electric Boat	1/12/63	2/1/64	8/14/64	S5W
634	*Stonewall Jackson*	1962	Mare Island	7/4/62	11/30/63	8/26/64	S5W
635	*Sam Rayburn*	1962	Newport News	12/3/62	12/20/63	12/2/64	S5W
636	*Nathanael Greene*	1962	Portsmouth	5/21/62	5/12/64	12/19/64	S5W

Nuclear Surface Ships

	Guided Missile Cruiser						
CGN 9	*Long Beach*	1957	Bethlehem-Quincy	12/2/57	7/14/59	9/9/61	C1W (2 reactors)
	Aircraft Carrier						
CVAN 65	*Enterprise*	1958	Newport News	2/4/58	9/24/60	11/25/61	A2W (8 reactors)
	Guided Missile Frigate						
DLGN 25	*Bainbridge*	1959	Bethlehem-Quincy	5/15/59	4/15/61	10/6/62	D2G (2 reactors)

* Lost at sea

Appendix 3

Financial Data

AEC Investment in the Naval Reactors Program, 1947–1963

(in millions)

AEC Research and Development Costs[1]

Submarine propulsion reactors	$476.4
Surface ship propulsion reactors	267.4
Supporting work and capital equipment	52.6
Central-station nuclear power reactor (Shippingport)	155.4
Total	$951.8

AEC Prototype Construction Costs

For Submarine Propulsion:

S1W *(Nautilus)*	$ 27.3
S1G *(Seawolf)*	27.9
S3G/S4G *(Triton)*	26.5
S1C *(Tullibee)*	13.3
S5G	12.3
Total	$107.3

For Surface Ship Propulsion:

A1W *(Enterprise)*	$ 34.8
D1G *(Bainbridge)*	34.2
Total	$ 69.0

AEC Costs for Construction of Central-Station Nuclear Power Reactor (Shippingport) $ 54.9

AEC Costs for Shipboard Nuclear Propulsion Plants[2]

Nautilus (SSN 571)	$ 16.3
Seawolf (SSN 575)	18.3
Total	$ 34.6

1. Includes original prototype operations, work related to startup and testing of the shipboard plant, development of new reactor components, and training of crews.
2. The only shipboard plants for which the AEC provided funds were for the *Nautilus* and *Seawolf*. These plants were transferred to the Navy on a nonreimbursable basis.

Navy-Funded Nuclear Propulsion Plant Research, Development, Test, and Evaluation, 1946–1963

	(in millions)
Submarine nuclear propulsion plant development	$129.5
Surface ship nuclear propulsion plant development	67.1
Nuclear propulsion plant general support	33.3
Nuclear propulsion plant application engineering	10.4
Total	$240.3

Costs for Nautilus (SSN 571) and Its Land-Based Prototype

	(in millions)
Prototype	
AEC research and development through start of prototype	$ 57.4
AEC cost for construction of prototype	27.3
Navy research and development through delivery of ship	18.6
Total	$103.3
Ship	
AEC cost for shipboard nuclear reactor plant	$ 16.3
Navy cost for construction of ship	58.2
Total	$ 74.5

Abbreviations of
Sources Cited in Notes

AAB	Papers of Admiral Arleigh A. Burke, Naval History Division, Department of the Navy, Washington, D. C.
AEC	Records of Headquarters, U. S. Atomic Energy Commission, Washington, D. C.
ANL	Records of the Argonne National Laboratory, Argonne, Illinois
BAPL	Records of the Bettis Atomic Power Laboratory, Westinghouse Electric Corporation, West Mifflin, Pa.
KAPL	Records of Knolls Atomic Power Laboratory, General Electric Company, Schenectady, New York
NAVS	Records of the Naval Ship Systems Command, Department of the Navy, Washington, D. C.
NHD	Records of the Naval History Division, Department of the Navy, Washington, D. C.
NRD	Records of the Division of Naval Reactors, U. S. Atomic Energy Commission, Washington, D. C.
PNR	Records of the Pittsburgh Naval Reactors Office, U. S. Atomic Energy Commission, West Mifflin, Pa.
TEM	Papers of Thomas E. Murray, Washington, D. C.
WAPD	Records of the Atomic Power Division, Westinghouse Electric Corporation, West Mifflin, Pennsylvania
WEC	Records of the Westinghouse Electric Corporation, Pittsburgh, Pennsylvania
WNRC	Washington National Records Center, Modern Military Records Division, National Archives and Records Service, Suitland, Maryland

Notes

The notes which follow are intended as a guide to the material we consulted and should not be considered a rigorous citation of all the documentary evidence available. Neither should the citation of specific documents be interpreted to mean that the materials are necessarily unclassified or available to the public. In fact, most of the material we consulted is closely linked to current technology and must remain classified. We have, however, in the source abbreviations, indicated where the records we used are located. Except for those materials cited as being in the files of the Atomic Energy Commission, none of the materials are now available to the historical staff, and requests for access should be directed to the organization cited in each note.

Chapter 1

1. Samuel E. Morison, *History of United States Naval Operations in World War II,* vol. 14, *Victory in the Pacific, 1945* (Boston: Little Brown, 1960), pp. 362–65.

2. *New York Times,* Oct. 3, 1945.

3. *New York Times,* Oct. 6, 1945. Nimitz's Washington statements were published in *Vital Speeches* 12 (Nov. 1, 1945): 39–41.

4. *New York Times,* Sept. 21, 1945. Nimitz's conversations with Forrestal are recorded in the unpublished Forrestal diaries and are reproduced in Edwin P. Hoyt, *How They Won the War in the Pacific: Nimitz and His Admirals* (New York: Weybright and Talley, 1970), p. 500. On the King-Nimitz relationship during the war, see the folder "Correspondence with FADM King, 1942–1945," Nimitz Papers, *NHD;* Hoyt, *How They Won the War,* pp. 40–45; King to Forrestal, Oct. 8, 1945, Nimitz Papers, *NHD.*

5. Hoyt, *How They Won the War,* p. 500.

6. COMINCH (Commander in Chief), U.S. Fleet, and CNO (Chief of Naval Operations) to COMINCH, U.S. Pacific Fleet, Aug. 30, 1945, encl. (A) of Report of The Board Convened by Order of the COMINCH, U. S. Pacific Fleet, to Report upon the Characteristics of Ships and Aircraft Types, issued Nov. 8, 1945, *NAVS,* hereafter cited as Board Report, Nov. 8, 1945.

7. Walter Millis, ed., *The Forrestal Diaries* (New York: Viking Press, 1951), p. 46; King to Forrestal, April 27, 1945, King Papers, *NHD.*

8. CNO to Distribution List, Subject: Basic Post-War Plan No. 1, May 7, 1945, King Papers, *NHD.* For King's views on the balanced fleet see Vincent Davis, *Postwar Defense Policy and the U. S. Navy, 1943–1946* (Chapel Hill: University of North Carolina Press, 1966), p. 195. See King to Forrestal, Aug. 19, 1945, and King to Vinson, Aug. 24, 1945, in King Papers, *NHD.*

9. Ship Characteristics Board Memorandum No. 48-45, Aug. 22, 1945, King Papers, *NHD.*

10. Charles O. Paullin, *History of Naval Administration, 1775–1911* (Annapolis: U. S. Naval Institute Press, 1968), pp. 201–5; Leonard D. White, *The Jacksonians: A Study in Administrative History, 1829–1861* (New York: Macmillan, 1954), pp. 213–31; Elting E. Morison, "Naval Administration in the United States," U. S. Naval Institute *Proceedings* 72 (Oct. 1946): 1303–7; Thomas W. Ray, "The Bureaus Go On Forever...," U. S. Naval Institute *Proceedings* 94 (Jan. 1968): 50–63.

11. Bradley A. Fiske, *From Midshipman to Rear-Admiral* (New York: Century, 1919), pp. 526–33, 540–89; Henry P. Beers, "The Development of the Office of The Chief of Naval Operations," *Military Affairs* 10 (Spring 1946): 40–68; ibid. 10 (Fall 1946): 10–38; ibid. 11 (Summer 1947): 88–89.

12. A good summary of the development of the CNO and the COMINCH, U. S. Fleet, is in Julius A. Furer, *Administration of the Navy Department in World War II* (Washington: Government Printing Office, 1959), pp. 102–94. An inadequate biography is Ernest J. King and Walter Muir Whitehill, *Fleet Admiral King, A Naval Record* (New York: W. W. Norton, 1952). On Roosevelt and the CNO, see Robert H. Connery, *The Navy and the Industrial Mobilization in World War II* (Princeton: Princeton University Press, 1951), pp. 27–28. The evolution of the position of Commander in Chief is traced in Richard W. Leopold, "Fleet Organization, 1919–1941" (unpublished ms., Naval History Division, 1945), pp. 1–6.

13. King and Whitehill, *Fleet Admiral King,* pp. 295–309, 628; Furer, *Administration of the Navy,* pp. 107–8, 166–67; Beers, "Development of the Office of Chief of Naval Operations," 10: 55. Navy Department Bulletin 45-275, Appointment of Ship Characteristics Board, March 15, 1945.

14. BuShips, "An Administrative History of the Bureau of Ships During World War II," 1: 17–41; a copy of the history, bound in manuscript form, is in the NAVSHIPS Library, Navy Dept., Washington. Connery, *The Navy and the Industrial Mobilization,* pp. 23–25. A good brief history of the bureau is in E. A. Wright, "The Bureau of Ships: A Study in Organization," *Journal of the American Society of Naval Engineers* 71 (Feb. 1959): 7–21. A dispute over responsibility for overweight destroyers was also a factor in the merger. See Furer, *Administration of the Navy,* pp. 217–19.

15. BuShips, "Administrative History," 2: 67–68, 163–67, and 3: 187–213.

16. For biographical data on Mills see *Army and Navy Journal,* Jan. 19, 1946. The most complete biography of Cochrane is in *Current Biography, 1951* (New York: H. W. Wilson, 1952), pp. 117–19.

17. Furer, *Administration of the Navy,* pp. 210–61; BuShips, "Administrative History," 2: 197–272.

18. For the work of the laboratory between the wars see L. S. Howeth, *History of Communications—Electronics in the United States Navy* (Washington: Government Printing Office, 1963), pp. 443–68; J. P. Baxter, 3rd, *Scientists Against Time* (Boston: Little, Brown, 1952), pp. 136–45; A. Hunter Dupree, *Science in the Federal Government: A History of Policies and Activities to 1940* (Cambridge: Harvard University Press, 1957), p. 333.

19. "The Evolution of the Office of Naval Research," *Physics Today* 14 (Aug. 1961): 30–35. Robert G. Albion and Robert H. Connery, *Forrestal and the Navy* (New York: Columbia University Press, 1962), pp. 241–42.

20. Furer, *Administration of the Navy,* pp. 754–56; Harold G. Bowen, *Ships, Machinery and Mossbacks* (Princeton: Princeton University Press, 1954), pp. 7–46, 59–77; Bowen, "Steam in Relation to Marine Engineering," *Journal of the American Society of Naval Engineers* 48 (Feb. 1936): 49–58.

21. Board Report, Nov. 8, 1945, *NAVS.*

22. A thoughtful analysis of World War II submarine operations is in Robert E. Kuenne, *The Attack Submarine, A Study in Strategy* (New Haven: Yale University Press, 1965), especially Part I, "The Conventional Submarine as a Weapons System in World War II," and Part II, "Formal Analysis of the United States and German Submarine Offensives of World War II." A good discussion of the fleet-type is in Andrew I. McKee's "Recent Submarine Design Practices and Problems," in the Society of Naval Architects and Marine Engineers, *Transactions* 67 (1959): 623–52. Data on German submarines are from Eberhard Roessler, *U-Boottyp XXI* (Munich: J. F. Lehmanns Verlag, 1967), p. 136.

23. King and Whitehill, *Fleet Admiral King,* pp. 629–37.

24. *New York Times,* Oct. 20, 24, 28, Nov. 2, 1945; Nimitz had testified before a JCS (Joint Chiefs of Staff) special committee, headed by Admiral J. O. Richardson, at Pearl Harbor on Dec. 8, 1944. Nimitz's testimony appears in Senate Committee on Military Affairs, *Department of Armed Forces, Department of Military Security, Hearings on S. 84 and S. 482, Oct. 17–Dec. 17, 1945* (Washington, 1945), pp. 411–34, hereafter cited as *Unification Hearings.* In April 1945, Nimitz cabled King that he did not have a copy of his testimony before the Richardson Committee, but he sent King the gist of it; see Nimitz to King, April 19, 1945, Nimitz Papers, *NHD.* For background on the unification fight, see Davis, *Postwar Defense Policy,* pp. 138–56, 225–34.

25. *Unification Hearings,* pp. 383–403. The quotations are taken from Nimitz's prepared statement, Nov. 17, 1945, filed in Nimitz Papers, *NHD.*

26. See note 3 of this chapter.

Chapter 2

1. Laura Fermi, *Atoms in the Family: My Life With Enrico Fermi* (Chicago: University of Chicago Press, 1954), pp. 154–56; R. B. Roberts et al., "Droplet Fission of Uranium and Thorium Nuclei," *Physical Review* 55 (Feb. 15, 1939): 416–17; Herbert L. Anderson et al., "The Fission of Uranium," *Physical Review* 55 (March 1, 1939): 511.

2. G. B. Pegram to Admiral S. C. Hooper, March 16, 1939, quoted in Fermi, *Atoms in the Family,* pp. 162–63; H. W. Graf, memorandum for file, March 17, 1939, *AEC*. Some information on early development of nuclear propulsion is in Carl O. Holmquist and Russell S. Greenbaum, "The Development of Nuclear Propulsion in the Navy," U. S. Naval Institute *Proceedings* 86 (Sept. 1960): 65–71.

3. Gunn to Bowen, June 1, 1939; Beams to Gunn, June 2, 1939; and Bowen, Memorandum on Sub-Atomic Power Sources for Submarine Propulsion, Nov. 13, 1939 (all in *AEC*).

4. A. Hunter Dupree, *Science in the Federal Government: A History of Policies and Activities to 1940* (Cambridge: Harvard University Press, 1957), pp. 344–68.

5. R. G. Hewlett and O. E. Anderson, Jr., *The New World, 1939–1946, Vol. I of A History of the U. S. Atomic Energy Commission* (University Park: Pennsylvania State University Press, 1962), pp. 14–20.

6. Harold G. Bowen, *Ships, Machinery and Mossbacks* (Princeton: Princeton University Press, 1954), pp. 137–38.

7. Bowen to Pegram, April 11, 1940; Pegram to Briggs, May 6, 1940; and Urey to Gunn, June 7, 1940 (all in *AEC*).

8. Briggs to Bowen, June 19, 1940, and Briggs to Bush, July 1, 1940, *AEC*.

9. Gunn to Bowen, Jan. 28, 1941; and J. A. Fleming to R. P. Briscoe, May 17, 1941, *AEC*.

10. Conant to Briggs, July 30, 1941; Bush to Gunn, Aug. 14, 1941; and Gunn to Bush, Aug. 18, 1941 (all in *AEC*).

11. R. P. Briscoe to Briggs, with encl., July 9, 1941, *AEC*; Hewlett and Anderson, *The New World,* pp. 66, 71–72.

12. Bowen, *Ships, Machinery and Mossbacks,* pp. 186–87. Gunn to Files, NRL, Dec. 10, 1942; Bush to W. R. Purnell, Dec. 31, 1942; and Bush to Conant, Jan. 14, 1943 (all in *AEC*).

13. Briggs et al. to Conant, Sept. 8, 1943; and Conant to Purnell, Sept. 15, 1943. J. R. Ruhoff to J. R. Dole, Oct. 11, 1943; A. H. Van Keuren to Groves, Nov. 10, 1943; and Groves to Van Keuren, Nov. 12, 1943 (all in *WNRC*).

14. U. S. Atomic Energy Commission, *In The Matter of J. Robert Oppenheimer* (Washington: Government Printing Office, 1954), pp.

164–65. Minutes, Military Policy Committee, May 10, June 21, 1944; Groves to K. D. Nichols, Sept. 1, 1945; and Monthly Report on DSM Project, Sept. 1945 (all in *WNRC*). J. E. Bigelow, The Contribution of the S-50 Plant to the Production of Uranium Contained in the First Weapon, June 30, 1961, *AEC.*

15. W. A. Shurcliff to R. C. Tolman, Nov. 13, 1944, *NRD;* Hewlett and Anderson, *The New World,* pp. 324–25.

16. Report of Committee on Postwar Policy, Dec. 28, 1944, *WNRC;* Mills, Probable Advantages of a Substitute of Fuel Oil as a Source of Power in Naval Vessels, encl., Tolman to Groves, June 12, 1945, and Solberg to Mills, Aug. 29, 1945, *NRD.* In mentioning refueling difficulties, Mills was probably referring to the loss of three destroyers which capsized and sank during a typhoon off the Philippines in December 1944. Later investigation indicated that difficulties in refueling and thus in maintaining sufficient ballast were the major causes of the losses. Hans C. Adamson and George F. Kosco, *Halsey's Typhoons* (New York: Crown, 1967); Samuel E. Morison, *History of United States Naval Operations in World War II,* vol. 13, *The Liberation of the Philippines, Luzon, Mindanao, The Visayas, 1944–1945* (Boston: Little, Brown, 1959), pp. 71–81.

17. Henry D. Smyth, *A General Account of the Development of Methods of Using Atomic Energy for Military Purposes under the Auspices of the United States Government* (Princeton: Princeton University Press, 1945).

18. Senate Special Committee on Atomic Energy, *Hearings Pursuant to Senate Resolution 179, Dec. 13, 1945* (Washington, 1946), pp. 365–67.

19. Navy Department, *Annual Report of the Secretary of the Navy, 1946* (Washington: Government Printing Office, 1947), pp. 32–37.

20. Official Report of the Chief of Naval Operations to the Secretary of the Navy Covering the Period 1 Oct. 1945 to 30 June 1946, p. 48, *NHD.*

21. For an excellent view of Navy activities on new ship designs beginning early in 1945, see CNO to Secretary of the Navy, Jan. 17, 1946, with its numerous enclosures; Cochrane to Chairman, General Board, Feb. 8, 1945; C. L. Brand, BuShips, to CNO, Feb. 8, 1946; COMINCH, Pacific Fleet, to CNO, Dec. 22, 1945 (all in CNO files, *NHD*).

22. "The Evolution of the Office of Naval Research," *Physics Today* 14 (Aug. 1961): 30–35.

23. H. A. Schade to Mills, Dec. 3, 1945, *NRD.*

24. Solberg to Mills, Dec. 29, 1945, *NRD.*

25. Hewlett and Anderson, *The New World,* pp. 408–55.

26. Russell S. Greenbaum, "Nuclear Power for the Navy: The First Decade (1939–1949)," (mimeographed, n.d.), pp. 48–67; Bowen to Forrestal, Jan. 25, 1946; Bowen to CNO, Feb. 7, 1946; W. H. P. Blandy to Bowen,

March 5, 1946; Bowen to Nimitz, March 8, 1946; and Bowen to Forrestal, March 12, 1946 (all in *NHD*).

27. E. L. Cochrane to Nimitz, March 29, 1946, and P. H. Abelson, R. E. Ruskin, and C. J. Raseman, Atomic Energy Submarine, March 28, 1946, *NRD*. Andrew I. McKee, "Recent Submarine Design Practices and Problems," Society of Naval Architects and Marine Engineers, *Transactions* 67 (1959): 637–39; Emil Kruska and Eberhard Roessler, *Walter-U-Boote* (Munich: J. F. Lehmanns Verlag, 1969), p. 81.

28. Greenbaum, "Nuclear Power for the Navy," pp. 75–76; Forrestal to Chairman, General Board, March 16, 1946; Chairman, General Board, to Chief, BuShips, March 20, 1946; and Chairman, General Board, to Secretary of the Navy, April 4, 1946 (all in CNO Files, *NHD*).

29. Cochrane to Chairman, General Board, May 16, 1946, CNO Files, *NHD.*

30. BuShips *Journal* 4 (June 1955): 8–9.

31. Most of the biographical information comes from Rickover's personal files. An interesting but rather episodic account is in Clay Blair, *The Atomic Submarine and Admiral Rickover* (New York: Henry Holt, 1954), pp. 34–80.

32. Rickover to Mills, June 4, 1947, *NRD*. For Rickover's recollections of early days at Oak Ridge see Senate Committee on Armed Services, Preparedness Investigating Subcommittee, *Inquiry Into Satellite and Missile Programs,* Jan. 6–22, 1958 (Washington, 1958), pp. 1382–88.

33. Examples of work by the group are found in the following AEC reports: M-3216, The Shielding of Power Piles, Aug. 14, 1946; M-3219, The Daniels Pile, n.d.; M-3535, Abstracts from Information Meeting Papers Presented at Argonne, April 21–23, 1947; and M-3553, Discussion of Design Factors for Gas Cooled Pile Steam Boilers, April 1, 1947 (all in *AEC*).

34. Rickover, Nuclear Energy Propulsion for Naval Vessels, Nov. 11, 1946, encl., Mills to Nichols, Nov. 12, 1946, *NRD*.

35. General Electric, Charts for Destroyer Project, May 13, 1946, *NRD;* Hewlett and Anderson, *The New World,* p. 629; Mills to Groves, Aug. 12, 1946, *WNRC;* Groves to Mills, Aug. 20, 1946, *NRD*.

36. Kendall Birr, *Pioneering in Industrial Research—The Story of the General Electric Research Laboratory* (Washington: Public Affairs Press, 1957), pp. 88–89, 160–65.

37. Price to Westinghouse Staff, Pittsburgh, Nov. 18, 1946, *NRD.*

38. Submarine Conference, Submarine Program, Jan. 8, 1947, encl., C. W. Styer to CNO, Jan. 9, 1947, *NHD*. A note in *NRD* indicates the CNO approved the program the next day.

39. BuShips, Administrative Order 47-1, Jan. 2, 1947, *NRD.*

40. R. G. Hewlett and F. Duncan, *Atomic Shield, 1947–1952, Vol. II of A History of the U. S. Atomic Energy Commission* (University Park: Pennsylvania State University Press, 1969), pp. 1–53.

41. K. H. Kingdon to C. W. LaPierre, Nov. 11, 1946, and H. W. Paige to E. S. Lee, Jan. 8, 1947, *KAPL.*

42. Report on Conference with GE, Jan. 20–22, 1947, *NRD.*

43. W. S. Parsons, Atomic Energy for Submarine Propulsion, Estimate of the Situation As Of 1947, n.d., *NRD.*

44. Rickover to Mills, June 4, 1947, *NRD.*

45. Nimitz to Mills, July 16, 1947, *NHD.*

46. General Electric Co., Survey of Atomic Plants for Naval Ship Propulsion, June 6, 1947, *NRD.*

47. Minutes, Committee on Atomic Energy, Joint Research and Development Board, July 30, 1947, *NRD.*

48. Williams Diary, July 29, 1947, *AEC.*

49. The original of Teller's letter to Hafstad, Aug. 19, 1947, has not been found, but it is printed in Joint Committee on Atomic Energy, *Atomic Power and Private Enterprise,* Dec. 1952 (Washington, 1953), p. 191. For Hafstad's reaction see Hafstad to W. A. Hamilton, Oct. 29, 1952, Records of the Joint Committee on Atomic Energy, Box 116, National Archives, Washington, D. C.

50. Rickover to Mills, Aug. 20, 28, 1947, *NRD.*

Chapter 3

1. Mills to CNO, Dec. 3, 1946, *NRD.* See also chapter 2 near note 22.

2. R. G. Hewlett and F. Duncan, *Atomic Shield, 1947–1952, Vol. II of A History of the U. S. Atomic Energy Commission* (University Park: Pennsylvania State University Press, 1969), pp. 1–58, 261–75.

3. Ibid., pp. 2–4. David E. Lilienthal, *Journals,* vol. 2, *The Atomic Energy Years, 1945–1950* (New York: Harper & Row, 1964), pp. 10–236.

4. Joint Committee on Atomic Energy, *Confirmation of Atomic Energy Commission and General Manager,* Jan. 27–31, Feb. 3–8, 10–12, 17–22, 24–26, March 3–4, 1947 (Washington, 1947), pp. 35–43 (hereafter cited as *Confirmation Hearings*); Lewis L. Strauss, *Men and Decisions* (Garden City: Doubleday, 1962), pp. 211–15.

5. *Confirmation Hearings,* pp. 65–73; R. G. Hewlett and O. E. Anderson, Jr., *The New World, 1939–1946, Vol. I of A History of the U. S. Atomic Energy Commission* (University Park: Pennsylvania State University Press, 1962), pp. 534–39, 563–66, 638–50; Hewlett and Duncan, *Atomic Shield,* pp. 6, 21.

6. Hewlett and Duncan, *Atomic Shield,* pp. 27–32.

7. Ibid., pp. 66–71, 76–79, 103–6. Farrington Daniels to File, Oct. 20, 1947, *ANL.*

8. Minutes, GAC [General Advisory Committee] Meeting 7, Nov. 22, 1947, *AEC.*

9. Development of Nuclear Reactors in the AEC, Nov. 12, 1947, *AEC.* This document was probably prepared for Fisk's use at the meeting. The meeting itself is reported in J. B. Fisk to C. L. Wilson, Dec. 30, 1947, *AEC.* For Mills's statement, see Outline of Talk at AEC, Nov. 17, 1947, *NRD.*

10. A. M. Weinberg and F. H. Murray, High Pressure Water as a Heat Transfer Medium in a Nuclear Power Reactor, April 10, 1946, AEC Report Mon P-93, *AEC.*

11. Nimitz to Sullivan, Dec. 5, 1947; Sullivan to Forrestal, Dec. 8, 1947; Sullivan to Bush, Dec. 8, 1947; and Sullivan to Mills and P. F. Lee, Dec. 8, 1947 (all in *NRD*).

12. Rickover, Memorandum Report of Travel, Dec. 15, 1947, *NRD.* Attached to this memorandum are three others describing the visit in detail. The last, dated Dec. 11, 1947, recognizes the importance of the new data on zirconium.

13. J. F. Fisk to C. L. Wilson, Jan. 7, 1948, and Minutes, Reactor Development Group, Dec. 13, 14, 1947, *AEC.*

14. Hewlett and Duncan, *Atomic Shield,* pp. 118–26.

15. Bush to AEC, Jan. 9, 1948, *AEC.*

16. Mills to AEC, Jan. 20, 1948, *AEC.*

17. Minutes, Committee on Atomic Energy, Research and Development Board Meeting 11, Feb. 5, 1948, and Conant to L. H. Brereton, Feb. 11, 1948, *AEC.*

18. Minutes, GAC Meeting 8, Feb. 6, 1948; Oppenheimer to Lilienthal, Feb. 11, 1948; and AEC Division of Military Application, Nuclear Propulsion of Ships, Feb. 19, 1948 (all in *AEC*).

19. J. McCormack to Rickover, April 9, 1948, *NRD.*

20. Solberg, Project To Be Accomplished in the AEC, March 3, 1948, *NRD.*

21. McCormack to Rickover, April 9, 1948, *NRD.*

22. A transcript of Mills's speech is in *NAVS.* A written version which Mills apparently prepared from the transcript shortly after the event is in *NRD.* For another version see Clay Blair, *The Atomic Submarine and Admiral Rickover* (New York: Henry Holt, 1954), pp. 108–12. Mills's speech was later printed in Joint Committee on Atomic Energy, *Atomic Power and Private Enterprise,* Dec. 1952 (Washington, 1953), pp. 203–6.

23. S. T. Pike to Mills, April 27, 1948, *AEC.*

24. Carpenter to Lilienthal, May 5, 1948, *AEC.*

25. Solberg, Report of Conference at Argonne, May 4, 1948, *AEC.*

26. Mills to Lilienthal, May 12, 1948, *AEC.*

27. G. L. Weil to Fisk, May 12, 1948, and Wilson to Winne, June 14, 1948, *AEC.*

28. Minutes, GAC Meeting 10, June 4, 1948, *AEC.*

29. Winne to Wilson, June 3, 1948, *AEC.*

30. Lilienthal, *Atomic Energy Years,* pp. 302–3; Walter Millis, ed., *The Forrestal Diaries* (New York: Viking Press, 1951), p. 387; Hewlett and Duncan, *Atomic Shield,* pp. 157–61.

31. Agenda for General Board Serial 315, "Study of Nature of Warfare within Next Ten Years and Navy Contributions in Support of National Security," encl., Chairman, General Board to Distribution, March 30, 1948, in General Board Files, *NHD.*

32. Presentation of Naval Force Requirements for the Joint Chiefs of Staff, April 20, 1948, encl. R. E. Libby to chairman, General Board, May 3, 1948, General Board Records, *NHD.*

33. R. A. Spruance, The Future of the Navy, speech presented at the National War College, Feb. 11, 1948, in General Board Records, *NHD.*

34. Commander, Operational Development Force to Chairman, General Board, April 30, 1948, with attachments, in General Board Records, *NHD.*

35. Hewlett and Duncan, *Atomic Shield,* pp. 155–61, 165–72.

36. Minutes of Meeting to Discuss Nuclear Energy for Propulsion of Submarines, June 16, 1948, *NRD.*

37. Carpenter to Lilienthal, May 5, 1948; Lilienthal to Carpenter, May 17, 1948; and McCormack to Wilson, July 23, 1948 (all in *AEC*).

38. Lilienthal to Solberg, June 15, 1948, and Mills to Lilienthal, July 16, 1948, *AEC.*

39. BuShips, Administrative Order 18-48, Aug. 4, 1948, *NRD.*

40. For background on the reorganization see Hewlett and Duncan, *Atomic Shield,* pp. 336–40.

41. Fisk, Reactor Development Program at Knolls Atomic Power Laboratory, July 22, 1948, *AEC.*

42. Wilson to Winne, July 27, 1948, and Wilson to Mills, July 28, 1948, *AEC.*

43. Mills to Lilienthal, Aug. 2, 1948, *AEC.*

44. Mills to Sullivan, Aug. 3, 1948; first endorsement, Denfeld to Sullivan, Aug. 3, 1948; and Sullivan to Forrestal, Aug. 4, 1948 (all in *NRD*).

45. I. Harter to J. W. Parker, Aug. 9, 1948, *NRD.*

46. Carpenter, Memorandum of Visit to Argonne Laboratory, Aug. 13, 1948, encl., Carpenter to Wilson, Aug. 14, 1948, *NRD.*

47. Carpenter, Memorandum of Discussion with Admiral Mills, Captain Rickover, and General Wilson, Aug. 13, 1948, *NRD.*

48. H. Burris to L. E. Johnston, Aug. 24, 1948, *NRD.*

49. Rickover, Memorandum of Telephone Conversation with George Bucher, Aug. 25, 1948, *NRD.*

50. Minutes of BuShips-MLC-AEC Conference on NEPS [Nuclear Energy Propulsion Systems], Aug. 25, 1948, *NRD.*

51. Carpenter to Wilson, Mills, and Parsons, Sept. 2, 1948, *NRD.*

52. Rickover, Memorandum Report of Travel, Sept. 8, 1948, *NRD.*

53. Rickover to Mills, Sept. 9, 1948, *NRD.*

54. R. C. Robin, Minutes of Meeting, Sept. 24, 1948, *KAPL.*

55. C. W. Pomeroy, Announcement of Appointments, Oct. 5, 1948, *NRD.*

56. Mills to Wilson, Sept. 20, 1948, *NRD;* Minutes, BuShips-AEC Meeting, Oct. 8, 1948, *AEC.*

57. Rickover, Report of Conference at Argonne, Oct. 26, 1948, *NRD.*

58. Zinn recognized these points in a letter to Shugg, Nov. 8, 1948, *ANL.*

59. W. P. Bigler to Zinn, Nov. 29, 1948, *ANL;* A. Tammaro to Westinghouse, Letter Contract, Dec. 10, 1948, *AEC.*

60. The meeting took place on July 26, 1948. See McCormack to Wilson, July 23, 1948, *AEC.*

61. C. H. Greenewalt to Wilson, Oct. 14, 1948; Wilson to R. P. Russell, Nov. 2, 1948; Wilson to the Program Council, Oct. 5, 1948; and General Manager's Bulletin 22, Division of Reactor Development, Sept. 15, 1948 (all in *AEC*).

62. Wilson to L. R. Hafstad, Jan. 12, 1949, and AEC Press Release 152, Jan. 16, 1949, *AEC.*

Chapter 4

1. R. G. Hewlett and F. Duncan, *Atomic Shield, 1947–1952, Vol. II of A History of the U. S. Atomic Energy Commission* (University Park: Pennsylvania State University Press, 1969), pp. 317–18, 339.

2. Ibid., pp. 29–30, 185.

3. W. P. Bigler to J. H. McKinley, Dec. 3, 1948, *ANL.*

4. Etherington et al., Study of Water-Cooled Pile for Naval Application, Sept. 1, 1948, ORNL-133, *AEC;* Etherington to Zinn, Dec. 10, 1948, ANL-HE-7, *ANL.* A second program differing from the first in minor

revision is Etherington, Program for Development of a Nuclear Reactor for Submarine Propulsion, Feb. 17, 1949, ANL-HE-27, *ANL.*

5. C. W. Pomeroy to Management Mailing List, Oct. 5, 1948, *BAPL.*

6. A brief biography of Weaver is in the *New York Times,* Oct. 7, 1948.

7. *New York Times,* Dec. 21, 1949.

8. Weaver to Chicago Operations Office, Dec. 16, 1948; Tammaro to Weaver, Jan. 8, 1949; Tammaro to Shugg, Jan. 11, 1949; and Minutes, Commission Meeting 235, Jan. 19, 1949 (all in *AEC*).

9. Draft Contract, March 15, 1949, encl., Ramey to Tammaro, March 16, 1949, and Tammaro to Hafstad, April 7, 1949, *PNR.*

10. The basis for fee and overhead payments is described in Hafstad to Wilson, Aug. 25, 1949, *AEC.*

11. H. B. Eversole, Memorandum of Conference, Jan. 18–19, 1949; Notes on Washington Meeting, April 20, 1949; Tammaro to Weaver, May 6, 1949; and N. D. Cole to Geiger, May 9, 1949 (all in *PNR*). The definitive contract, AT-11-1-GEN-14, dated July 15, 1949, is in *AEC.*

12. P. N. Ross, Report on Conference at Argonne . . . , Jan. 24, 1949, *WAPD;* Operating Program Under Contract AT-11-1-GEN-14, Feb. 21, 1949, *BAPL.*

13. On the gas-cooled approach see Brown to Rickover, Jan. 8, 1949 and encls., *ANL;* Rickover to File, Dec. 1, 1949, and Roddis to File, Sept. 6, 1950, *NRD.*

14. Explicit references to Rickover's pressure are in Etherington to Distribution, March 9, 1949; Dietrich to Etherington, March 14, 1949; Etherington to Zinn, March 15, 1949; and Zinn to Hafstad, March 21, 1949, *NRD.*

15. Rickover, Nuclear Powered Submarine Development at Westinghouse, June 28, 1949, *BAPL.*

16. Rickover, Nuclear Powered Submarine Development at Argonne, July 15, 1949, and Rickover to File, July 30, 1949, *NRD.*

17. Weaver to all supervisors, Atomic Power Division, Sept. 12, 1949, *NRD;* T. P. Evans, Report of Trip to BuShips . . . , July 12, 1949, *BAPL;* R. H. Armstrong to Etherington, Aug. 9, 1949, *ANL;* Rickover to Etherington, Aug. 8, 1949, *BAPL;* Etherington to Zinn, Aug. 16, 1949, *NRD.* The first naval reactor division newsletter was dated July 18, 1949. *ANL* has a complete file of the letters.

18. Etherington to Weaver, Aug. 1, 1949, and Weaver to Etherington, Aug. 31, 1949, *ANL.* A good summary of work and personnel strengths of Argonne and Westinghouse is in Rickover to File, Aug. 29, 1949, *NRD.*

19. Naval Reactor Coordinating Committee, Minutes of Meeting, Bettis Field, Sept. 1, 1949, *BAPL;* Minutes of Meeting, Argonne National Laboratory, Oct. 6, 1949, *ANL.*

20. Gray to Rickover, Oct. 11, 1949, *NRD.*

21. Rickover to File, Oct. 4, 1949, *NRD.*

22. Zinn to Hafstad, Oct. 13, 1949, *AEC.*

23. Rickover to File, Oct. 15, 1949, *NRD.*

24. Zinn to Etherington, Oct. 28, 1949, encl., Rough Draft–Agenda for 31 October 1949 Meeting–Argonne National Laboratory, *ANL.*

25. Dick, Nuclear Powered Submarine Meeting . . . , 31 October, 1949, *NRD.*

26. General Electric Nucleonics Project, Nucleonics Committee Minutes No. 17, Feb. 15, 1949, *KAPL;* Suits and Kingdon, Preliminary Feasibility Report on the KAPL Reactor, Jan. 12, 1949, KAPL-116, *AEC.*

27. Hewlett and Duncan, *Atomic Shield,* pp. 213–14, 217–19.

28. D. Cochran et al., Nuclear Powered Submarine, A Comparison of Development Programs, April 15, 1949, KAPL-132, *NRD.*

29. Director of Reactor Development, Site Development Authorization for Intermediate Reactor at Schenectady, Aug. 2, 1949, and Commission Meeting 298, Aug. 3, 1949, *AEC.*

30. The only reference to this visit and its impact is a remark by Suits recorded in General Electric Nucleonics Project, Nucleonics Committee Minutes, No. 19, Aug. 30, 1949, *KAPL.*

31. Winne to Shugg, Aug. 22, 1949, *AEC.*

32. General Electric Nucleonics Project, Nucleonics Committee Minutes No. 19, Aug. 30, 1949, *KAPL;* Rickover to File, Aug. 31, 1949, *NRD.*

33. Suits to Winne, Sept. 8, 1949, and General Electric Nucleonics Project, Nucleonics Committee Minutes No. 20, Sept. 20, 1949, *KAPL.*

34. Shugg to Winne, Nov. 9, 1949, *AEC.*

35. Technical Feasibility Report for the KAPL West Milton Reactor, Feb. 14, 1950, KAPL-238; West Milton Construction Feasibility, Jan. 19, 1950, KAPL-239; and "Estimate of Capital Investment and Operating Cost Requirements for the KAPL West Milton Reactor, Feb. 14, 1950, KAPL-240 (all in *AEC*); Laney, Minutes of SIR Conference, Feb. 28, 1950, *NRD;* Kingdon, Conference on Relation of WMA to SIR, Feb. 28, 1950, *KAPL;* Rickover to Shugg, March 16, 1950, *NRD;* James C. Stewart, Conference Held in Carroll Wilson's Office, March 17, 1950 . . . , *AEC.*

36. Langmuir, Notes on a Meeting to Discuss the Program of the Knolls Atomic Power Laboratory, April 13, 1950, *AEC.*

37. Evidence for the March 21 meeting with McMahon is from notes prepared by Hafstad and Rickover, Meeting between Chief, Naval Reactors Branch, and Chairman, Joint Committee on Atomic Energy, *AEC.*

38. Joint Committee on Atomic Energy, Hearing Held Before Reactor Subcommittee of the Joint Committee on Atomic Energy to Discuss Deferral of West Milton Reactor Construction, April 3, 1950, *AEC.*

39. Winne to Manager, Schenectady Operations Office, April 6, 1950, *AEC.*

40. Schenectady *Gazette,* June 6, 21, 1950.

41. Report of Conference, KAPL, Oct. 27, 1950, *NRD.*

42. Hewlett and Duncan, *Atomic Shield,* pp. 200–212, 216–19, 495–96.

43. AEC Press Release No. 165, April 4, 1949, Bulletin GM-131, effective April 4, 1949, issued July 28, 1949, *AEC.*

44. Johnston to Hafstad, Jan. 13, 1950; Hafstad to Shugg, Jan. 18, 1950; and Rickover to File, Feb. 16, 1950 (all in *NRD*).

45. Rickover to File, Feb. 20, 1950, *NRD;* Memorandum of Agreement Between the Manager, Chicago Operations Office and the Manager, Idaho Operations Office . . . , Feb. 7, 1950, *WEC.*

Chapter 5

1. On the decentralized organization of the Bureau of Ships, see General Board of the Navy, *Shipbuilding and Conversion Program, Fiscal Year 1949,* July 23, 1947, 2:397, *NHD;* BuShips Instruction 4700.3, Sept. 8, 1954, *NRD.* On the decentralized Commission, see Richard G. Hewlett and Francis Duncan, *Atomic Shield, 1947–1952, Vol. II of A History of the U. S. Atomic Energy Commission* (University Park: Pennsylvania State University Press, 1962), pp. 19–21.

2. Laney, Kelley, and LaSpada to Rickover, Nov. 21, 1949, *NRD.*

3. Roddis and Dick, Report of Trip to MIT, Jan. 13, 1949; Rickover to Adm. Wheelock, Jan. 13, 1949; and Dunford to File, Aug. 22, 1951 (all in *NRD*); Joint Committee on Atomic Energy, *Hearings on the Shortage of Scientific and Engineering Manpower,* April 18, 1956 (Washington, 1956), pp. 131–32.

4. Rickover to RADM D. H. Clark, Sept. 2, 1949; Code 390 to Code 101, May 18, 1950; Dunford to Rickover, Nov. 10, 1950; Rickover to Moore, Dec. 7, 1950; Code 390 to Code 200, Mar. 2, 1951; Kintner to Rickover, June 6, 1951; and Roddis to Rickover, Oct. 10, 1951 (all in *NRD*).

5. Dick, Draft Staff Paper on Oak Ridge School, Sept. 28, 1949; Code 200 to Codes 400, 330, and 800, April 11, 1950; and Rickover to Weinberg, Dec. 12, 27, 1950 (all in *NRD*).

6. Directors of Reactor Development and Organization and Personnel, The Reactor Development Training School at ORNL, Nov. 7, 1949; Minutes, Commission Meeting 332, Nov. 10, 1949; and AEC Press Release 218, Nov. 28, 1949 (all in *AEC*); Joint Committee on Atomic Energy, *Hearings on the Shortage of Scientific and Engineering*

Manpower, p. 133; Philip N. Powers, "The History of Nuclear Engineering Education," *Journal of Engineering Education* 54 (June 1964): 364–68.

7. Code 390 to other Codes, Dec. 2, 1949; Code 390 to Code 300, April 5, 1951; Roddis to Rickover, April 25, 1951; Kintner to Rickover, May 22, 1951; Dunford, List of People Attending Second Lecture Course, Dec. 27, 1950; Roddis to Dunford, Feb. 27, 1951; Rickover to Distribution, April 4, 1951; G. W. Faurot to S. A. Goudsmit, May 8, 1951; Rickover to All Personnel, Oct. 31, 1951, and Nov. 13, 1951; and Kintner to All Personnel, Feb. 12, 1952 (all in *NRD*).

8. Organization of Naval Reactors Branch, Jan. 16, 1950, *NRD.*

9. Rickover to Members of Naval Reactors Branch, April 18, 1950, and Rickover to Code 390, Aug. 9, 1950, *NRD.*

10. The best description of the pressurized-water reactor in early 1950 is in Etherington et al., Interim Report on Water-Cooled Water-Moderated Reactor for Naval Application, March 1, 1950, ANL-4393, *NRD.* A good nontechnical summary is in Naval Reactors Branch, The Nuclear Powered Submarine, Oct. 1, 1951, *NRD.*

11. The problems of boiler design are discussed in Bryan, First Rough Draft, Feb. 6, 1953, *NRD.*

12. A good description of the sodium-cooled reactor is in General Electric Technical Manual, Nuclear Reactor U.S.S. *Seawolf,* SSN 575, Jan. 1957, NAVSHIPS 389-0018, *KAPL.* SIR Project Preliminary Design Report, Oct. 16, 1950, KAPL-409, *AEC,* contains a good summary of the plant as it appeared on that date. The advantages of the sodium-cooled, intermediate reactor for submarine propulsion are found in Naval Reactors Branch, The Nuclear Powered Submarine, Oct. 1, 1951, *NRD.*

13. NRD files contain scores of letters and memorandums on water-corrosion studies. For a general description see D. J. DePaul, ed., *Corrosion and Wear Handbook For Water Cooled Reactors,* AEC Report TID-7006 (Washington: USAEC, 1957).

14. Carey B. Jackson, ed., *Liquid Metals Handbook, Sodium-NaK Supplement* (Washington: USAEC and BuShips, Department of the Navy, 1950). Revisions were published in 1952 and 1955.

15. Naval Group, Fundamentals of the Shielding Problem, April 2, 1947, M-3551, *NRD.*

16. Rickover, The Shielding of Power Piles, Aug. 14, 1946, M-3216, *NRD.* H. J. Muller, "Changing Genes," Public Lecture sponsored by Biology Division . . . , Oak Ridge . . . , April 8, 1947. The lecture notes taken by a member of the Navy group are filed in a binder, Project Survey and Biology, *NRD.* For a historical summary of the development of radiation protection standards see statement by Dr. Lauriston S. Taylor before

the JCAE, June 3, 1957, in Joint Committee on Atomic Energy, *The Nature of Radioactive Fallout and Its Effects on Man, Hearings,* May 27–June 3, 1957 (Washington, 1957), pp. 827–52.

17. Oak Ridge Shielding Symposium, Sept. 27–30, 1948, CF-48-10-44, *AEC;* Dunford, Shielding Conference of Nov. 14, 1949, *NRD;* Theodore Rockwell, ed., *Reactor Shielding Design Manual,* AEC Report TID-7004 (Washington: USAEC, 1956). The manual was also published commercially both in the United States and the Soviet Union.

18. Radkowsky's role in the development of the seed-and-blanket concept is described in chapter 8. The scope of reactor physics undertaken for the Navy program is suggested in A. Radkowsky, *Naval Reactors Physics Handbook,* vol. 1, *Selected Basic Techniques* (Washington: Government Printing Office, 1964).

19. A. Radkowsky and R. Brodsky, *A Bibliography of Available Digital Computer Codes for Nuclear Reactor Problems,* AEC Report AECU-3078 (Washington: Naval Reactors Branch, USAEC, 1955).

20. B. Lustman and Frank Kerze, Jr., eds., *The Metallurgy of Zirconium* (New York: McGraw-Hill, 1955); D. E. Thomas and Earl T. Hayes, *The Metallurgy of Hafnium* (Washington: Government Printing Office, n.d.); and D. W. White and J. E. Burke, eds., *The Metal Beryllium* (Cleveland: American Society for Metals, 1955).

21. For an extensive treatment of zirconium production methods see S. M. Shelton, ed., "Zirconium Production Methods," and Z. M. Shapiro, "Iodide-Decomposition Process for Production of Zirconium" from Lustman and Kerze, *Metallurgy of Zirconium,* pp. 59–215. A very good contemporary account of the status of zirconium is S. Naymark et al., Feasibility of Naval Reactor Fuel Elements Made of Zirconium-Uranium Alloy Clad with Zirconium, March 31, 1950, ANL-4436, *NRD.*

22. C. L. Wilson, Establishment of Office of New York Directed Operations, June 19, 1947, with encl., *AEC.* Rickover to director, Argonne National Laboratory, Oct. 7, 1949; Weaver to Geiger, March 21, 1950; Rickover to Hafstad, March 27 and April 10, 1950; Hafstad to Williams, Feb. 6, 1950; Dick to Hafstad, April 10, 1950; and Rickover to Hafstad, May 1, 1950 (all in *NRD*).

23. E. B. Roth et al., Zirconium Program for the Mark I Naval Reactor, July 11, 1950; Etherington and Slack to Rickover, July 14, 1950; and Rickover to File, July 17, 1950 (all in *NRD*).

24. The general tenor of internal Navy documents and some published statements by naval officers imply the system of passive review described in this paragraph (e.g., Chief, BuShips, SCAP Recommendation No. 4, May 21, 1959, *NRD*). Although the authors have found no explicit statement of the idea either in documents or books on management, we are convinced from our experience in

government and from our observations of research and development projects that the distinction is both real and fundamental.

25. This paragraph is based on interviews with a number of Rickover's representatives in the field.

26. Rickover often warned his staff with comments on pinks, such as: "You are telling them how to do the job." For example, see Code 390 to Code 500, April 24, 1950, *NRD*.

27. J. W. Crawford, Fuel Element Development at Westinghouse, Jan. 16, 1951; Crawford, Status of Fuel Element Development at Battelle, Feb. 6, 1951; Crawford, Report of the Development Status of the STR Control Rod Drive Mechanism, March 26, 1951; Dick to Weaver, March 17, 1951; and Dick to Geiger, March 29, 1951 (all in *NRD*).

28. Naymark et al., Feasibility of Naval Reactor Fuel Elements Made of Zirconium-Uranium Alloy Clad with Zirconium, March 31, 1950, ANL-4436, *NRD*. Bettis Materials and Metallurgy Dept. Memorandum for the Mark I Fuel Assembly Board, March 18, 1950, and Report of the Mark I Fuel Assembly Board, March 31, 1950, *BAPL*.

29. Etherington et al., Interim Report on Water-Cooled, Water-Moderated Reactor for Naval Application, March 1, 1950, ANL-4393, *NRD;* Minutes of Subdivision Managers' Meeting, Sept. 13, 1950, *BAPL*.

30. Wilson to Kerze, Dec. 5, 1950; Kerze to Kyger, Dec. 16, 1950; and Wilson, Hafnium Tip for Control Rod Test in ZPR (Conference held at Westinghouse), March 1, 1951 (all in *NRD*).

31. Tentative Specifications Concerning Reactor Core and Integral Control Mark I STR, June 1, 1950, ANL-HE-507, *ANL;* Roddis to File, Conference held at Westinghouse Atomic Power Division . . . , July 5, 1950, *PNR.*

32. Roddis, Interim Report on STR Core Cartridge Project, April 12, 1951; Rickover to Price, April 7, 1951; Dick, draft letter to Price, May 3, 1951; and Roddis, Trip Report on STR Individual Rod Drives, April 30, 1951 (all in *NRD*).

33. STR Project Review Board, Interim Report on Control Rod Drive Development, June 27, 1951, *BAPL*.

34. This and the following paragraphs are based on STR Project Review Board, Interim Report on STR Schedule, July 13, 1951, *NRD.*

35. Dick to Laney, Oct. 15, 18, 1950, and Rickover to Stewart, Oct. 19, 1950, and encl., Conference Report, *NRD.*

36. Report of Conference at KAPL, Jan. 26, 1951, *NRD.*

37. Laney to Rickover, undated (but sometime in January 1952), *NRD.*

38. Rickover, Conference with GE, Jan. 8, 1952, *NRD.*

39. Schenectady *Union-Star,* May 29, 1952.

Chapter 6

1. The request of March 28, 1949, and Rickover's views are cited in the ad hoc committee's reply to Admiral Momsen, April 27, 1949, *NRD*.

2. A transcript of the May 18 conference is attached to Chairman, Submarine Conference, to CNO, June 7, 1949, *NAVS*.

3. CNO to Distribution, Aug. 19, 1949, transmitting Operational Requirement SW-07601, *NRD*.

4. Oppenheimer to Carpenter, Aug. 18, 1948, and encl., Draft of the Long Range Military Objectives in Atomic Energy, *AEC*. For examples of Oppenheimer's views on military reactors see R. G. Hewlett and F. Duncan, *Atomic Shield, 1947–1952, Vol. II of A History of the U. S. Atomic Energy Commission* (University Park: Pennsylvania State University Press, 1969), pp. 17–18, 189–90, 192, 211. Oppenheimer was chairman of a group which reported in December 1950 that it saw no reason for changing its 1948 assessment on a naval reactor. See Military Objectives in the Use of Atomic Energy, Dec. 29, 1950, *AEC*.

5. CNO to Distribution, Aug. 19, 1949, *NRD*.

6. The *Guppy* and *Tang* developments can be found in E. S. Arentzen and Philip Mandel, "Naval Architectural Aspects of Submarine Design," *Transactions of the Society of Naval Architects and Marine Engineers* 68 (1960): 622–70, and A. I. McKee, "Recent Submarine Design Practices and Problems," *Transactions of the Society of Naval Architects and Marine Engineers* 67 (1959): 623–36. Dunford to Rickover, draft, Dec. 9, 1949, *NRD,* summarizes bureau work on hull development.

7. Information on closed-cycle development has been drawn from an unpublished paper by John V. Flynn, "U. S. Navy Closed-Cycle Development," on file in the Historian's Office, *AEC*. A useful article on submarine propulsion is J. H. Reinertson, L. E. Alsager, and Morely, "The Submarine Propulsion Plant—Development and Prospects," *Naval Engineers Journal,* May 1963, pp. 349–64.

8. CNO to Distribution, Aug. 19, 1949, transmitting Operational Requirement SW-07601; Barker, Naymark, and Turnbaugh to Rickover, draft, Oct. 17, 1949; Dick, Report to Naval Reactor Policy Board by the Scheduling Committee, Nov. 14, 1949; and Naval Reactor Policy Board, Meeting 1, Nov. 18, 1949, in binder: R. H. Dick, Naval Reactor Policy Board Meetings (all in *NRD*).

9. Dunford to File, Nov. 22, 1949, *NRD*.

10. Rickover to File, Dec. 6, 1949; biographical data are from New London *Day,* Feb. 27, 1956, *NRD*.

11. Rickover to File, Dec. 6, 1949, *NRD*.

12. John Niven, Courtlandt Canby, and Vernon Welch, *Dynamic America: A History of General Dynamics Corporation and Its Predecessor Companies* (New York: General Dynamics and Doubleday, n.d.), p. 283.

13. Roddis, Memorandum of Conference held . . . Jan. 6, 1950, *NRD.*

14. There are several accounts of Rickover's early contacts with Electric Boat. None are satisfactory. See Clay Blair, Jr., *The Atomic Submarine and Admiral Rickover* (New York: Henry Holt, 1954), pp. 140–42; Niven, Canby, Welch, *Dynamic America,* p. 349; and "So They Named It General Dynamics," *Fortune,* April 1953.

15. For Rickover's contract philosophy see Rickover to Stewart, Oct. 19, 1950; and encl., Rickover to File, Feb. 16, 1950, *NRD.* The contractual relations are from AEC Monthly Status and Progress Reports, Jan., Feb., Dec., 1950, *AEC.*

16. Chief, BuShips to Distribution, Nov. 22, 1949, *NRD.*

17. Chief, BuShips to CNO, Jan. 20, 1950, *NRD.*

18. Ship Characteristics Board Memo 15-50, March 23, 1950, and Dunford to Kyger, Roddis, and Dick, March 27, 1950, *NRD.*

19. Joint Committee on Atomic Energy, Hearing of the Subcommittee on Reactor Development, Feb. 9, 1950, *NRD.*

20. General Board Hearing, March 28, 1950, General Board Files, *NHD;* Dunford to File, April 10, 1950, *NRD.*

21. House Committee on Armed Services, *To Authorize the Construction of Modern Naval Vessels, and For Other Purposes. Hearings on H.R. 7764,* April 25, 1950 (Washington, 1950); P. L. 674, 81 Cong., 64 *Stat.* 420; Chief, BuShips to CNO, Aug. 3, 1950, *NRD.*

22. Oppenheimer to Carpenter, Aug. 18, 1948, and encl., Draft of Long Range Military Objectives in Atomic Energy, *AEC;* D. Cochran et al., Nuclear Powered Submarine, A Comparison of Development Programs, April 15, 1949, KAPL-132, *NRD.* For the view of the bureau chief see: Statement of Rear Admiral Homer N. Wallin, USN, Chief of the Bureau of Ships, Before the Senate Armed Services Committee on 3 March, 1953, *NRD.*

23. Concurrent development is stressed in "Mark I Equals Mark II," in Jean F. Brennan, *The Elegant Solution* (Princeton: Van Nostrand Reinhold, 1967), pp. 25–53. Concurrent development had been a common practice in the Manhattan Project during World War II. Rickover used concurrent development with the sodium-cooled as well as the water-cooled approach.

24. Code 410 to Code 390, May 29, 1950, *NRD;* Code 410 to Code 401, June 27, 1950, *NAVS.*

25. Code 410 to Code 500 to Code 400, June 29, 1950, *NAVS.*

26. Rickover to Files, July 7, 1950, *NRD.*

27. Chief, BuShips to CNO, Aug. 3, 1950, *NRD.*

28. The responsibilities of the various groups are described in several conference reports. M. J. Lawrence, Confidential Memorandum of Conference 430–94, March 8, 1950; Dunford to Files, March 7, 1950; Dunford, Report of Conference . . . Held at Electric Boat . . . on 31 Oct. 1950, *NRD.* The letter contract between Westinghouse and Electric Boat was signed by Weaver and John Jay Hopkins, President of Electric Boat, dated Feb. 28, 1950, *WEC.* The Westinghouse-Electric Boat contract for the Mark I plant layout is dated May 19, 1950, *AEC.*

29. Rickover to File, March 15, 1950, *NRD.*

30. McGaraghan, Report of Conference, Mark I Reactor Priority and Means of Speeding Its Completion, Feb. 18, 1952, *NRD.*

31. For examples of O'Grady's work see O'Grady to Dignan, Mark I Completion, May 28, 1952, O'Grady to Rickover, Trip Report, Foster Wheeler Co. . . . , June 12, 1952; and O'Grady to File, Removal of STR Components from Strike-Bound Babcock & Wilcox Plant, July 19, 1951 (all in *NRD*).

32. Dick to Rickover, March 3, 1952; Rickover to McGaraghan, March 14, 1952; and I. H. Mandil, Report of Trip . . . to Inspect the Cleanliness of Steam Generator #4 . . . , June 11, 1952 (all in *NRD*).

33. Mark I STR Design Report, Section 13, Bettis Mock-up, Aug. 2, 1952, WAPD-40-Sect-13, *AEC;* a photograph of the Electric Boat mock-up is in Mark I STR Design Report, Section 15, Plant Layout, April 20, 1952, WAPD-40-Sect-15, *AEC;* McGaraghan, Report of Conference [held] Feb. 18, 1952 . . . , *NRD.*

34. For earlier work see chapter 2 of this work, between nn 34 and 36. AEC, Monthly Status and Progress Reports, April, Oct., Nov. 1950, *AEC.*

35. Report on Reactor Site Study for KAPL, July 30, 1948, SNY-LEJ-1; Report by Division of Engineering, Acquisition of Land for Location of Intermediate Reactor, Aug. 30, 1948, Minutes, Commission Meeting 191, Sept. 10, 1948; and Commission Meeting 298, Aug. 3, 1949 (all in *AEC*).

36. E. Teller, Statement on Danger Area Regulations and on Schenectady Intermediate Reactor, Nov. 17, 1948; G. S. Mikhalapov, Minutes of Meeting on the Evaluation of the Structural Integrity of the SIR Power Plant Building, Dec. 26, 1951; encl., W. H. Milton, Jr., to J. D. Anderson, Jan. 9, 1952; and AEC Press Release 412, Feb. 21, 1952 (all in *AEC*).

37. KAPL, SIR Program Progress Report, Dec. 15, 1952–Jan. 31, 1953, KAPL-861, pp. 11, 62, *AEC.*

38. Department of Defense Press Release 1636-51, Dec. 13, 1951, and Roddis to Distribution, Jan. 4, 1952, both in *NRD.*

39. Secretary of the Navy to Chief, BuShips, Oct. 25, 1951, *NRD.*

40. Rickover briefed Truman on Feb. 9, 1952. See Schedule for Saturday, 9 Feb. 1952, *NRD.*

41. For Truman's speech see Address in Groton, Conn., at the Keel Laying of the First Atomic Energy Submarine, June 14, 1952, *Public Papers of Harry S. Truman, 1952–1953* (Washington: Government Printing Office, 1966), pp. 425–29. The ceremony is described in the New London *Evening Day,* June 14, 1952.

42. AEC Press Release No. 337, Jan. 17, 1951, *AEC.*

43. Rickover to File, Aug. 23, 1952, *NRD.*

44. For the changes Robinson was making see Scheel to Robinson, Aug. 29, 1952, and Rickover to File, Sept. 9, 1952, *NRD.* First news of the Groton meeting is in Rickover to File, Sept. 25, 1952, *NRD.*

45. Rickover to File, Sept. 25, 1952, and Turnbaugh to Kintner, Sept. 26, 1952, *NRD.*

46. Panoff to Rickover, Oct. 6, 1952, *NRD.*

47. General Dynamics Corporation Executive Order No. EB-5, Nov. 1, 1952, *NRD.*

48. Kintner to Rickover, Nov. 14, 1952, *NRD.*

49. Sylvester, Code 400 Memorandum to File, Dec. 10, 1952, *NRD.* Memorandum Code 515 to Code 500, Dec. 22, 1952, *NAVS.*

50. Code 400 to 500, 600, 490, 410, Feb. 3, 1953, *NAVS.*

51. Two good brief articles on the Mark I startup are E. E. Kintner, "Admiral Rickover's Gamble, The Landlocked Submarine," *The Atlantic Monthly* 203 (Jan. 1959): 31–35, and E. E. Kintner, "The First Days of the Mark I," *Journal of the American Society of Naval Engineers* 72 (Feb. 1960): 9–13. Kintner gives the criticality date as March 31, 1953. The log book of the Mark I located at the Naval Reactor Facility, gives the time as 11:17 p.m., March 30, 1953, Records of the Naval Reactor Facility, National Reactor Testing Station, Idaho. The discrepancy is between time zones.

52. Kintner, "The First Days of the Mark I."

53. Officer Personnel Act of 1947, P.L. 381, 80 Cong., 1 sess., 34 U.S.C. 306(a). A good summary of the act and promotion procedures is in Rexford G. Wheeler and Sheldon H. Kinney, "The Promotion of Career Officers II—Operation of the Promotion System," U. S. Naval Institute *Proceedings* 80 (July 1954): 761–71.

54. Julius A. Furer, *Administration of the Navy Department in World War II* (Washington: Government Printing Office, 1959), pp. 295–97.

55. Clay Blair, Jr., "The Dawn of Atomic Plenty: U. S. Contracts for a New Submarine," *Life* 31 (Sept. 3, 1951): 18–21; *Time* 58 (Sept. 3, 1951):

23–24, and 60 (Aug. 4, 1952): 18; C. B. Palmer, "SSN-571—Making of the Atomic Sub," *New York Times Magazine,* Oct. 26, 1952; New York *Sunday Mirror,* Nov. 16, 1952.

56. Although none of the naval officers, including Rickover, interviewed by the authors would acknowledge that religious prejudice was an influence in the Navy during these years, the authors concluded on the basis of all the oral evidence that Rickover's Jewish origins did in some instances fuel the antagonism between Rickover and his fellow officers. Religious prejudice was in our opinion a reinforcing but not a controlling factor in Rickover's unpopularity in the Navy.

57. Admiral Arleigh Burke, Chief of Naval Operations, expressed this view in a memo to file, Feb. 13, 1960, *AAB.* See also Burke to Vice Admiral Smedberg, Aug. 20, 1960, *AAB.*

58. Statement by Rear Admiral Homer N. Wallin Before Senate Armed Services Committee, March 3, 1953, *NRD.*

59. Michael Amrine, "Fifteen-Day Runaround For Atom Sub Story," *Editor and Publisher,* Feb. 7, 1953; John A. Giles in the Washington *Star,* Feb. 22, 1953.

60. *Congressional Record,* 83 Cong., 1 sess., pp. 1024–29, 1220–21, 1553–62.

61. Michael Amrine, article for North American Newspaper Alliance, Feb. 20, 1953, *NRD.*

62. Rickover to the Files, March 3, 1953, *NRD.*

63. *New York Times,* Feb. 27, 1953.

64. Anderson to Saltonstall, March 6, 1953, *NRD.*

65. "The Reminiscences of James L. Holloway, Jr." (mimeographed, Oral History Research Office, Columbia University, 1963), pp. 141–42. A copy is on file in *NHD.*

Chapter 7

1. Roddis to Etherington, Weaver, et al., Feb. 9, 1950, *NRD.*

2. CNO to Chief, BuShips, Oct. 22, 1951, *NRD.*

3. Ibid.; a draft of the memo prepared by Roddis on Oct. 12, 1951, is also in *NRD.* J. W. Logan, Minutes of Meeting at Knolls, Nov. 26, 1951, and A. Radkowsky to J. A. Kyger, Jan. 5, 1952, *NRD;* K. H. Kingdon to C. G. Suits, Nov. 30, 1951, *KAPL.*

4. H. V. Erben to W. J. Williams, Nov. 1, 1951; Erben to M. W. Boyer, Feb. 4, 1952; Minutes, Preliminary Program Conference with G. E., March 3, 1952; Director of Reactor Development, Summary of AEC Criticism of G. E., June 6, 1952, Erben to Boyer, May 29, 1952; Minutes, Commission Meeting 713, June 18, 1952; and Boyer to Erben, June 23, 1952 (all in *AEC*).

5. S. Naymark to Rickover, July 28, 1952; Roddis, Memorandum Report of Conference, Aug. 8, 1952; R. L. Schmidt and T. Trocki, SAR Progress Review, March 20, 1953; and J. W. Crawford to Rickover, April 16, 1953 (all in *NRD*).

6. Roddis to NRB, Feb. 13, 23, 1950; Roddis to File, March 23, 1950; Dunford to NRB, March 3, 1950; and Dick to NRB, Feb. 28, 1950 (all in *NRD*).

7. Sherman to Chief, BuShips, Aug. 15, 1950, *NHD*. On the effects of the Korean War, see David L. McDonald, "Carrier Employment Since 1950," U. S. Naval Institute *Proceedings* 90 (Nov. 1964): 26–33; "Floating Bases Show Their Value," *U. S. News & World Report,* Sept. 8, 1950, pp. 16–17; A. W. Jessup, " 'Mobile Bases' Carry Navy Punch," *Aviation Week* 53 (Oct. 9, 1950), 14–15. James A. Field, *History of United States Naval Operations: Korea* (Washington: Government Printing Office, 1962), pp. 60–62.

8. Sherman to JCS, Dec. 5, 1950, *NRD*. The laboratory studies completed in Oct. and early Nov. 1950 were: ORNL-50-10-114, KAPL-425, and ANL-4529, *AEC*. For correspondence leading up to Sherman's action see: D. H. Clark to CNO, Nov. 30, 1950, and L. R. Hafstad to Clark, Nov. 20, 1950, *NRD;* and Division of Reactor Development, Study of a Large Mobile Nuclear Power Plant, Nov. 17, 1950, WASH-27, *AEC.*

9. R. G. Hewlett and F. Duncan, *Atomic Shield, 1947–1952, Vol. II of A History of the U. S. Atomic Energy Commission* (University Park: Pennsylvania State University Press, 1969), pp. 417–20, 427–32, 489–95.

10. Minutes, AEC-MLC Joint Meeting 56, March 27, 1951, and Dean to R. LeBaron, April 26, 1951, *AEC;* Weaver to File, Aug. 3, 1951, *PNR;* G. L. Weil to J. J. Flaherty, Aug. 8, 1951, *NRD.*

11. Murray to Dean, Sept. 23, 1951, and Boyer to Murray, Oct. 4, 1951, *AEC;* Rickover to Hafstad, Oct. 22, 1951, *NRD;* Murray to Boyer, Oct. 26, 1951, and Boyer to Murray, Oct. 30, 1951, *AEC;* McMahon to Dean, Dec. 5, 1951, Records of Joint Committee on Atomic Energy, National Archives.

12. Weil to Geiger, Nov. 7, 1951, and Westinghouse, CVR Design Study, WAPD-50, Jan. 31, 1952, *NRD;* Weaver to Geiger, Feb. 5, 1952, *PNR;* Boyer to LeBaron, Feb. 19, 1952, and LeBaron to Boyer, Feb. 29, 1952, *AEC.* Rickover to Hafstad, Feb. 20, 1952; Rickover, Memorandum of Discussion with Admiral Withington, Feb. 28, 1952; Rickover, Memorandum of Conference, March 12, 1952 (all in *NRD*). Director of Reactor Development, Naval Aircraft Carrier Reactor Project, March 31, 1952, and Minutes, Commission Meeting 678, April 2, 1952, *AEC;* Roddis to File, April 4, 1952, *NRD.*

13. *Public Papers of Dwight D. Eisenhower, 1953* (Washington: Government Printing Office, 1960), pp. 19–20; J. M. Dodge to Dean, Feb. 3, 1953, *AEC.*

14. LeBaron to Secretary of the Navy, April 3, 1953, and J. S. Lay, Jr., to Dean, April 3, 1953, *AEC.*

15. Dean Diary, April 22, 1953; Dean, Memorandum Concerning NSC Meeting, April 22, 1953; and Lay to the Commission, April 24, 1953 (all in *AEC*).

16. Ship Design Coordinating Committee to Chief, BuShips, July 30, 1952, and Wallin to Secretary of the Navy, Aug. 19, 1952, *NRD.*

17. W. D. Leggett to Code 400, Dec. 1, 1952; Leggett to Duncan, Dec. 2, 1952; and Rickover to File, Dec. 24, 1952 (all in *NRD*).

18. Minutes, Commission Meeting 847, April 3, 1953, *AEC.*

19. The reorganization plan is briefly summarized in Carl W. Borklund, *The Department of Defense* (New York: Praeger, 1968), pp. 65–67. Dwight D. Eisenhower, *White House Years: Mandate for Change, 1953–1956* (Garden City: Doubleday, 1963), explains the reorganization and changes in the JCS, pp. 445–49.

20. The figures on ships are from the Secretary of the Navy's semiannual reports which are a part of the Secretary of Defense reports. See Department of Defense, *Semiannual Report to the Secretary of Defense, January 1 to June 30, 1950* (Washington, 1950), p. 109, and *Semiannual Report . . . , January 1 to June 30, 1953* (Washington, 1953), pp. 173–74. For the Navy in Korea see Field, *History of United States Naval Operations: Korea.*

21. Borklund, *The Department of Defense,* p. 153.

22. House Committee on Armed Services, *To Authorize the Construction of Modern Naval Vessel, and For Other Purposes, Hearings on H.R. 7764, April 25, 1950* (Washington, 1950); Ship Characteristics Board Memo 15-50, March 23, 1950, *NRD;* CNO to Secretary of the Navy, Feb. 24, 1950; CNO to Distribution List, May 29, 1950, and encl., Approved Characteristics for Conversion of Fleet Type Submarines to Antisubmarine Submarine (SSK), Shipbuilding Project No. 58, Ship Characteristics for Conversion of Fleet Type Submarine to Radar Picket Submarine (SSR) Shipbuilding Project 12A, *NAVS;* Raymond V. B. Blackman, ed., *Jane's Fighting Ships, 1969–1970* (New York: McGraw-Hill, 1969), pp. 406–12, contains photographs of the various types of submarines.

23. The evolution of the American aircraft carrier and British contributions can be traced in the annual editions of *Jane's Fighting Ships.* Another good summary is Norman Polmar, *Aircraft Carriers: A Graphic History of Carrier Aviation and Its Influence on World Events* (New York: Doubleday, 1969).

24. "New Class 'Attack' Submarines Launched" and "New Diesel to Power Submarines," both in *Marine Engineering and Shipping Review* 56

(Aug. 1951): 40–42. Edward L. Beach, *Around The World Submerged, The Voyage of the Triton* (New York: Holt, Rinehart & Winston, 1962), pp. xv–xvi; DCNO (Operations) to DCNO (Logistics), June 26, 1953, *NRD*.

25. Carney, Address Before the National Convention of the Military Order of the World Wars, Pittsburgh, Oct. 27, 1953, DOD Press Release 1016-53; Report of Conference, Naval Nuclear Propulsion Plant Development Held by the Secretary of the Navy, Dec. 6, 1953, p. 14, *NRD*.

26. Wheelock to Wallin, June 11, 1953, and Chief, BuShips, to CNO, June 26, 1953, *NRD*.

27. References to the preference for small nuclear-powered submarines are in Code 490 (H. G. Rickover) to Code 400 (E. W. Sylvester), July 29, 1953, *NRD*. Assistant CNO (Undersea Warfare) to CNO, Oct. 13, 1953, *NRD,* refers to a conference between the submarine force commanders of the Atlantic and the Pacific which called for the development of nuclear power to be concentrated on small attack submarines. Rickover later recalled the Navy's interest in small submarines in Joint Committee on Atomic Energy, *Naval Nuclear Propulsion Program, 1967–1968,* Feb. 8, 1968 (Washington, 1968), p. 164.

28. R. V. Laney, Memorandum of Meeting, Aug. 5, 1953, and Chief, BuShips, to CNO, Improved Submarine Designs, June 26, 1953, *NRD*.

29. Code 490 (Rickover) to Code 100 (Wallin), Aug. 10, 1953, *NRD*.

30. Rough Draft, Information Requested Verbally by Mr. Zuckert and Mr. McCarthy, Aug. 12, 1953, and B. E. Manseau to Secretary of the Navy, Aug. 19, 1953, *NRD*. R. B. Anderson to R. LeBaron, Aug. 22, 1953; Director of Reactor Development, Advanced Naval Reactor Project, Sept. 1, 1953; Minutes, Commission Meeting 908, Sept. 2, 1953; and Commission Meeting 912, Sept. 9, 1953 (all in *AEC*).

31. Rough Draft, Fleet Submarine Reactor, Sept. 3, 1953, WAPD-STR-524, and Rickover to Hafstad, Sept. 30, 1953, *NRD*.

32. Sylvester's reasoning is derived from Rickover's arguments. Director, Nuclear Power Division (Code 490) to Assistant Chief, BuShips (Code 400), Sept. 30, 1953, *NRD*.

33. Code 436 to Code 430, and encl., Sept. 29, 1953, and Sylvester, Memorandum for File, Oct. 13, 1953, *NRD*.

34. Rickover to Sylvester, Sept. 30, 1953, *NRD*.

35. Rickover to Code 400, Nov. 19, 1953, *NRD*.

36. Roddis, Report on Sycamore Meeting, Sept. 10–11, 1953, *NRD*.

37. Naval Reactors Branch, Partial Organization (SIR and SAR only), July 1, 1954, and Personnel Roster, Naval Reactors Branch, Aug. 24, 1954, *NRD*.

38. BuShips Notice 5430, July 11, 1955, *NRD*. The reorganization abolished the positions of assistant chiefs of bureau for research and development

(Code 300) and for ships (Code 400), creating in their stead the assistant chiefs for ship design and research (Code 400) and for shipbuilding and fleet maintenance (Code 500). Thus the principal effect of the reorganization was to move the design function from the shipbuilding to the research code.

39. Chief, BuShips to CNO, Jan. 25, 1954, *NRD;* House Subcommittee on Appropriations, 83 Cong., 2 sess., *Hearings on Department of the Navy Appropriations for 1955* (Washington, 1954), pp. 3, 41, 52.

40. Notes for Carney Meeting, undated but probably Jan. 12, 1954, and Roddis to File, Feb. 23, 1954, *NRD.*

41. Naval Message from CNO to Fleet Commanders, Type Commanders LANT, Type Commanders, PAC, Feb. 11, 1954; Rickover to Code 400, Feb. 12, 1954; and Chief, BuShips, to Chairman, Standing Committee, Long Range Shipbuilding and Conversion Plan, Feb. 17, 1954 (all in *NRD*).

42. Ship Design Coordinating Committee to Chief, BuShips, March 12, 1954, and first endorsement, June 3, 1954, *NRD.*

43. CNO to Chief, BuShips, April 20, 1954, and Rickover, Memorandum of Conference in Adm. Leggett's Office, April 26, 1954, *NRD.*

44. Chief, BuShips, to CNO, May 5, 1954, *NAVS.*

45. Chief, BuShips, to CNO, draft, June 17, 1954, and J. C. Cochran, Report of Trip, May 20, 1954, *NRD;* Lay to Secretary of Defense and Chairman, AEC, July 26, 1954; Director of Reactor Development, Large Ship Reactor Program, Aug. 12, 1954; Minutes, Commission Meeting 1020, Aug. 18, 1954; and Joint DOD-AEC Press Release 573, Oct. 15, 1954 (all in *AEC*).

46. Joint Committee on Atomic Energy, *Naval Nuclear Propulsion Program, 1967–1968,* Feb. 8, 1968 (Washington, 1968), p. 164.

47. Nuclear Submarine Propulsion School, WAPD-82, June 15, 1953, *NRD;* Dean L. Axene, " 'School of the Boat' for the *Nautilus,*" U. S. Naval Institute *Proceedings* 81 (Nov. 1955): 1229–35. On the launching see "USS Nautilus, The United States' First Nuclear Powered Submarine Is Launched," BuShips *Journal* 2 (Feb. 1954): pp. 18–21.

48. The pipe incident is described in Record of Proceedings . . . To Investigate Casualty in the Boiler and Steam Piping System on Board USS Nautilus (SSN-571) Which Occurred on 16 Sept. 1954, 4 vols., *NRD.* See also chapter 10 in this vol., between nn 49 and 53. On startup see Davis to Nichols, Jan. 4, 1955, *AEC,* and Code 500 to Code 100, Jan. 4, 1955, *NRD.*

49. Official descriptions of the trials are in Chief, Naval Reactors Branch to General Manager, AEC, Jan. 19, 24, 1955, *NRD,* and Rickover to CNO, Jan. 19, 1955, *NAVS.* Other accounts are in New London *Evening Day,*

Jan. 17, 1955; and Rickover, "Nuclear Power and the Navy," *Journal of the American Society of Naval Engineers* 68 (1956): 17–21.

50. Scrapbooks in *NRD* contain press clippings, advertisements, and excerpts from the *Congressional Record* describing the Joint Committee trip.

51. The first ships in the *Skate* class had been authorized in the 1955 shipbuilding program (P.L. 458, 83 Cong., 68 *Stat.,* 344). For the allocation to Portsmouth, see Chief, BuShips, to Commander, Portsmouth Naval Shipyard, July 30, 1954. On Rickover's views see Rickover, Memorandum of Conference Held in Leggett's Office, April 25, 1954, *NRD*. Mare Island's reactions to the assignment were reported in Oakland *Tribune,* Sept. 17, 1954.

52. Carney's statement appeared in the *Congressional Record,* 84 Cong., 1 sess., March 10, 1955, pp. 2675–77. See also "Fiscal Year 1956 Navy Shipbuilding and Conversion Program," *House Report,* 84 Cong., 1 sess., no. 209 (March 15, 1955), pp. 4–7.

53. Data on the shakedown cruise are reported in an internal status report, the Navy Nuclear Propulsion Program, a Joint AEC-Navy Program, Feb., 1966, *NRD*.

54. J. F. Calvert to Rickover, June 20, 1955, *NRD;* Commander, Antisubmarine Force, Atlantic Fleet to CNO, Sept. 22, 1955, *NAVS*.

55. COMSUBLANT (Commander, Submarine Force, Atlantic Fleet) to CNO, June 23, July 25, 1955, *NRD*.

56. D. A. Paolucci, "The Development of Navy Strategic Offensive and Defensive Systems," U. S. Naval Institute, *Proceedings* 96 (May 1970): 210–11; Harvey M. Sapolsky, "Creating the Invulnerable Deterrent: Programmatic and Bureaucratic Success in the Polaris System Development," lithographed (Mass. Institute of Technology, 1971), pp. 267–72. Most of the historical material in the manuscript was not included in Sapolsky's book *The Polaris System Development: Bureaucratic and Programmatic Success in Government* (Cambridge: Harvard University Press, 1972).

57. Price to Secretary of Defense, June 14, 1955, and Anderson to Secretary of the Navy, June 22, 1955, *NRD*.

58. Carney to Thomas, Aug. 1, 1955, *NRD*.

59. Rickover to Mumma, Aug. 15, 1955, *NRD;* Chairman, Military Liaison Committee to Chairman, AEC, Aug. 17, 1954; Director of Reactor Development, Selection of Contractor for the SRS Project, July 11, 1955; and Minutes, Commission Meeting 1092, June 2, 1955, *AEC*.

60. CINCLANTFLT (Commander in Chief, Atlantic Fleet) to CNO, Oct. 11, 1955, *NAVS*.

Chapter 8

1. Rickover to LeBaron, Feb. 13, 1951, *NRD.*

2. R. G. Hewlett and F. Duncan, *Atomic Shield, 1947–1952, Vol. II of A History of the U. S. Atomic Energy Commission* (University Park: Pennsylvania State University Press, 1969), pp. 512–14; Director of Reactor Development, Program for Development of Non-Mobile Reactors, March 11, April 8, 1952, and Zinn to File, May 13, 1952, *AEC.*

3. Hewlett and Duncan, *Atomic Shield,* pp. 437–38, 517–18. Director of Reactor Development, Industrial Participation Program, Jan. 14, 1952; Dow-Detroit Edison Proposal for Reactor Development, April 8, 1952; A. N. Anderson to File, Sept. 5, 1952; Willis Gale, Memorandum Regarding Discussion of Atomic Energy at Conference of Presidents, Sept. 9, 1952; and Director of Reactor Development, Additional Proposals for Industrial Participation in Reactor Program, Oct. 9, 1952 (all in *AEC*).

4. Director of Reactor Development, Site for Large Ship Reactor, Aug. 26, 1952, *AEC;* Hafstad to Tammaro, Aug. 21, Sept. 29, 1952, *NRD.*

5. For evidence of widespread industry interest in late 1952 see Joint Committee on Atomic Energy, *Atomic Power and Private Enterprise,* Dec., 1952 (Washington, 1952).

6. Rickover to Gordon Dean, May 15, 1953, and Rickover to File, April 30, 1953, *NRD.*

7. Murray to Dean, April 6, 1953, cited in Thomas E. Murray and the PWR Project (unpublished ms.), pp. 1, 6, 7, *TEM.* For a brief published description, see Thomas E. Murray, *Nuclear Policy for War and Peace* (Cleveland: World Publishing Co., 1960), pp. 158–59.

8. H. D. Smyth to Eisenhower, April 29, 1953, and Smyth to NSC, May 6, 1953, *AEC.*

9. Joint Committee on Atomic Energy, Transcript of Hearing on CVR Project, May 6, 1953, *AEC.*

10. W. S. Cole to John Phillips, May 20, 1953, *AEC.*

11. Joint Committee on Atomic Energy, Transcript of Hearings on FY 1954 Budget, May 18, 1953, *AEC; New York Times,* May 19, 1953.

12. Both Rickover's and McLain's approaches are included in Hafstad's report, Pressurized Water Reactor Program, May 19, 1953, *AEC.*

13. Murray to Dean, May 21, 1953, quoted in Murray and the PWR Project, p. 14, *TEM.*

14. Minutes, Commission Executive Session, June 16, 1953, and Hafstad to Tammaro, June 23, 1953, *AEC.*

15. Hafstad, Reorientation of the CVR Program, July 3, 1953; Minutes, Commission Meeting 884, July 7, 1953; Hafstad to Boyer, July 8, 1953; and Minutes, Commission Meeting 885, July 9, 1953 (all in *AEC*).

16. Dunford, who was then serving as Murray's assistant, recalled the July 9 meeting in a note in Murray and the PWR Project, p. 36, *TEM;* Cole to Strauss, July 8, 1953, *AEC*.

17. Minutes, Commission Meeting 885, July 9, 1953, *AEC*. See also Dunford's note cited in note 16 above.

18. Murray, Diary Memo 279, July 13, 1953, quoted in Murray and the PWR Project, pp. 22–23, *TEM;* Minutes, Commission Meeting 896, July 27, 1953, and Strauss to Cole, Aug. 5, 1953, *AEC*.

19. Minutes, GAC Meeting 36, Aug. 17–19, 1953, *AEC;* Murray, Diary Memo, Aug. 20, 1953, quoted in Murray and the PWR Project, p. 24, *TEM*.

20. Minutes, Commission Meeting 906, Aug. 20, 1953, *AEC;* Murray, Diary Memo, Aug. 20, 1953, quoted in Murray and the PWR Project, p. 24, *TEM*.

21. Rickover to Hafstad, July 28, 1953, *AEC*.

22. Supplement No. 18 to Contract AT-11-1-Gen-14, Oct. 9, 1953, Westinghouse Technical Progress Reports, WAPD-MRP-41, 42, 43, covering the period Aug. 15, 1953 to March 1, 1954, *AEC*.

23. Rickover to Geiger, April 22, 1952, and Weaver to All Supervisors, Atomic Power Division, May 23, 1952, *NRD*.

24. Simpson to All Supervisors, Feb. 1, March 23, 1954; Weaver to Rickover, Nov. 18, 1954; and Westinghouse, Organization Chart for Atomic Power Division, Jan. 1, 1955 (all in *NRD*). C. H. Weaver to L. E. Lynde, Report of Operations, Atomic Power Division, April 20, Aug. 24, 1954, *PNR*.

25. Murray, Diary Memo, Aug. 20, 1953, and Diary Memo 322, Oct. 5, 1953, quoted in Murray and the PWR Project, pp. 24–25, 28, *TEM;* Murray to Strauss, Oct. 12, 1953, and AEC Press Release, Oct. 22, 1953, *AEC*. The Soviet detonation was announced in AEC Press Release 495, Aug. 20, 1953, *AEC*.

26. The group consisted of the following companies: American Gas & Electric Service Corp., Bechtel Corp., Commonwealth Edison Co., Pacific Gas & Electric Co., and Union Electric Co. of Missouri.

27. Rickover, Report of Conference with South Carolina Electric & Gas Co., Nov. 14, 1953, *NRD*. Proposal from Nuclear Power Group, Nov. 17, 1953; S. C. McMeekin to Rickover, Dec. 4, 1953; Murray to the Commissioners, Dec. 1, 1953; and AEC Press Release 509, Dec. 7, 1953 (all in *AEC*).

28. Fleger to Hafstad, Feb. 12, March 6, 8, 1954, *AEC*.

29. The proposal was submitted by letter, Fleger to Dean, April 24, 1953, *AEC*. Hafstad submitted a revised proposal to the Commission on June 2, 1953. It was considered at Commission Meeting 875 on June 10,

1953, and the agreement was announced in AEC Press Release 501, Oct. 6, 1953, *AEC.*

30. Fleger later summarized some of these views in a speech before the Edison Electric Institute in Los Angeles on June 15, 1955. A copy of the speech, which Mr. Fleger provided, is in *AEC.*

31. Hafstad, PWR Participation and Site Selection, March 10, 1954; Minutes, Commission Meeting 966, March 11, 1954; and AEC Press Release 526, March 14, 1954, *AEC.*

32. Cole to Strauss, Nov. 23, 1953, and Strauss to Cole, Dec. 15, 1953, *AEC.*

33. A good example of the selection board's activities may be found in the file on the selection of Dravo, which Ramey submitted to Hafstad by memorandum, Sept. 1, 1955, *NRD.*

34. The technical aspects of the design process are covered in detail in the volume prepared by AEC for the 1958 Geneva Conference, *The Shippingport Pressurized Water Reactor* (Reading, Mass.: Addison-Wesley, 1958), hereafter cited as *Shippingport;* the material in this and the following paragraphs comes from pp. 5–24. Individual items were not listed together in a single document, but appeared in various forms of internal communications. For example, see: Rickover to Strauss, Aug. 21, 1953; Geiger, Report of Conference, April 19, 1954; and R. Kay, Report of Conference, April 4, 1955 (all in *NRD*).

35. Westinghouse prepared a series of reports covering its work on the project, e.g., Technical Progress Report, PWR Program, Aug. 15 to Nov. 1, 1953, WAPD-MRP-41, *AEC.*

36. Shaw to Geiger, Sept. 15, 1954; T. Fahrner, PWR Pressure Vessel, Recommended Outline and Principal Dimensions, Nov. 16, 1954; Shaw to Geiger, Aug. 13, 1957; G. L. Rogers, Report of Conference, Jan. 24, 1956 (all in *NRD*). For a general description of development see *Shippingport,* pp. 63–76, 381–412, 471–500.

37. Mandil, Report of Conference, May 25, 1956, *NRD; Shippingport,* pp. 76–116.

38. A. Radkowsky to T. Rockwell, Nov. 11, 1954, *NRD;* A. Radkowsky and R. T. Bayard, "The Physics Aspects of Seed and Blanket Core With Examples from PWR," P/1067, in United Nations, *Proceedings of the Second United Nations International Conference on the Peaceful Uses of Atomic Energy,* vol. 13, *Reactor Physics and Economics* (Geneva: UN, 1958), pp. 128–45; *Shippingport,* pp. 43–55, 60–63.

39. *Shippingport,* pp. 121–40; E. T. DiBerto to J. H. Barker, Jr., Dec. 13, 1954, and Report of Conference, Dec. 8, 1954, *NRD.*

40. Z. M. Shapiro, Minutes of Special Uranium-Alloy Panel Meeting, April 19, 1955; T. J. Iltis and W. H. Wilson, Report of Trip to Bettis, April 21–22, 1955; and I. H. Mandil to Geiger, June 2, 1955 (all in *NRD*).

41. AEC Press Release, Sept. 6, 1954; Strauss to K. E. Fields, May 25, 1955; and W. K. Davis to Strauss, June 7, 1955 (all in *AEC*).

42. Minutes, PWR Schedule Committee, Feb. 10, 15, 23, 1955, *NRD;* USAEC, Monthly Report to the General Advisory Committee, March, 1955, *AEC.*

43. Minutes of the Coordinating Committee are on file in *NRD* and *PNR.*

44. Minutes, Coordinating Committee, Jan. 11, 1956, *NRD.*

45. Minutes, Coordinating Committee, Aug. 24, Sept. 29, Oct. 26, 1955, July 18, 1956, *NRD.*

46. Rickover later summed up his differences with Dravo in a memorandum to W. K. Davis, the AEC director of reactor development, March 27, 1957, *AEC.* On the overtime request, see W. K. Davis to K. E. Fields, July 25, 1956, and K. E. Fields to the Commission, July 27, 1956, *AEC.*

47. Kenneth Jay, *Calder Hall* (New York: Harcourt, Brace, 1956), pp. 15–79.

48. Minutes, Coordinating Committee, Nov. 20, 1956, *NRD;* W. K. Davis to K. E. Fields, Feb. 1, 1957; and J. W. Simpson to Manager, PAO, Jan. 9, 1957, encl., K. E. Fields, Revised Cost Estimate, PWR, Feb. 14, 1957, *AEC.*

49. J. J. Flaherty to Davis, March 15, 1957, and Rickover to Davis, March 27, 1957, encls., W. K. Davis, Revised Cost Estimate, PWR, April 25, 1957, *AEC.*

50. Review Committee on PWR Project to W. K. Davis, Feb. 14, 1957, encl., W. K. Davis to K. E. Fields, Feb. 26, 1957, *NRD.* The General Accounting Office reviewed the Commission's administration of the Shippingport project in a report to the Congress, "Review of Atomic Energy Commission Shippingport Atomic Power Station, . . . June, 1958," reproduced by GAO in March 1959; copy in *AEC.*

51. Minutes, Coordinating Committee, Jan. 30, 1957, *NRD;* General Manager's Monthly Report to the Commission, March 1957, and Minutes, Commission Meeting 1278, April 30, 1957, *AEC.*

52. R. V. Laney, Report of Conference, May 2, 1957, *NRD.* Minutes of the Operations Committee on file in *NRD.*

53. Committee on Interior and Insular Affairs, House of Representatives, *Hearings on Fuel and Energy Resources,* Part II, April 14, 17, 18, 19, 1972 (Washington, 1972), p. 685.

54. Westinghouse, Technical Progress Report, Oct. 24 to Dec. 23, 1957, WAPD-MRP-71, pp. 28–37. A useful chronology of early operation is in Duquesne Light Co. and BAPL, "Shippingport Operations From Start-Up to First Refueling," DLCS-364, pp. I-36–I-39, *AEC.*

55. The Commission declassified virtually all technical information on the project at Commission Meeting 1082, May 18, 1955. C. E. Teeter, Jr., Notes on PWR Forum, Bettis Field, Aug. 27, 1954; AEC Press Release

738, Dec. 1, 1955; USAEC, "The Pressurized Water Reactor Forum Held Dec. 2, 1955, at Mellon Institute, Pittsburgh," TID-8010 (all in *AEC*). The Duquesne training course was described in DLCS-364 (cited in note 54), pp. I-21–I-35. See also AEC Press Release A-331, Dec. 11, 1958, Summary of Distribution of Students in Nuclear Power Station Training Program, Dec. 30, 1970, *AEC*.

56. DLCS-364, pp. II-38–II-67. An excellent summary of early operations appeared in "PWR: The Significance of Shippingport," *Nucleonics* 16 (April 1958):53–72.

57. USAEC, "Nuclear Reactors Built, Being Built, or Planned in the United States as of Dec. 31, 1968," TID-8200 (19th rev.), p. 7, *AEC*.

Chapter 9

1. Rickover later discussed the expansion of the project in Joint Committee on Atomic Energy, *Hearings on the Tour of the "U.S.S. Enterprise" and Report on Joint AEC-Naval Reactor Program,* March 31, 1962 (Washington, 1962), p. 51. On Navy projects see C. E. Slonim, "Project-Type Organization for Ships," *Naval Engineers Journal* 79 (Oct. 1967): 833–38. For a generic description of project management see D. I. Cleland and W. R. King, *Systems Analysis and Project Management* (New York: McGraw-Hill, 1968), pp. 151–57.

2. Carney to Thomas, Aug. 1, 1955, *NRD.*

3. Burke, Informal Notes in Binder Marked "Return to Op-00's Office," ca. May 10, 1955, *AAB.*

4. BuShips Notice 5430, July 11, 1955, *NRD.*

5. Panoff to Rickover, June 15, 1955, *NRD.*

6. For a roster of Code 1500, see W. K. Davis to Staff, June 27, 1955, *NRD.*

7. Kintner to Rickover, June 17, 1955, and Panoff and Rockwell to Rickover, June 23, 1955, *NRD.*

8. Leighton described his reasons for leaving the Navy in Senate Committee on Armed Services, *Hearings on U. S. Submarine Program,* March 13, 1968 (Washington, 1968), pp. 91–94. Rickover attacked the Navy's rotation policy in the same hearing (pp. 94–98) and in Senate Committee on Armed Services, *Hearings on Missiles, Space, and Other Defense Matters,* Feb. 3, 1960 (Washington, 1960), pp. 171–73; Joint Committee on Atomic Energy, *Naval Nuclear Propulsion Program,* Jan. 26, 1966 (Washington, 1966), pp. 22–25.

9. Senate Armed Services Committee, *Hearings on U. S. Submarine Program,* March 13–27, 1968 (Washington, 1968), p. 97.

10. Chief, BuShips, to CNO, Sept. 7, 1955, and Summary Sheet, Sept. 13, 1955, *NRD.*

11. Rickover to Mumma, Sept. 17, 21, 1955, *NRD.*

12. Burke to D. B. Duncan, Oct. 19, 1955; Burke, Memo for the Record, Oct. 21, 1955; Burke to J. E. Clark and W. F. Raborn, Dec. 2, 1955 (all in *AAB*). W. D. Miles, "The Polaris," *Technology and Culture* 4 (1963): 478–80; D. A. Paolucci, "The Development of Navy Strategic Offensive and Defensive Systems," U. S. Naval Institute, *Proceedings* 96 (May 1970): 214–15.

13. Statement of Admiral Burke Before the House Armed Services Committee, Jan. 18, 1956, encl., OPNAVNOTE 5720, Jan. 26, 1956, *NRD;* House Appropriations Committee, *Department of the Navy Appropriations for 1957,* March 9, 1956 (Washington, 1956), pp. 505–6.

14. Chairman, Ship Characteristics Board to Chief, BuShips, Aug. 24, 1955, *NRD;* Chairman, Military Liaison Committee to Chairman, AEC, Oct. 20, 1955, *AEC.*

15. E. S. Arentzen and P. Mandel, "Naval Architectural Aspects of Submarine Design," *Transactions of the Society of Naval Architects and Marine Engineers* 68 (1960): 622–70; A. I. McKee, "Recent Submarine Design Practices and Problems," *Transactions . . .* 67 (1959): 623–36; W. D. Roseborough, Jr., "The Evolution of the Attack Submarine," reprinted from *Sperryscope* in *Naval Engineers Journal* 74 (Aug. 1962): 425–29; Norman Polmar, *Atomic Submarines* (Princeton: Van Nostrand Reinhold, 1963), pp. 140–43.

16. Harry Jackson, "USS *Albacore:* The 'New Look' in Submarine Design," BuShips *Journal* 2 (March 1954): 2–4. David Taylor Model Basin, USS *Albacore* Standardization Trials and a Comparison of Trial Instrument Performance, April 1956, DTMB Report C-770, and R. P. Metzger to W. C. Barnes, July 11, 1955, *NRD.*

17. V. A. Lascara to NR Staff, Oct. 14, 1955, *NRD.*

18. House Appropriations Committee, *Hearings on Department of Defense Appropriations, Fiscal Year 1962,* Part 6, May 11, 1961 (Washington, 1961), p. 6; Joint Committee on Atomic Energy, *Tour of USS "Enterprise" and Report on Joint AEC-Naval Reactor Program,* March 31, 1962 (Washington, 1962), pp. 52–53.

19. Rickover to R. J. Cordiner, GE president, Nov. 5, 1954; J. D. Anderson to K. R. Van Tassel, Feb. 14, 1955; and F. K. McCune to Rickover, Jan. 11, 1957, *NRD.*

20. KAPL Chronological History, encl., J. G. Shaw to General Manager, KAPL, Aug. 24, 1959, *KAPL;* AEC Press Release 663, July 13, 1955, *AEC.*

21. K. E. Fields to W. P. Brobeck, Aug. 17, 1955, *AEC;* Rickover to File, Aug. 17, 1955, and Rickover to Chief, BuShips, Sept. 5, 1955, *NRD.*

22. KAPL Chronological History, pp. 22, 24; P. K. Taylor, Code 525, BuShips to File, Sept. 5, 1956; A. E. Francis to Rickover, Sept. 12, 1956; and SIR Review Committee to Rickover, Sept. 27, 1956 (all in *NRD*).

23. Rickover to W. K. Davis, Nov. 2, 1956; Davis to K. E. Fields, Nov. 29, 1956; and Rickover to CNO, First Sea Trials of the *Seawolf,* Jan. 22, 1957 (all in *AEC*).

24. D. T. Leighton to Manager, Schenectady Office, Revised SAR Power Plant Specifications, Jan. 27, 1955; Rickover to F. K. McCune, Feb. 25, 1955; McCune to Rickover, March 2, 1955; Rickover to File, Responsibility for SAR Project, March 11, 1955; and Leighton to Rickover, April 13, 1955 (all in *NRD*).

25. E. E. Kintner to Rickover, SCB Working Level Meeting on SAR, Jan. 20, 1955; Chief, BuShips to CNO, June 10, 1955; and Chairman, Ship Characteristics Board to Chief, BuShips, Aug. 3, 1955 (all in *NRD*).

26. Conference on SAR Project, July 14, 1955; D. T. Leighton, Conference Report on SAR Project, Sept. 15, 1955; J. D. Anderson to C. Shugg, Oct. 19, 1955; and Leighton, Conference on S3G Project, Dec. 27, 1955, July 11, 1956 (all in *NRD*).

27. KAPL Monthly Report, April 1953; Rickover to Van Tassel, Nov. 3, 1953; McCune to Rickover, Dec. 21, 1953; and J. D. Anderson to Rickover, May 21, 1954 (all in *NRD*).

28. Rockwell to Rickover, July 11, 1955; Rickover to Mumma, July 18, Oct. 17, 1955; and Rickover to McCune, July 18, 1955, May 2, 1956 (all in *NRD*).

29. McCune to Rickover, April 23, 1956; W. K. Davis to K. E. Fields, Aug. 23, 1956; and McCune to J. D. Anderson, Aug. 23, 1956 (all in *NRD*). Also in Joint Committee on Atomic Energy, *Hearings on Nuclear Propulsion for Naval Surface Vessels,* Oct. 30–Nov. 13, 1963 (Washington, 1964), pp. 224–29.

30. Strauss to Secretary of the Navy, Aug. 9, 1956, *AEC;* Rockwell, Report of Conference on Study, July 24, 1957, *NRD.* Code 1500's critique of the proposal appears as an appendix in Report of Conference on Naval Nuclear Propulsion Plant Development Held by the Secretary of the Navy, Dec. 6, 1963, *NRD.*

31. On the efforts of the three companies, see the booklet issued by the Bath Iron Works, General Electric, and Gibbs & Cox, Letter to the Honorable Secretary of Defense, "Design and Construction of Nuclear Powered Destroyer/Frigate Type Ships," June 30, 1958, *NRD.*

32. C. H. Weaver to L. E. Lynde, Report of Government Operations at Atomic Power Division, Jan. 21, 1955, and Robert Heller & Assoc., Inc., Management Study of Atomic Power Division, March 9, 1955, *PNR.*

33. A. Squire to Weaver, Monthly SFR Report, Jan. 21, 1955, March 24, 1955, and SFR Core Section, Conceptual Design of SFR Core I, Report WAPD-SFR-R-141, April 1, 1955, *PNR.*

34. Squire to Weaver, Monthly SFR Report, April 21, 1955; M. Shaw, Report of Conference on SFR Reactor Compartment Arrangements, Feb. 10, 1955; and Weaver to Lynde, April 21, 1955 (all in *PNR*).

35. J. C. Cochran to L. D. Geiger, LSR Plant Specifications and Development Objectives, Jan. 3, 1955, *NRD;* J. C. Rengel, LSR Project Manager to Weaver, Monthly LSR Reports, Jan. 21, March 23, 1955, *PNR.*

36. J. T. Stiefel to J. W. Simpson, Monthly A1W Management Report, Oct. 21, Dec. 23, 1955, March 23, 1956, *PNR.* Chief, BuShips, to CNO, Sept. 13, 1955; Rickover, Conference with Bethlehem Steel on C1W Project, April 11, 1956; P. D. Foote to L. L. Strauss, Oct. 8, 1957; Strauss to Foote, Nov. 13, 1957; and V. A. Lascara, Action Taken to Reprogram Work on the C1W and A1W Projects and to Re-Designate the F1W Project, Jan. 17, 1955 (all in *NRD*).

37. Simpson to Weaver, Report of Operations, Bettis Plant, Oct. 21, 1955, and J. E. Mealia, Report of Conference on S5W Nuclear Reactor Design, Nov. 18, 1955, *PNR.*

38. For a comprehensive description of the S5W core and its development, see Bettis Report, S5W Core Design Report, WAPD-S5W-R-150, May 1956, *WAPD.*

39. Simpson to Weaver, Report of Operation, Bettis Plant, Feb. 24, 1956, and D. C. Spencer to Simpson, Monthly S5W Reports, Feb. 19, May 22, 1957, Jan. 21, June 17, 1958, *PNR;* E. J. Takitch to R. V. Laney, S5W Weekly Progress Reports, Oct. 3, 1956, March 6, 1957, *NRD.*

40. Weaver to L. E. Osborne, Report on Operations, Atomic Power Division, Sept. 25, 1953, and Weaver to L. E. Lynde, Report on Operations, Atomic Power Division, Jan. 23, 1954, *PNR.*

41. Weaver to Osborne, Report of Operations, Atomic Power Division, July 24, 1953, *PNR.*

42. R. V. Laney to Rickover, Aug. 13, 1956, *NRD.*

43. Although literally thousands of documents in NRD files illustrate the activities of PAD, virtually all of them are concerned with engineering details rather than with the general mission of the organization. This and the following paragraph were based on general reading in the files and discussions with those involved.

44. Borden to Weaver, Dec. 22, 1958, *NRD.*

45. This section is based extensively on an informal *PNR* report written in 1968 and entitled "Summary of the Supply of Zirconium." PNR also has detailed files on all zirconium contracts. The most useful documents are USAEC, Invitation for Bids on the Supply of Zirconium and Hafnium, Nov. 5, 1951, and J. J. Flaherty to W. K. Davis, June 28, 1955, with enclosures, *PNR.*

46. Weaver to Osborne, Report of Operations, July 24, 1953, *PNR.*

47. Weaver to Geiger, draft, July 26, 1954; Geiger to H. R. Kelly et al., Aug. 16, 1954; and Kelly to Geiger, Sept. 7, 1954 (all in *PNR*).

48. Simpson to Weaver, Report of Operations, Aug. 19, 1955, *PNR.*

49. E. S. Wolslegel to Laney et al., SFR Core Vendor Justification, Dec. 13, 1955, *PNR.*

50. Geiger, Report of Conference on Core Procurement, Aug. 9, 1956, *PNR.*

51. Simpson to Weaver, Report of Operations, Jan. 20, 1956, and Geiger to J. J. Flaherty, Procurement of S5W Cores, March 16, 1957, *PNR.*

52. W. K. Davis to K. E. Fields, Procurement of Navy Reactor Cores for the Navy by the Commission, April 11, 1957, *PNR.*

53. Geiger to Flaherty, March 5, Aug. 10, 1956; Simpson to Weaver, March 23, 1956; Eisenschmidt to Simpson, Nuclear Core Department Monthly Report, May 22, 1957; Conference Report on S5W Cores, June 6, 1957; and Geiger to D. Saxe, Award of Contract for S5W Cores, June 27, 1957 (all in *PNR*).

54. Davis to Rickover, Aug. 7, 1957, *PNR.*

55. Rickover to Davis, Aug. 8, 1957, and Laney to Rickover, Sept. 5, 1957, *PNR.*

56. As for PAD operations, the documentation of core production activities is both too detailed and voluminous to cite comprehensively. The following documents illustrate the kinds of materials available: Simpson to Weaver, Report of Operations, Feb. 18, 1958, *PNR;* and I. H. Mandil to Geiger, Evaluation of Fuel Element Bond Defects, April 21, 1960; Rickover to L. S. Wilcoxson, Oct. 6, 1960; E. Hartshorne to Rickover, Oct. 12, 1960; and G. L. Williams to Rickover, Oct. 14, 1960 (all in *NRD*). The attempt to build a commercial core industry is discussed in Laney to Files, Sept. 4, 1957, *PNR.*

57. AEC Press Release A-67, March 27, 1958, *AEC.*

Chapter 10

1. Rickover to Chief, BuShips, Oct. 2, 1955, *NRD.* One example of a follow yard's dissatisfaction with the system is illustrated in BuShips Code 400 to Code 525, July 29, 1960; J. F. Fagan, Jr., to Personal File, July 30, 1960; and Fagan to RADM J. M. Farrin, Aug. 2, 1960 (all in *NRD*).

2. Shugg to Rickover, May 27, 1954, and J. S. Bethea, Memorandum of Conference at Portsmouth, June 15, 1954, *NRD.*

3. D. Anderson to T. W. Dunn, Dec. 8, 1955, and F. J. Horne, Jr., to Panoff, Jan. 20, 1959, *NRD.* Mare Island Naval Shipyard, Planning Dept. Training Program for Installation and Repair of Submarine Nuclear Propulsion Plants, Nov., 1955, Records of the Mare Island Naval Shipyard, California.

4. Commander, Portsmouth Naval Shipyard to Chief, BuShips, Nov. 1, 1955, Feb. 24, 1959, and Panoff to Code 1500 Staff, July 11, 1955, *NRD.*

5. Chief, BuShips to Commander, Portsmouth Naval Shipyard, Feb. 10, 1955; Commander, Portsmouth Naval Shipyard to Chief, BuShips, Dec. 22, 1954. Chief, BuShips to Commander, Mare Island Naval Shipyard, March 23, 1955; and Functional Statements, Nuclear Power Division, April 22, 1960 (all in *NRD*). E. A. Wright, "The Bureau of Ships: A Study in Organization," *Journal of the American Society of Naval Engineers* 71 (Feb. 1959): 7–21; Bureau of Naval Personnel, *The Engineering Duty Officer (General),* NAVPERS 10814-B (Washington, 1963), pp. 94–108.

6. Hinchey to Rickover, Nov. 9, 1959, *NRD.*

7. "1952 Hits Peak in Shipbuilding," *Marine Engineering,* Feb. 1953, pp. 67–68, 81. "Gloom in the Shipyards," *Fortune,* July 1954, pp. 96–99, contains a good survey of the ship construction industry. Also useful is the testimony of Anderson and Leggett of March 8, 9, 10, 1954, in House Subcommittee on Appropriations, *Department of the Navy Appropriations for 1955* (Washington, 1954), pp. 521–633, and Leggett's testimony on Feb. 26, March 22, 29, 1954, in House Subcommittee on Defense Activity of the Committee on Armed Services, *Award of Noncompetitive Negotiated Contract by the Navy Department to Build Destroyers at Bethlehem Yard, Mass.* (Washington, 1954).

8. Rickover, Report of Conference, Sept. 10, 1954, and Blewett to Leggett, Oct. 22, 1954, *NRD.* For earlier Newport News activity see Roddis, Report of Conference, July 28, 1952, and Roddis to Rickover, Sept. 10, 1952, *NRD;* Newport News Shipbuilding and Dry Dock Co., CVR Project, Summary Report 3892-V/10633, Aug., 1953, *BAPL.* For the 1956 program see John D. Alden, "A New Fleet Emerges: Combat Ships," *Naval Review, 1964* (Annapolis, 1963), pp. 315–24. On Ingalls see Monro B. Lanier to Mumma, May 28, 1955, *NAVS.*

9. Rickover, Report of Conference, Sept. 10, 1954; W. E. Blewett to Leggett, Oct. 22, 1954; Rickover to Mumma, Oct. 22, 1955; D. B. Strohmeier to Rickover, July 29, 1955; and A. D. Huff to Rickover, Sept. 2, 1955 (all in *NRD*).

10. M. B. Lanier to Mumma, May 28, 1955, *NAVS.*

11. Rickover to Mumma, May 24, 1956; Blewett to Mumma, April 2, 1956; and Mumma to Blewett, April 9, 1956 (all in *NRD*).

12. "Naval Appropriation Bill," *House Report,* 57 Cong., 1 sess., no. 1792 (April 28, 1902), p. 19; P. L. 234, 57 Cong., 32 Stat., 662–91. There is an abundance of literature comparing the private and Navy yards. One of the best is a cost study by Arthur Andersen & Co., "Report on Survey and Analysis of Differences Between U. S. Navy Shipbuilding Costs at Naval and Private Shipyards: Shipbuilding Cost Study," Nov. 30, 1962, *NAVS.* Less technical but useful are John D. Alden, "The Case for Naval Shipyards?" *Naval Engineers Journal* 77 (Aug. 1965): 660–64, and J. J. Meyer, "Our Nation's Shipyards," United States Naval Institute

Proceedings 90 (Nov. 1964): 34–35. On the history of shipyards see E. A. Wright, "The Bureau of Ships: A Study in Organization," *Journal of the American Society of Naval Engineers* 71 (Feb. 1959): 7–21, and Holden A. Evans, *One Man's Fight For a Better Navy* (New York: Dodd, Mead, 1940). The perspective of the private shipbuilder is presented in "Naval Construction, Conversion and Repair in Private Yards," Shipbuilders Council of America, *Annual Report,* April 1, 1958, pp. 16–18.

13. BuShips, Request for Proposals, C-P.R 525-220(1712), Aug. 13, 1956, and Blewett to Mumma, Nov. 28, 1956, *NRD.*

14. Wyndham D. Miles, "The Polaris," *Technology and Culture* 4 (Fall 1963): 478–80; Harvey M. Sapolsky, "Creating the Invulnerable Deterrent: Programmatic and Bureaucratic Success in the Polaris System Development" (unpublished Ms., Massachusetts Institute of Technology, 1971), pp. 273–77.

15. Burke, Memorandum for the Record, Dec. 2, 1955, and Burke to D. B. Duncan, Oct. 19, 1955, Burke Papers, *AAB.*

16. Miles, "The Polaris," pp. 480–81; Sapolsky, pp. 278–81.

17. H. A. Jackson and E. R. Lacey, "Milestone in the Development of the New Navy," BuShips *Journal* 7 (Sept. 1958): 2–3.

18. Miles, "The Polaris," pp. 481–82; CNO to Chairman, AEC, Sept. 14, 1956, *AEC;* P. J. Sloyan, "Polaris—How Red Tape Can Be Cut," Baltimore *News-American,* Nov. 15–17, 1970; Sapolsky, pp. 282–86.

19. Ship Characteristics Board Memo No. 63-57, SCB Project No. 180, March 29, 1957, with covering note to Rickover, April 16, 1957, *NRD.*

20. F. J. Callahan to Rickover, March 26, 1956; D. T. Leighton to Rickover, March 27, 1956; and D. P. Brooks to Rickover, April 12, 1957 (all in *NRD*).

21. Code 1500 to Code 300, April 22, 1957; Chief, BuShips, to CNO (Chairman SCB), June 7, 1957; Chief, BuShips, to Assistant Chiefs, June 11, 1957; and CNO to Distribution, June 17, 1957 (all in *NRD*).

22. *New York Times,* Aug. 27, 1957; R. L. Shifley to Burke, Aug. 28, 1957, *AAB;* Joint Committee on Atomic Energy, Executive Hearings, Military Applications Subcommittee, Aug. 29, 1957, pp. 95–102, 146, Joint Committee Files.

23. Burke to Op-09, Aug. 31, 1957, *AAB.*

24. Burke, Memo of NSC Meeting, July 25, 1957, *AAB.*

25. Burke to Op-03, Feb. 2, 1957, and Burke to Secretary of the Navy, Sept. 27, 1957, *AAB.*

26. Rickover to W. K. Davis, March 17, 1956, and Davis to J. D. Anderson, March 30, 1956, *NRD;* Davis to W. B. McCool, Feb. 6, 1957, *AEC.*

27. Gates to Strauss, Sept. 30, 1957; W. F. Libby to Gates, Oct. 2, 1957; and Director of Reactor Development, Destroyer Nuclear Propulsion Plant, Feb. 13, 1957 (all in *AEC*); Rickover to Davis, Oct. 10, 1957, *NRD.*

28. Burke, Debriefing on NSC Meeting, Nov. 7, 1957, *AAB.*

29. Presentation by the Secretary of the Navy to NSC, Nov. 12, 1957, *AAB.*

30. Adm. Libbey, Debriefing on JCS Meeting, Nov. 16, 1957, Burke, Memo for the Record, Nov. 18, 1957, *AAB.*

31. Burke, Memo for the Record, Nov. 18, 20, 1957, *AAB.*

32. Burke, Item for Flag Officers' Dope, Nov. 23, 1957, and Burke, Memo for the Record, Nov. 22, 1957, *AAB;* D. A. Quarles to Strauss, Dec. 12, 1957, *AEC.*

33. R. L. Shifley to Burke, Nov. 26, 1957, *AAB;* G. S. Patrick to File, Nov. 27, 1957, *NRD.*

34. This section on shipbuilding status is based on summaries and chronologies in *NRD* and the following documents: Op-03C22, Memorandum for Record, Dec. 12, 1957, with attachs., and Mumma, Memorandum for File, Dec. 27, 1957, *NRD.*

35. This section is based on the following three hearings before the House Committee on Appropriations: *Department of the Navy Appropriations for 1958,* Jan., 1957 (Washington, 1957), pp. 632–33; *Supplemental Defense Appropriations for 1958,* Jan. 13, 1958 (Washington, 1958), pp. 196–99; *Department of Defense Appropriations for 1959,* Feb. 19, 1958 (Washington, 1958), pp. 384, 391, 395–402.

36. Rickover to Code 500 (undated, but sometime in Dec. 1957), *NAVS.*

37. R. B. Laning, Conference Report, Feb. 6, 1959; Code 1500 to Code 100, Feb. 16, 1959; Panoff, Memo to File, Feb. 19, 1959; and BuShips Contract NObs-4268 with New York Shipbuilding Corp., March 3, 1959 (all in *NRD*). Later New York Shipbuilding also built two nuclear-powered surface ships.

38. A. M. Morgan and Rickover, Memorandum of Conference, April 4, 1956; A. C. Smith, Sup-Ship-INSORD Instruction 4355.2, April 23, 1956; and Supervisor of Shipbuilding and Naval Inspector of Ordnance to Electric Boat, July 22, 1957 (all in *NRD*).

39. For examples of Shor's reports, see Shor to Rickover, March 22, April 9, 1956, *NRD.* Electric Boat was also required to submit detailed written reports. See Rickover to Shugg, Comments on First Monthly Progress Report Received June 12, 1956, *NRD.*

40. Francis to Rickover, June 9, 1956, and Dec. 22, 1958, *NRD.*

41. J. W. Crawford to Panoff, Feb. 28, 1958, and Crawford to Rickover, March 3, 14, 1958, *NRD.*

42. Rickover to Lanier, Nov. 20, 1957, and Lanier to Rickover, Nov. 29, 1957, *NRD.*

43. Panoff, Memo to File, Feb. 19, 1959; Teale to Rickover, March 18, 1959; and Organization for Reactor Plant and Overall Propulsion Plant Control at New York Ship, March 20, 1959 (all in *NRD*).

44. For examples of continuing troubles at all the private yards, including Electric Boat, see: D. T. Leighton to Rickover, Feb. 21, 1956; J. R. Byrd, Jr., to Rickover, June 12, 1959; Francis to Rickover, July 20, 1959; Code 1500 to Code 100, Oct. 1, 1959; Rickover to A. B. Homer, Oct. 16, 1959; Rickover to F. J. Mayo, Nov. 21, 1959; and R. W. Bass to Rickover, Sept. 19, Oct. 26, Nov. 14, 1961 (all in *NRD*).

45. Rickover to Distribution, Responsibilities of NR Representatives at Field Offices, March 27, 1962, with attachment, *NRD.*

46. D. C. Spencer to Simpson, Reports of Operations, July 22, Oct. 18, 1958, *PNR;* V. H. Hayden to BuShips Technical Representative, Bettis, Dec. 26, 1958, *NRD.*

47. The duties of the Bettis resident engineer are described in Conference Report, Bettis Atomic Power Division, Clairton Site, Nov. 18–20, 1958, WAPD-S5W-A(S)f-1791, *NRD.*

48. Spencer to Simpson, Reports of Operations, March 16, 1959, Feb. 15, March 18, April 18, May 16, 1960, *PNR.*

49. For a brief historical sketch of the Navy's inspection system see Julius A. Furer, *Administration of the Navy Department in World War II* (Washington: Government Printing Office, 1959), pp. 870–79; Robert H. Connery, *The Navy and the Industrial Mobilization in World War II* (Princeton: Princeton University Press, 1951), pp. 124–28. A few paragraphs on inspection in Navy shipyards in 1956 are in Bureau of Naval Personnel, *Naval Shipyard Duty for Engineering Specialists,* NAVPERS 10815-A (Washington, 1956), pp. 60–61. The duties of the supervisor of shipbuilding are described briefly in Bureau of Naval Personnel, *The Engineering Duty Officer (General)* NAVPERS 10814-B (Washington, 1963), p. 59. Of some use is Edward D. Maissan, "The Bureau's Machinery Inspection Service," BuShips *Journal* 2 (June 1956):11–13.

50. Statement by LCDR S. W. W. Shor at Electric Boat, Oct. 2, 1954, NRD Annotated draft press release approved Sept. 18, 1954, *NRD.*

51. Rickover to Chief, BuShips, Nov. 4, 1954, and Officer in Charge [E. P. Wilkinson] of the *Nautilus* to CNO, Sept. 28, 1954, *NAVS;* Shugg to Supervisor of Shipbuilding, Sept. 30, 1954, *NRD.*

52. Rickover pointed out that federal and military specifications governing pipe marking permitted painted markings, one marking per length of pipe for large pipe, and no markings at all on the surface of small pipe. Rickover to Chief, BuShips, Nov. 4, 1954, *NAVS.* Electric Boat reactions are in R. A. Hawkins, Memorandum of Meeting [at Groton on Sept. 26, 1954] . . . , Oct. 15, 1954, and J. J. Hopkins to L. L. Strauss, undated but about Oct. 1, 1954, *NRD;* Frank Pace, Jr., to Thomas S. Gates, Jr., Sept. 29, 1954, *NAVS.*

53. *Record of Proceedings of An Investigation Conducted at Groton, Connecticut . . . ordered on 12 Oct. 1954,* 5 vols., *NRD.* Vol. II contains the "Statement on 19 November 1954 by Carlton G. Lutts . . . Head of Laboratories," which gives examples of wrong material use. Date of circulation is from a Route and Office Memo Form, NAVS, with an inked date of Dec. 8, 1954, and which transmitted the report within the bureau. More piping incidents, mistakes on propeller shafting, along with the *Nautilus* pipe error, led to an investigation summarized in "Report on Quality of Work U. S. Naval Shipyards, prepared by Inspector General, Bureau of Ships," June 13, 1958, NAVSHIPS Library. The quality control engineering office was established by BuShips Instruction 5430.29, Oct. 22, 1959, *NRD.* The absence of standard procedures in Navy shipyards is described in Virgil C. Johnston, "Quality Assurance in Naval Shipyards," *Naval Engineers Journal* 75 (Oct. 1963): 731–34. Besse B. Day, who became the bureau's expert on quality control delivered a speech on March 2, 1955, calling for bureau action. The talk is listed in the Inspector General's Report and printed in "Principles of Quality Control," BuShips *Journal* 7 (May 1958): 2–8, 14.

54. Joseph M. Juran, *Quality Control Handbook* (New York: McGraw-Hill, 1951), contains several representative essays on various aspects of quality control. Robert F. Hart, "The Path to Quality Control," *Naval Engineers Journal* 76 (Oct. 1964): 691–96, deals with application to the bureau.

55. Establishing a quality control department at Electric Boat is referred to in Operations Manager and Manager of Quality Control to Distribution, March 6, 1962, *NRD.*

56. Undated handwritten letter from M. Shaw to Rickover, probably written around Aug. 21, 1958, *NRD.*

57. Crawford to Rickover, Sept. 30, 1958, *NRD.* Broad's duties are in Rawlings to File, Oct. 10, 1958, *NRD.*

58. N.S.S. & D.D. Co. Quality Inspection for Nuclear Propulsion Plants Organization Chart, Dec. 29, 1958, *NRD.* A candid description of the division was made by the New York Shipbuilding Corp. See J. A. Sweeney, Memorandum for File, Jan. 11, 1960, *NRD.*

59. NR files on quality control reports and shipyard audits are voluminous. For Rickover's speech see H. G. Rickover, "The Never-Ending Challenge," Oct. 29, 1962, Appendix 3, Joint Committee on Atomic Energy, *Hearings . . . on the Loss of the U.S.S. Thresher* (Washington, 1965), pp. 135–44.

60. This section describes the general pattern of trials for all nuclear submarines rather than one particular ship. Despite obvious differences, trials for nuclear submarines and surface ships were fundamentally the same. This account is based heavily on Naval Ship Systems

Command, *Manual for the Control and Testing and Plant Conditions,* NAVSHIPS 0980-0280-5000, Washington, Dec. 1969, and on direct observation by the authors on three submarine trials and one surface ship trial. Some comparison between conventional and nuclear ships can be found in Harley F. Cope and Howard Bucknell, III, *Command at Sea,* 3d ed. (Annapolis: U. S. Naval Institute Press, 1966), pp. 35–93.

61. Reference to qualified personnel is in OPNAV INSTRUCTION 9080.2D, April 19, 1960, *NRD.*

62. The first references to the joint test group appear in Agenda—Joint Test Group Meeting No. 1, Nov. 15, 1955, *PNR,* which was drawn up for the *Skate,* the first production model of a nuclear submarine. An example of tests set by the group for the *Skate* is in Memorandum from A. E. Francis, et al., to Distribution, June 12, 1957, *PNR.*

63. Naval Ship Systems Command, *Manual for the Control and Testing and Plant Conditions,* pp. 3–8.

64. Cope and Bucknell, *Command at Sea,* pp. 70–72. A description of a fast cruise is in Edward L. Beach, *Around the World Submerged, The Voyage of the Triton* (New York: Holt, Rinehart & Winston, 1962), pp. 15–16.

65. This section is based on the authors' direct observation. A newspaper account of a trial is in Washington *Sunday Star,* Oct. 31, 1971.

66. Beach, *Around the World Submerged,* pp. 31–36.

67. Rickover missed only two of sixty trials between 1955 and 1966, and those for reasons of illness. See Joint Committee on Atomic Energy, *Hearings on Naval Nuclear Propulsion Program, Jan. 26, 1966* (Washington, 1966), p. 33.

Chapter 11

1. Naval Personnel Act of 1899, Chap. 413, 55 Cong., 3 sess., 30 *Stat.* 1004–09.

2. Naval Appropriations Act of 1916, P. L. 241, 64 Cong., 39 *Stat.* 556–619. The history of personnel in the Navy is a complex subject. This brief summary is based on the following articles in the *Journal of The American Society of Naval Engineers:* "Designated Engineering Duty Only," 63 (Nov. 1951): 751–60; R. E. Bassler, "The Origin of Engineering Duty Only," 65 (Nov. 1953): 771–74; and R. B. Madden, "The Bureau of Ships and Its E. D. Officers," 66 (Feb. 1954):12–14. A good account of the competition between the line and engineers is in Edward W. Sloan, III, *Benjamin Franklin Isherwood Naval Engineer: The Years as Engineer in Chief, 1861–1869* (Annapolis: U. S. Naval Institute Press, 1965), pp. 193–212. A brief summary of the distinction between unrestricted and

restricted line and staff is in James Calvert, *The Naval Profession,* rev. ed. (New York: McGraw-Hill, 1971), pp. 84–90.

3. For the legal basis see: Sec. 3(a), The Atomic Energy Act of 1946 (P. L. 585, 79 Cong., 60 *Stat.* 758–59) and Sec. 161, The Atomic Energy Act of 1954 (P. L. 703, 83 Cong., 68 *Stat.* 948).

4. Rickover described these early procedures in some detail in Summary of Events and Correspondence Concerning Operation of Nuclear Powered Naval Ships Into Ports, encl., Rickover to G. T. Seaborg, AEC Chairman, Aug. 31, 1962, *AEC.*

5. Director of Reactor Development, Policy for Operation of Military Power Reactors, Feb. 15, 1954, and Minutes, Commission Meeting 962, Feb. 17, 1954, *AEC.*

6. Statement of Policy for Operation of Military Power Reactors, Nov. 8, 1954, encl., L. L. Strauss to H. B. Loper, Dec. 2, 1954, and Memorandum of Understanding concerning the USS *Nautilus,* encl., C. S. Thomas to Strauss, May 10, 1954, *AEC.* President Eisenhower authorized the transfer of fissionable material to the Navy in his memorandum to Strauss, April 2, 1954, *AEC.* Examples of Rickover's operating specifications are: Summary Report on Reactor Hazards Associated with Operation of the USS *Nautilus,* Part I, Initial Dockside Operation, Feb. 1, 1954, *NRD,* and Chief, BuShips, to Officer in Charge, USS *Nautilus,* Aug. 24, 1954, *NAVS.*

7. On the *Seawolf* see: Strauss to Loper, Sept. 29, 1955, and Loper to Strauss, Nov. 9, 1955, *AEC.* On the views of the reactor safeguards committee see: C. R. McCullough, chairman, to K. E. Fields, June 6, 1957, *AEC.*

8. Joint Committee on Atomic Energy, *Hearings on Governmental Indemnity and Reactor Safety,* March 25–27, 1957 (Washington, 1957), pp. 6–8. P. L. 256, 85 Cong., 1 sess., (71 *Stat.* 576).

9. McCullough to Strauss, Sept. 19, 1957, and Burke to Strauss, Jan. 21, 1958, *AEC.*

10. The basic directive was CNO to Distribution, Operation of Nuclear Powered Ships, OPNAV 03000.5, Feb. 6, 1958, *NRD.* The other directives were: Chief of Naval Personnel to Distribution, Personnel and Training Aspects of the Nuclear Propulsion Program, BuPers Instruction 1540.38, Dec. 31, 1957, and Chief, BuShips, to Distribution, Repair and Maintenance of Nuclear Propulsion Plants for Naval Ships, BuShips Instruction 9890.4, Feb. 25, 1958, *NRD.*

11. The reader should note that important issues concerning fleet operations have been omitted from the following sections for reasons of national security. A classified version of this chapter is on file in the Historian's Office, *AEC.*

12. Edward L. Beach, *Around the World Submerged: The Voyage of the Triton* (New York: Holt, Rinehart & Winston, 1962), pp. xv–xvi.

13. Theodore Roscoe, *United States Submarine Operations in World War II* (Annapolis: U. S. Naval Institute Press, 1949), p. 493. The book is a good example of the spirit of the submarine force. Destroyermen had their own esprit de corps, and Roscoe wrote a similar book on destroyer operations.

14. Bureau of Naval Personnel, *Personnel Administration in the Navy,* NAVPERS 10848-B (Washington, 1955), p. 83, contains a brief summary of officer assignment and training. For a nostalgic comparison of nuclear and prenuclear submarines see: Tom B. Thamm, "The Quiet Crisis in the Silent Service," U. S. Naval Institute *Proceedings* 97 (Aug. 1971): 50–58.

15. James F. Calvert, *Surface at the Pole: The Extraordinary Voyages of the USS Skate* (New York: McGraw-Hill, 1960), pp. 50–51, describes the initial reaction to Rickover's request.

16. Bruton describes the assignment in his Memorandum for Admiral Leggett, July 14, 1954, *NRD.*

17. Laning to Rickover, Aug. 9, 1954, *NRD.*

18. Roddis to Distribution List, Aug. 19, 1954, and encl., Conference Report, *NRD.*

19. COMSUBLANT to CNO, Aug. 12, 1954, *NRD.* Attached to this document is an endorsement by CINCLANTFLT, Aug. 26, 1954.

20. Report of the Board to Study the Personnel Aspects of Nuclear Power Utilization in the Navy, Aug. 20, 1954, pp. 32–33, 40–42, and C. H. Andrews, Conference Report, Nov. 1, 1954, *NRD.*

21. John H. Farson to Head, Field Administration Branch, July 13, 1955; Officer in Charge, U. S. Naval Submarine School, New London to COMSUBLANT, Dec. 21, 1955; Summary of Nuclear Propulsion Crew Training, June 12, 1959; and Rickover to W. K. Davis, Feb. 9, 1956 (all in *NRD*).

22. Effects of Accelerated SSGN(FBM) Construction Schedule on Submarine and Nuclear Power Training Programs—Tentative, Revised Feb. 24, 1958, Chief of Naval Personnel to CNO, Jan. 30, 1956, *NRD.*

23. Rickover, Memorandums for VADM H. P. Smith . . . , April 15, Oct. 16, Nov. 4, 1958; C. S. Carlisle to Rickover, March 31, 1960; Carnahan to Rickover, Aug. 3, 1961; and Rickover, Memorandum for VADM W. R. Smedberg, III, Aug. 18, 1961 (all in *NRD*).

24. The authors spent several days at the Naval Reactor Facility at the National Reactor Testing Station observing prototype training.

25. Report of Meeting, March 5, 1957, BUPERS INSTRUCTION 1540.33A, Aug. 29, 1957, *NRD.*

26. Chief of Naval Personnel to Officer in Charge, U. S. Naval Submarine School . . . , March 4, 1960, *NRD.*

27. William R. Anderson with Clay Blair, Jr., *Nautilus 90 North* (Cleveland: World Publishing Co., 1959), pp. 20–29; Calvert, *Surface at the Pole,* pp. 11–15. George P. Steele, *Seadragon: Northwest Under The Ice* (New York: Dutton, 1962), pp. 21–29, notes preliminary interviews by Rickover's staff.

28. Anderson, *Nautilus 90 North,* pp. 31–40; Calvert, *Surface at the Pole,* pp. 17–19; Steele, *Seadragon,* pp. 30–41. Calvert described the prototype training in an article, "What We Don't Know Can Hurt Us," U. S. Naval Institute *Proceedings* 85 (Jan. 1959), 55–59.

29. Rickover expressed this opinion in: Subcommittee of House Committee on Appropriations, *Department of Defense Appropriations for 1971,* May 13, 1970 (Washington, 1971), p. 30.

30. The quotation is from Rickover to VADM H. P. Smith, Chief of Naval Personnel, Jan. 8, 1959, *NRD.* He expressed these same thoughts in House Committee on Science and Astronautics, *Hearings on Scientific Manpower and Education, May 18, 1959* (Washington, 1959), pp. 407–12.

31. The Commission discussed the mounting pressure for Rickover's promotion at Commission Meeting 1338, Feb. 28, 1958, *AEC.* Congressional pressure was reported in Washington *Evening Star* and Washington *Post,* Aug. 12, 1958.

32. Secretary Gates apologized for the oversight in DOD Press Release 759-58, Aug. 12, 1958. Two days later twenty-one Senators sponsored a bill creating a special medal for Rickover. Washington *Evening Star,* Aug. 14, 1958; Senate Joint Resolution 201, 85 Cong., 2 sess. President Eisenhower also designated Rickover as his personal representative to welcome the *Nautilus* in New York on Aug. 25, 1958. New York *Journal American,* Aug. 19, 25, 1958; New York *Times,* Aug. 20, 1958; Washington *Daily News,* Aug. 20, 1958. J. T. Ramey described the Joint Committee's role in forcing Rickover's promotion to vice admiral in a speech at Los Alamos honoring Senator Anderson, April 8, 1972, AEC Press Release, April 8, 1972, *AEC.*

33. VCNO to CNO, Operation of Nuclear Powered Ships, Sept. 2, 1958, *NRD;* OpNav Instruction 03000.5A, Operation of Nuclear Powered Ships, Nov. 25, 1958, encl., Burke to McCone, Nov. 25, 1958, *AEC.*

34. Some discussion of interpretation of the Act appears in Report of U. S. Atomic Energy Commission on Section 91 of the Atomic Energy Act of 1954, as amended, in Joint Committee on Atomic Energy, *Hearing on Amending the Atomic Energy Act and Authorization of Stanford Accelerator Project,* Aug. 26, 1959 (Washington, 1959), pp. 4–8. Hereafter cited as *Amendment Hearings.*

35. For Commission discussions of weapon custody see: Minutes, Commission Meeting 1393, July 29, 1958; Meeting 1413, Oct. 14, 1958; and Meeting 1464, Feb. 4, 1959 (all in *AEC*).

36. For DOD views see: H. B. Loper to Senator Anderson, Aug. 28, 1959, with encls., in *Amendment Hearings,* pp. 15–16.

37. Joint Committee on Atomic Energy, *Hearings on Review of Naval Reactor Program and Admiral Rickover Award,* April 11, 15, 1959 (Washington, 1959).

38. The proposed amendment was introduced as S.2569 by Senator Anderson and as H.R. 8754, 86 Cong., 1 sess., by Congressman Carl T. Durham. The amendments and McCone's testimony appear in *Amendment Hearings,* pp. 1–9. On subsequent AEC actions, see: Directors of Licensing and Regulation, Military Application, and Reactor Development, Draft Amendment to Section 91 of the Atomic Energy Act, Feb. 11, 1960; Minutes, Commission Meeting 1591, Feb. 19, 1960; Rickover to F. K. Pittman, Director of Reactor Development, Feb. 18, 1960; and McCone to Senator Anderson, June 6, Dec. 2, 1960 (all in *AEC*).

39. VADM Wallace M. Beakley to G. T. Seaborg, June 5, 1961, transmitting the following BUSHIPS INSTRUCTIONS: 9890—Control and Disposition of Radioactive Equipment and Material from Naval Nuclear-Powered Ships, June 5, 1961; 9890.10—Accountability for Special Nuclear Material Utilized in Connection with Naval Nuclear Propulsion Plants, Nov. 30, 1960. Also enclosed was a draft SECNAV INSTRUCTION 4555, Disposal of Radioactive Wastes. All documents are in *AEC*.

40. The directive is attached to J. F. Kennedy to Chet Holifield, Sept. 23, 1961, *AEC*. This directive was still in effect more than a decade later.

41. The bulletins are in Naval Reactors Technical Bulletin Vol. A., *NRD*.

42. In this context fleet commanders were those in charge of numbered fleets and had operational command over all ships of whatever type assigned to that fleet. Type commanders had authority for the detailed procedures required to keep ships of that type operational. The assignment of operational command when a ship was not assigned to a numbered fleet was a complex matter depending on ship type and area of operations, but in general terms submarines were more often under the type commander in this situation than were surface ships.

43. For a general description of Navy overhaul procedures in the mid-1950s see: John N. Gross, "A Ship's Overhaul," BuShips *Journal* 4 (Jan. 1955): 2–7.

44. F. J. Callahan to Rickover, June 13, 1956, and Chief, BuShips, to Inspector of Naval Material, Pittsburgh, Sept. 25, 1956, *NRD*.

45. *Nautilus* Core Replacement Highlights, April 12, 1957, *NRD.*

46. Chief, BuShips, to Commander, Portsmouth Naval Shipyard, Oct. 22, 1958, and Chief, BuShips, to COMSUBLANT, Oct. 29, Dec. 23, 1958, *NRD.*

47. Commanding Officer, *Nautilus,* to COMSUBLANT, Nov. 28, 1959; Draft Letter, Commanding Officer, *Nautilus,* to COMSUBLANT, undated but about June 30, 1960; and Draft Summary of *Nautilus* Overhaul, Oct. 4, 1960 (all in *NRD*).

48. CNO to Chief, BuShips, Oct. 26, 1959; Study of Assignment of Polaris Submarine Overhaul Workload in Pacific, Nov. 5, 1959; Code 1500 to Code 100, Expansion of Facilities for the Overhaul of Nuclear Powered Ships, Nov. 5, 1959; and Chief, BuShips to Commanders, Pearl Harbor and Charleston Naval Shipyards, Feb. 24, 1959 (all in *NRD*).

49. This and the following paragraphs are based on the authors' direct observation of refueling operations and on Bettis Technology Proceedings, "A Symposium on Refueling Naval Reactors," Nov., 1967, WAPD-X-3906, *PNR.*

50. The number of ships in commission and building has been derived from The Naval Nuclear Propulsion Program: A Joint Atomic Energy Commission—Navy Program, Feb., 1966, *NRD.*

51. There are several popular accounts of early nuclear submarine voyages. Anderson with Blair, *Nautilus 90 North;* Beach, *Around the World Submerged;* Calvert, *Surface at the Pole;* Steele, *Seadragon: Northwest Under the Ice.* A popularly written and inadequate account of the effort is James Baar and William E. Howard, *Polaris: The Concept and Creation of a New and Mighty Weapon* (New York: Harcourt, Brace, Jovanovich, 1960). Baar and Howard state that the *George Washington* departed on patrol secretly. However, preparations for the patrol were followed by several newspapers, among them the *New York Times,* Jan. 2, Oct. 28, Nov. 14, 1960. The latter story gave the planned departure ceremony. The Washington *Evening Star,* Nov. 16, 1960, noted that messages on departure had been sent by Eisenhower, Gates, and Burke.

52. CNO to Distribution, Jan. 13, 1958, *NRD.*

53. Among the purposes of Sea Orbit were to demonstrate the unique capabilities of nuclear surface ships, to demonstrate the mobility of American military force, and to visit the Indian Ocean area. See Office of the CNO, Memorandum for the Joint Chiefs of Staff, April 11, 1964, *NRD.* The Assistant Secretary of Defense (Public Affairs) issued several press releases on the progress of Sea Orbit, among them No. 564-64, July 31, 1964; No. 613-64, Aug. 22, 1964; No. 684-64, Sept. 23, 1964; and No. 713-64, Oct. 2, 1964 (all in *NRD*).

54. The advantages of surface nuclear propulsion are described in: "A Treatise on Nuclear Propulsion in Surface Ships," Appendix I, Joint

Committee on Atomic Energy, *Hearings on Nuclear Propulsion for Naval Surface Vessels,* Oct. 30, 31, and Nov. 13, 1963 (Washington, 1964), pp. 197–201. For combat see: Rear Admiral Henry L. Miller, "Advantages of Nuclear Power and Its Utilization in a Combat Environment," Appendix I, Joint Committee on Atomic Energy, *Hearings on Naval Nuclear Propulsion Program,* Jan. 26, 1966 (Washington, 1966), pp. 37–42.

55. Hearings on the cost differential between nuclear and conventional ships were frequent and too numerous to cite. A good example is Joint Committee on Atomic Energy, *Hearings on Nuclear Propulsion for Naval Surface Vessels,* Oct. 30, 31, Nov. 3, 1963 (Washington, 1964). See also: Joint Committee on Atomic Energy, *Hearing on Nuclear Propulsion for Naval Warships,* May 5, 1971–Sept. 30, 1972 (Washington, 1972).

Chapter 12

1. One example is the Experimental R&D Incentive Program directed by the National Science Foundation and the National Bureau of Standards. See Deborah Shapley, "Technology Incentives: NSF Gropes for Relevance," *Science* 179 (March 16, 1973): 1105–7.

2. In April 1973 there were 101 submarines and 4 surface ships in the nuclear fleet. The total investment in these ships through fiscal year 1973 was $20.4 billion.

3. G. Geoffrey Smith, *Gas Turbines and Jet Propulsion* (New York: Philosophical Library, 1955), pp. 38–41, 399–401; Otis E. Lancaster, ed., *Jet Propulsion Engines,* vol. 12, *High Speed Aerodynamics and Jet Propulsion* (Princeton: Princeton University Press, 1959), pp. 32–37; Ray Wagner, *American Combat Planes* (Garden City: Doubleday, 1968), pp. 144–48, 252–54, 412–13.

4. R. G. Hewlett and F. Duncan, *Atomic Shield, 1947–1952, Vol. II of A History of the U. S. Atomic Energy Commission* (University Park: Pennsylvania State University Press, 1969), pp. 185–88, 193–96, 214–20. Cost data on the MTR are from Joel W. Chastain, Jr., *U. S. Research Reactor Operation and Use* (Reading, Mass.: Addison-Wesley, 1958), p. 326.

5. On the costs of nuclear ships, see appendix 3. Total procurement costs for all Navy ships in fiscal year 1962 were $2.7 billion. U. S., *The Budget of the United States Government, Fiscal Year 1964* (Washington, 1963), Appendix, p. 278.

6. Harvey M. Sapolsky, *The Polaris System Development: Bureaucratic and Programmatic Success in Government* (Cambridge: Harvard University Press, 1972), pp. 44–52, 153–59.

7. Sapolsky evaluates PERT in some detail on pp. 110–30. The quotation is on p. 125.

8. House Subcommittee on Military Operations, Committee on Government Operations, *Organization and Management of Missile Programs,* March 20, 1959 (Washington, 1959), pp. 620–22; Joint Committee on Atomic Energy, *Naval Nuclear Propulsion Program, 1970,* March 19–20, 1970 (Washington, 1970), pp. 4–5; Joint Committee on Atomic Energy, *Loss of the U.S.S. Thresher,* July 23, 1963 (Washington, 1963), pp. 86–89.

9. Senate Committee on Armed Services, *Hearings on U.S. Submarine Program,* March 13, 1968 (Washington, 1968), pp. 91–94. Rickover attacked the Navy's rotation policy in the same hearing (pp. 94–98) and in Senate Committee on Armed Services, *Hearings on Missiles, Space, and Other Defense Matters,* Feb. 3, 1960 (Washington, 1960), pp. 171–73; Joint Committee on Atomic Energy, *Naval Nuclear Propulsion Program,* Jan. 26, 1966 (Washington, 1966), pp. 22–25.

10. Rickover to Leighton, Dec. 9, 1958, and Lascara and Leighton to Rickover, Dec. 12, 1958, *NRD;* Joint Committee on Atomic Energy *Naval Nuclear Propulsion Program, 1969,* April 23, 1969 (Washington, 1969), pp. 64–73.

Sources

In writing this study of technological innovation, we could not begin to capture the complexities of naval nuclear propulsion simply by describing the development of reactor systems or shipbuilding techniques. As in most government projects, the direction of technological development was often influenced by political, budgetary, or bureaucratic pressures in Washington. At times technical accomplishments or difficulties influenced policy decisions at the highest levels in the Commission and the Department of Defense. With these relationships in mind, we approached our research in the broadest possible context and sought a wide variety of primary sources which carried us far beyond the records of Admiral Rickover's office.

Primary Sources

The primary sources most accessible to us were the Commission's own official files in the Office of the Secretary. Because we had already used many of the pertinent records in preparing two volumes of the Commission's history, we could quickly exploit this source. Many types of records were available in the Secretary's files but the most useful to us were internal correspondence between the division of reactor development (including Rickover's group) and the Commission, official correspondence between the Navy Department and the Commission, Commission staff papers, and minutes of Commission meetings. These files, which are remarkably complete and well organized, provide an excellent view of the naval reactors project from the Commission's perspective.

For a broad view of policy development in the Navy we relied on the files of the Office of Chief of Naval Operations which have been transferred to the Naval History Division in Washington. Because the files are organized by operational unit, it was relatively easy to isolate the pertinent documents. The Naval History Division also holds the personal papers of many high-ranking naval officers. We were able to obtain permission to consult the papers of Fleet Admiral Ernest J. King, Fleet Admiral Chester W. Nimitz, and Admiral Arleigh A. Burke. We also used the transcript of an interview with Admiral James L. Holloway, Jr., for the Columbia University Oral History Project. All these materials helped us to understand the complex of forces which come to bear on high officials in the Navy and how these can influence their attitudes toward technical projects like nuclear propulsion. We are grateful to Vice Admiral Edwin B. Hooper, USN (Ret.), Curator of the Navy Department and director of Naval History, and Dean C. Allard, head of the Naval History Division's operational archives branch, for guiding us to these sources.

By far the largest and most valuable documentary source for this book was found in the files of the Division of Naval Reactors at its offices in the National Center in Arlington, Virginia. With a keen sense of both the administrative and historical value of records, Admiral Rickover from the beginning of the project saw to it that his staff prepared summaries of meetings and filed copies of correspondence involving the project. Each project officer and senior member of the technical staff maintained files of his activities. From time to time these files were retired in the normal manner to the Federal Records Center for the Washington area, now located in Suitland, Maryland. Fortunately the successive librarians for the naval reactors project have maintained indexes to most of these retired materials so that it is possible to locate them among the hundreds of thousands of boxes stored in the center. The indexes, however, are largely by reactor type or construction project and do not always indicate the group within the organization in which they originated. Thus it is often difficult to discover without examining the boxes whether the records deal with policy conferences and correspondence of interest to the historian or with minute engineering design details which only the originator could fully appreciate. Even working with the available indexes we found it necessary to examine in detail several hundred linear feet of records scattered through the center.

As a check on our research we were able to use the substantial number of historical records which Admiral Rickover and his staff have retained in their office files. Conscious of the historical significance of much of the program, Rickover and his senior staff over the years have collected copies of many key documents. Most of these materials are duplicated in the official files, but we found these special collections useful, particularly in the early stages of our research when we were attempting to gain a general understanding of the project. The unique documents of this type are copies of correspondence between Rickover and other high officials in the Commission, the Navy, or the Department of Defense. The records of the Division of Naval Reactors were not only valuable but essential to our task. Without them it would be impossible to write an adequate history of the project.

The records of the Bureau of Ships, of which the nuclear power division was a part, were an obvious source of materials for this book, but these records were not easy to use. Now held by the Naval Ship Systems Command, the bureau's records have long since been retired to the Suitland records center and for the most part forgotten. We found no useful index to these records, which were apparently filed chronologically only by ship number. By combing the files for ships of interest to us, we were able to find many helpful documents scattered through voluminous files of technical or administrative documents. Unfortunately we discovered no general policy files which documented the positions of the bureau chief or his principal advisors. Documents reflecting bureau positions on policy issues, therefore, had to come from other sources. For a general understanding of the bureau organization and procedures we relied on the Bureau of Ships *Journal,* organization charts, and telephone books held in the Naval Ship Systems Command's technical library at the National Center.

To gain a broader perspective of the project beyond Admiral Rickover's Washington headquarters, we systematically mined the record repositories of the principal laboratories and field offices. Because the Bettis Laboratory developed most of the reactors used in the nuclear fleet during the period covered by this book, the records of that laboratory and the Pittsburgh Naval Reactors Office were particularly important. Not only did these records give us a close-up view of technical problems, but they also revealed the impact of technical activities on laboratory organization, the Shippingport project, the establishment of the Plant Apparatus Department, and relations with the shipyards. The records of the Pittsburgh office richly document the special responsibilities which this office exercised for the Commission, notably in directing zirconium procurement and production, managing construction and operation of the Shippingport plant, and administering contracts with manufacturers of fuel elements.

The records of the Knolls Atomic Power Laboratory were essential in documenting the technology of the sodium-cooled reactor and the struggles of the General Electric Company to find a workable relationship with the Navy and Rickover's organization. In addition to the usual files of technical memorandums and reports, the library at the laboratory has assembled a valuable collection of records which document the origins of the Navy project at Knolls.

The records of the Argonne National Laboratory are indispensable for any study of the Commission's reactor development program in the 1950s. Designated the center for reactor development in 1948, Argonne, under Walter Zinn's direction, was involved in policy decisions and technical activities extending far beyond that one laboratory. Thus we found much useful material, not only in the files of the Argonne naval reactors division but also in Zinn's files and other laboratory records. The Argonne collections are all the more important because they became the best single source on the Commission's reactor program in the 1950s after the destruction of the files of the division of reactor development in Washington about 1957.

Like all historians, we depended heavily upon administrative officials, research specialists, and librarians in many of the organizations which made records available: Velma E. Lockhart and Lester C. Koogle, Jr., of the Commissions' staff; Ann L. Buck, Rose V. Gayle, Theresa Leone, Isabel Lovell Moore, Jean Scroggins, and Barbara J. Whitlark of Admiral Rickover's staff; Linda Nunly Carl and Ferda K. Muzzi of the Naval Reactors Library; Lucille Achauer of the Naval Ship Systems Command Library; Rita L. Halle and Fred S. Meigs of the Naval History Library; Charles W. Flynn, Raymond E. Denne, Charles E. Doria, Helen L. Russell, and Janet C. Stuler of the Commission's Pittsburgh office; Helen S. Brown, John H. Martens, and E. Newman Pettitt of the Argonne National Laboratory; Madeline T. Barringer, Stuart Sturges, and Adelaide B. Oppenheim of the Knolls Laboratory; William L. Kabler of the Bettis Laboratory; Mack C. Corbett of the Commission's Idaho office; and Howard R. Canter of the Idaho Naval Reactors Facility.

Secondary Sources

For background on the Atomic Energy Commission, its organization, and activities the reader should consult Richard G. Hewlett and Oscar E. Anderson, Jr., *The New World, 1939–1946* (University Park: Pennsylvania State University Press, 1962), and Richard G. Hewlett and Francis Duncan, *Atomic Shield, 1947–1952* (University Park: Pennsylvania State University Press, 1969), the first two volumes in the published history of the Commission. The Commission has also sponsored the writing and publication of many books on reactor technology. John F. Hogerton, *The Atomic Energy Deskbook* (New York: Reinhold, 1962), is a valuable reference guide for the general reader. Much of the fundamental technology of water-cooled reactors has been set forth in a number of handbooks prepared by the naval reactors branch and cited in notes 13–20 of chapter 5.

On the Navy side, the serious reader will find relatively few secondary sources that even begin to provide an adequate background for the naval nuclear propulsion project. Traditionally histories of the United States Navy have concentrated on combat operations rather than on high policy, organization, administration, and technology. The Navy's experience in all of these areas during World War II influenced the origins of the nuclear propulsion project. Yet in no place, even in brief summary, is there a general account of the full scope of the Navy's activities during World War II. Samuel E. Morison's fifteen volumes cover naval operations, while Julius A. Furer, *Administration of the Navy Department in World War II* (Washington: Government Printing Office, 1959), gives some insight into the organization and administrative procedures used in the Navy. Robert H. Connery, *The Navy and the Industrial Mobilization in World War II* (Princeton: Princeton University Press, 1951), is useful within its self-imposed limits. The library of the Naval Ship Systems Command in Arlington, Virginia, has a manuscript history of the Bureau of Ships during World War II which helps explain the technical activities of ship procurement during the period.

The lack of secondary sources is even more evident in dealing with the early phases of the nuclear propulsion program. The few biographies of senior naval officers throw little light on the nuclear power project. Vincent Davis, *Postwar Defense Policy and the U. S. Navy, 1943–1946* (Chapel Hill: University of North Carolina Press, 1966), focuses upon the role of the Navy in the postwar military establishment but has little to say about the technological problems of the period. The best published accounts of the nuclear propulsion project have been books by commanders of nuclear submarines: William R. Anderson with Clay Blair, Jr., *Nautilus 90 North* (Cleveland: World Publishing Co., 1959); Edward L. Beach, *Around the World Submerged: The Voyage of the Triton* (New York: Holt, Rinehart & Winston, 1962); James F. Calvert, *Surface at the Pole: The Extraordinary Voyages of the USS Skate,* (New York: McGraw-Hill, 1960); and George P. Steele, *Seadragon: Northwest Under the Ice* (New York: E. P. Dutton, 1962). Although these books give the reader a sense of the discipline and technical excellence which nuclear power brought to the fleet, they deliberately avoid any discussion of the new technology or its impact on the Navy as an institution. Even if superficial in some respects, these accounts are far superior to the book by Clay Blair, Jr., *Admiral Rickover and the Atomic Submarine* (New York: Henry

Holt, 1954), which is a popularized and partisan account centering on the promotion struggle of 1953. Students of the development of ship types will find valuable information in Norman Polmar's two volumes: *The Atomic Submarine* (Princeton: Van Nostrand Reinhold Company, 1963) and *Aircraft Carriers, A Graphic History of Carrier Aviation and Its Influence on World Events* (New York: Doubleday, Inc., 1969).

Thus the student of technological development in the postwar period must rely for secondary materials on professional journals and other serial publications. The reader may glean some insights about the impact of technology on the postwar Navy from articles in the United States Naval Institute *Proceedings* and, since 1962, in the *Naval Review*. For information on naval engineering we found helpful articles in the annual *Transactions* of the Society of Naval Architects and Marine Engineers and the *Journal of the American Society of Naval Engineers*. A few articles in these journals throw light on the organization and evolution of the Bureau of Ships. Many more give the reader a sense of the incredible complexities of ship design, engineering, and construction.

The most voluminous printed sources of information about the nuclear Navy are transcripts of hearings before Congressional committees—mainly the Joint Committee on Atomic Energy, the Armed Services Committees, and the Appropriations Committees of both houses. Admiral Rickover has testified many times before Congressional committees on the Navy's need for ships, on training and education, reactor development and safety, and relations with industry. The easy give-and-take between Rickover and members of Congress should not be allowed to disguise the extreme care which he and his staff take in preparing for hearings. The transcripts reveal not only his mastery of the Congressional hearing forum but also an extraordinary amount of information about the nuclear propulsion program. Rickover frequently includes in the record substantial extracts of unclassified information from highly sensitive documents. For the student who is willing to dig through hundreds of pages of fine print the published transcripts provide a wealth of information on the project.

Interviews

Like all contemporary historians, we supplemented our documentary research with conversations with many of those who participated in the events we were describing. Following procedures established in writing the first two volumes of the Commission's history, we used interviews more as a supplemental than as the primary source of evidence. Most of our interviews occurred only after we had carefully studied the pertinent documents and prepared precise questions for each person to be interviewed. As in preparing the earlier volumes, we did not use a tape recorder because we believe that recording devices inhibit the frank expression of opinions, particularly when the persons being interviewed are discussing controversial subjects involving their living, and often still active, associates. Nor in our text do we quote from interviews or attribute what we have written to specific individuals. We have tried to base our conclusions on a judicious weighing of all the evidence from both written and oral sources.

Over a period of several years we discussed the naval nuclear project with more than 150 individuals, including former Commissioners and other Commission officials, admirals, and other high-ranking naval officers and civilian officials in the Navy Department, company presidents, laboratory directors, project officers, technical group leaders, scientists, engineers, technicians, and men in the fleet. We also made use of interviews with many individuals whom we saw in writing the first two volumes of the Commission's history. The following list can include only those whose names we recorded. Many others in casual conversations provided valuable insights and the flavor of authenticity.

Atomic Energy Commission, Washington: Robert F. Bacher, W. Kenneth Davis, James B. Fisk, Lawrence R. Hafstad, Robert E. Hollingsworth, David E. Lilienthal, Woodford B. McCool, John L. McGruder, James T. Ramey, Leonard F. C. Reichle, Glenn T. Seaborg, Cyril S. Smith, Lewis L. Strauss, Edward R. Trapnell, George L. Weil, Walter J. Williams.

Atomic Energy Commission Field Offices: Jon D. Anderson, John J. Flaherty, Charles W. Flynn, Lawton D. Geiger, Stanley W. Nitzman, David Saxe.

Bureau of Ships: David H. Clark, Wilson D. Leggett, Jr., Earle W. Mills, Albert G. Mumma, Homer N. Wallin, Charles D. Wheelock.

Fleet Operations: Edward L. Beach, Marvin S. Blair, Arleigh A. Burke, James F. Calvert, Robert B. Carney, David W. Cockfield, James H. Doyle, Jr., Paul J. Early, Elton W. Grenfell, James L. Holloway, Jr., Robert L. J. Long, Charles B. Momsen, Jr., John H. Nicholson, Forrest S. Petersen, Nils R. Thunman, Eugene P. Wilkinson.

Navy Technical Bureaus: Philip H. Abelson, John M. Fluke, Franklin C. Knock, George H. Main, Chad J. Raseman, Robert K. Reed, Frank G. Scarborough.

Naval Reactors Branch: Joseph H. Barker, Jr., Willis C. Barnes, Richard W. Bass, Edward J. Bauser, Robert S. Brodsky, Philip R. Clark, John W. Crawford, Jr., John F. Drain, James M. Dunford, Arthur E. Francis, William L. Givens, Merwin C. Greer, Jack C. Grigg, Souren Hanessian, Tom A. Hendrickson, William M. Hewitt, William S. Humphrey, Donald G. Iselin, Frank Kerze, Jr., Edwin E. Kintner, Robert V. Laney, David T. Leighton, Theresa Leone, Miles A. Libbey, John M. Maloney, I. Harry Mandil, Howard K. Marks, Robert P. Metzger, Murray E. Miles, Robert Panoff, Alvin Radkowsky, Hyman G. Rickover, Theodore Rockwell, III, Louis H. Roddis, Jr., Rachel J. Sarbaugh, Milton Shaw, Samuel W. W. Shor, Karl E. Swenson, James R. Vaughn, Thomas J. Walters, William Wegner, Steven A. White.

Polaris Project: Levering Smith.

Westinghouse Electric Corporation, including Bettis Laboratory and Plant Apparatus Department: Nicholas A. Beldecos, William L. Borden, George H. Cohen, Paul A. Cohen, Ralph F. Costa, William R. Ellis, William H. Hamilton, Vernon F. Hayden, William L. Kabler, Edward J. Kreh, Bernard F. Langer,

William H. Linton, Eli F. Lohr, Benjamin R. Lustman, Raymond C. Mairson, Wilfred D. Miller, Gwilym A. Price, Leonard B. Prus, Joseph C. Rengel, Philip N. Ross, John W. Simpson, Joseph J. Squilla, Alexander Squire, Harold E. Thomas, Charles H. Weaver, John E. Zerbe.

General Electric Company, including Knolls Atomic Power Laboratory: Donald J. Anthony, Ralph J. Cordiner, Earl B. Haines, Henry Hurwitz, Jr., W. Rudolph Kanne, Kenneth A. Kesselring, Kenneth H. Kingdon, Cramer W. LaPierre, William H. Milton, C. Robert Stahl, Harry E. Stevens, Henry E. Stone, Stuart Sturges, C. Guy Suits, Leonard B. Vandenburg, Volney C. Wilson.

Argonne National Laboratory: Alfred Amorosi, Norman H. Hilberry, John H. Martens, E. Newman Pettitt, Walter H. Zinn.

Electric Boat Company: William G. Atkinson, Robert B. Chappell, Thomas W. Dunn, John S. Leonard, Andrew I. McKee, Owen O'Neil, Joseph D. Pierce, Carleton Shugg.

Newport News Shipbuilding and Dry Dock Company: Lennis C. Ackerman, Richard S. Broad, R. Spencer Plummer.

Oak Ridge National Laboratory: Alvin M. Weinberg, Eugene P. Wigner.

Naval Reactor Facility, Idaho, including Westinghouse and General Electric personnel: John Armenta, Edwin M. Baldwin, Robert L. Cage, Howard R. Canter, Robert W. Chewning, Donald H. Krueger, C. Ray Lockard, Benjamin J. Rencher, Henry D. Ruppel, Emil H. Schoch.

Duquesne Light Company: Philip A. Fleger, John E. Gray.

Physical Evidence

One advantage of writing about the recent past is that the historian can often explore the physical setting of the events he is describing while the sites still retain some of their original appearance and atmosphere. In our research for this book we were able to visit and even work for extended periods in the very buildings where most of the events we were studying occurred.

From July 1969 until August 1970 we occupied an office only a few steps from Admiral Rickover's in the Main Navy Building along Constitution Avenue in Washington. There we could not help but observe the Rickover system in operation. Also during those months we were within fifty feet of the same offices which Rickover and some of his Oak Ridge group occupied when they returned to the Bureau of Ships in the autumn of 1946. Through old telephone books we were able to find each of those offices before the old building was demolished in the summer of 1970.

Before that we had worked for nine months in the ramshackle, decaying N Building behind Main Navy, which Rickover's group had occupied since 1955. Walking down the musty, dark corridors with their dirty yellow walls of crumbling plasterboard, it was hard to believe that such quarters could house one of the most important technical projects in the Navy.

Going back even further, one of the authors could recall numerous visits to the offices of the naval reactors branch in the T-3 Building, a few blocks east of Main Navy on Constitution Avenue. Both of the authors worked for a time in the Commission's headquarters building further west on the same street and frequently attended meetings in the conference room where the Commissioners made many of the decisions described in this book. Thus we were able to picture in our mind's eye the exact physical setting of much of our narrative.

We enjoyed similar advantages in writing about the laboratories and reactor installations described in this book. During several visits to the Bettis and Knolls laboratories, we worked in the offices and explored the plant facilities built for the project in the late 1940s. We spent a week at the naval reactor facility in Idaho to observe the training of new crews of officers and men on the Mark I prototype, which still looks much as it did at the time of initial startup in 1953.

We also had several opportunities to visit shipyards, naval installations, and ships in the nuclear fleet. At the Electric Boat Division of the General Dynamics Corporation at Groton, Connecticut, and at the Newport News Shipbuilding and Dry Dock Company in Newport News, Virginia, we spent many hours clambering through submarines under construction or observing the intricacies of refueling and overhaul operations. Both company officials and naval officers were available to answer our questions and to explain the fine points of shipbuilding and fleet operations. In addition to these private yards we visited the naval bases at New London, Connecticut; Mare Island, California; and Norfolk, Virginia. We accompanied Admiral Rickover on sea trials of the submarines *Spadefish* (SSN-668), *Hawkbill* (SSN-666), and *Drum* (SSN-677) and voyaged from New London to Norfolk on the *Bluefish* (SSN-675). On board these ships we studied the propulsion plants and witnessed training exercises. We were also aboard the aircraft carrier *Enterprise* (CVAN-65) in January 1971 for trials following refueling. We appreciated the courtesy and assistance of the officers and crews of these ships.

All of these experiences in the working world of the nuclear Navy gave us an insight into the project that we could never have attained in our Washington offices from documents or interviews.

Index